MYMATHWORKBOOK

to accompany

BASIC MATHEMATICS & ALGEBRA

with contributions from

James J. Ball
Indiana State University

Steve Oulette

Beverly Fusfield

Kevin Bodden
*Lewis and Clark
Community College*

Randy Gallaher
*Lewis and Clark
Community College*

Prentice Hall
is an imprint of

Prentice Hall
is an imprint of

www.pearsonhighered.com

Table of Contents

Solved Examples (SE)

Worksheets (WS)

Welcome to your MyMathWorkbook. This Workbook is designed to be used in conjunction with your MyMathLab course to help you succeed in math. Below you will find some additional tips to help you successfully complete your course in math this semester.

MyMathWorkbook
Eight Study Skills to Help You Succeed

1. **Take Notes**. Whether you are taking this course online, in a lab, or in a lecture hall, taking notes is a great way to improve your comprehension of the material and an even better way to study for an exam. You should take notes, whether you are writing down examples or writing down questions, when you are listening to a live lecture or when you are watching the videos in MyMathLab. After the class or video is complete, review your notes to make sure you understand the concepts. If anything is unclear, contact your professor for help—and later, update your notes with your newfound knowledge. Then, when it is time to review for your exam, you will have everything you need in front of you.

2. **Use This Workbook**: Solved Examples allow you to see a problem worked out, from beginning to end, and give you a place to take notes about key ideas or questions to ask your instructor. Practice Exercises can be worked in the lab and classroom or at home to reinforce the skills and concepts presented in the Solved Examples. By using this workbook, you can really see that practice makes perfect!

3. **Do your Homework**: Whether your instructor has assigned online homework or work for you to write out and hand in, mathematics is a discipline where you learn by doing. Only by actively participating in your class can you really practice and assess your comprehension of the material. And, since understanding builds step-by-step in math, you need to master the material in your first homework before you can be successful in your second homework.

4. **Know Where to Get Help**: If you are struggling with your homework, there are many different resources available to help you. Within MyMathLab, you can access a sample problem, a guided solution, or even a video to explain a difficult concept in more depth. You can contact the Pearson Tutor Center to talk live to a math instructor about your question. And you can always contact your professor, either in person or on the phone during office hours, or by email.

5. **Use the Study Plan**: MyMathLab's study plan helps you monitor your own progress, letting you see at a glance exactly which topics you need to practice. MyMathLab generates a personalized study plan for you based on your test results and the study plan links directly to interactive, tutorial exercises for topics you haven't yet mastered. So you can test, review your errors, and test again—allowing you to perfect your score and build confidence before going into your exams.

6. **Learn from your Mistakes**: If allowed by your professor, review past quizzes and tests. See where you made mistakes the first time, review those concepts, and quiz yourself again. Once you have really grasped a concept, you won't forget it when you are learning another topic down the road.

7. **Ask Questions**: If you don't understand a concept or an example, chances are others in your class are having trouble too. Ask your instructor for clarification. If you don't feel comfortable bringing up a question in class, try to catch your instructor in a one-on-one moment or go to the lab—and you can always contact the Pearson Tutor Center.

8. **Stay Positive**: Success is just around the corner—and the feeling that success brings is unmatchable!

1.1 Reading and Writing Whole Numbers

Learning Objectives:

1. Identify place value in whole numbers.
2. Write whole numbers in words.
3. Write whole numbers using digits.
4. Interpret tables using whole numbers.

Solved Examples:

1. Identify the place value of each digit in the numbers.

    a) 3,654   b) 265,812   c) 56,203,411

    Solution

    a) Use the place value chart and work from left to right. There are 3 thousands, 6 hundreds, 5 tens, and 4 ones.

    b) Use the place value chart and work from left to right. There are 2 hundred thousands, 6 ten thousands, 5 thousands, 8 hundreds 1 ten, and 2 ones.

    c) Use the place value chart and work from left to right. There are 5 ten millions, 6 millions, 2 hundred thousands, 0 ten thousands, 3 thousands, 4 hundreds, 1 ten, and 1 one.

2. Write a word name for each number.

    a) 325   b) 60,448

    Solution

    a) We want a word name for 325. There are 3 hundreds, 2 tens, and 5 ones. The answer is *three hundred twenty-five*.

    b) We want a word name for 60,448.

        60,448

        sixty thousand, ➤

        four hundred forty-eight

        The answer is *sixty thousand, four hundred forty-eight*.

3. Write a number for each word name.

    a) seven thousand, ninety-eight

    b) three hundred forty million, one hundred thirty-two

    Solution

    a) Use the place value chart. The answer is 7098.

    b) Use the place value chart. The answer is 340,000,132. Zeros indicate there are no thousands.

4. What is wrong with the following: *"three hundred and sixty-four"*

   Solution

   The word *and* is never used when writing whole numbers.

5. When do we use a hyphen when writing whole numbers?

   Solution .

   A hyphen is used when writing the numbers 21 – 99, except for numbers ending in zero.

1.2 Adding Whole Numbers

Learning Objectives:

1. Add when carrying is not required.
2. Add with carrying as necessary.
3. Solve application problems.

Solved Examples:

1. Add quickly.

    a) $5 + 3$ b) $4 + 7$ c) $8 + 9$ d) $6 + 2$ e) $3 + 3$

    Solution

    a) $\begin{array}{r} 5 \\ + 3 \\ \hline 8 \end{array}$ b) $\begin{array}{r} 4 \\ + 7 \\ \hline 11 \end{array}$ c) $\begin{array}{r} 8 \\ + 9 \\ \hline 17 \end{array}$ d) $\begin{array}{r} 6 \\ + 2 \\ \hline 8 \end{array}$ e) $\begin{array}{r} 3 \\ + 3 \\ \hline 6 \end{array}$

2. Add.

    a) $6 + 4 + 3 + 7$ b) $8 + 8 + 0 + 5$

    Solution

    a)

    b)

3. Add with no carrying required.

    a) $\begin{array}{r} 53 \\ + 12 \\ \hline \end{array}$ b) $\begin{array}{r} 1123 \\ + 345 \\ \hline \end{array}$ c) $\begin{array}{r} 40,001 \\ 32,442 \\ + 15,333 \\ \hline \end{array}$

    Solution

    a) $\begin{array}{r} 53 \\ + 12 \\ \hline 65 \end{array}$ b) $\begin{array}{r} 1123 \\ + 345 \\ \hline 1468 \end{array}$ c) $\begin{array}{r} 40,001 \\ 32,442 \\ + 15,333 \\ \hline 87,776 \end{array}$

4. Add with carrying required.

a)   96
    + 47

b)   5678
    + 3574

c)   6505
     173
     7044
    + 168

Solution

a)   1
     96
    + 47
    ───
    143

b)   111
     5678
    + 3574
    ─────
    9252

c)   1 2
     6505
     173
     7044
    + 168
    ──────
    13,890

5. A plane is flying at an altitude of 5932 ft. It then increases its altitude by 7384 ft. Find its new altitude.

Solution

   1 1
   5932
  + 7384
  ──────
  13,316

The new altitude is 13,316 feet.

1.3 Subtracting Whole Numbers

Learning Objectives:

1. Subtract when no borrowing is needed.
2. Subtract, borrowing as necessary.
3. Solve application problems.

Solved Examples:

1. Write two subtraction problems for each addition problem.

   a) $7 + 1 = 8$ b) $5 + 2 = 7$ c) $3 + 11 = 14$

   Solution

   a) $8 - 1 = 7$ or $8 - 7 = 1$ b) $7 - 2 = 5$ or $7 - 5 = 2$ c) $14 - 11 = 3$ or $14 - 3 = 11$

2. Change each subtraction problem to an addition problem.

   a) $9 - 5 = 4$ b) $11 - 3 = 8$ c) $15 - 3 = 12$

   Solution

   a) $9 = 5 + 4$ or $9 = 4 + 5$ b) $11 = 3 + 8$ or $11 + 8 + 3$ c) $15 = 3 + 12$ or $15 + 12 + 3$

3. Subtract with no borrowing.

   a) $98 - 51$ b) $54 - 22$

   c) $\begin{array}{r} 664 \\ -\ 51 \\ \hline \end{array}$

   Solution

   a) $\begin{array}{r} 98 \\ -51 \\ \hline 47 \end{array}$ Check $\begin{array}{r} 47 \\ +51 \\ \hline 98 \end{array}$
   b) $\begin{array}{r} 54 \\ -22 \\ \hline 32 \end{array}$ Check $\begin{array}{r} 32 \\ +22 \\ \hline 54 \end{array}$
   c) $\begin{array}{r} 664 \\ -51 \\ \hline 613 \end{array}$ Check $\begin{array}{r} 613 \\ +51 \\ \hline 664 \end{array}$

4. Subtract with borrowing.

   a) $\begin{array}{r} 51 \\ -12 \\ \hline \end{array}$ b) $\begin{array}{r} 1123 \\ -\ 345 \\ \hline \end{array}$ c) $\begin{array}{r} 40,001 \\ -15,333 \\ \hline \end{array}$

   Solution

   a) $\begin{array}{r} {}^{4\ 11} \\ \cancel{5}\cancel{1} \\ -1\ 2 \\ \hline 3\ 9 \end{array}$

   b) $\begin{array}{r} {}^{10\ 11} \\ {}^{0\ \cancel{0}\ \cancel{1}\ 13} \\ \cancel{1}\cancel{1}\cancel{2}\cancel{3} \\ -\ 3\ 4\ 5 \\ \hline 7\ 7\ 8 \end{array}$

   c) $\begin{array}{r} {}^{9\ \ 9\ \ 9} \\ {}^{3\ \cancel{10}\ \cancel{10}\ \cancel{10}\ 11} \\ \cancel{4}\cancel{0},\cancel{0}\cancel{0}\cancel{1} \\ -1\ 5,\ 3\ 3\ 3 \\ \hline 2\ 4,\ 6\ 6\ 8 \end{array}$

5. Earl has $729 in his checking account. After he writes a check to the bookstore for $249, how much is remaining in his account?

Solution

$$
\begin{array}{r}
\overset{6\ \ 12}{\$7\!\!\!/\,2\!\!\!/\,9} \\
-\ 2\ 4\ 9 \\
\hline
\$4\ 8\ 0
\end{array}
$$

Earl will have $480 remaining in his account.

1.4 Multiplying Whole Numbers

Learning Objectives:

1. Do chain multiplication.
2. Multiply by single-digit numbers.
3. Use multiplication shortcuts for numbers ending in zero.
4. Multiply by numbers having more than one digit.
5. Solve application problems.

Solved Examples:

1. Multiply.

   a) $5 \cdot 6 \cdot 0$ b) $(2)(3)(7)$

   Solution

   a) $5 \cdot 6 \cdot 0 = (5 \cdot 6) \cdot 0 = 30 \cdot 0 = 0$ b) $(2)(3)(7) = (2 \cdot 3)(7) = 6(7) = 42$

          or $5(6 \cdot 0) = 5 \cdot 0 = 0$ or $2(3 \cdot 7) = 2(21) = 42$

2. Multiply.

   a) $\begin{array}{r} 51 \\ \times\ 2 \\ \hline \end{array}$ b) $\begin{array}{r} 1123 \\ \times\ 5 \\ \hline \end{array}$ c) $\begin{array}{r} 40,001 \\ \times\ 3 \\ \hline \end{array}$

   Solution

   a) $\begin{array}{r} 51 \\ \times\ 2 \\ \hline 102 \end{array}$ b) $\begin{array}{r} {}^{11} \\ 1123 \\ \times\ 5 \\ \hline 5615 \end{array}$ c) $\begin{array}{r} 40,001 \\ \times\ 3 \\ \hline 120,003 \end{array}$

3. Multiply by powers of 10.

   a) 2×10 b) 2×100 c) 2×1000

   Solution

   a) $2 \times 10 = 20$ (one zero)   b) $2 \times 100 = 200$ (two zeros)   c) $2 \times 1000 = 2000$ (three zeros)

4. Multiply.

a) 18
    × 22

b) 534
    × 54

c) 4302
    × 107

d) 160
    × 200

Solution

a) 18
    × 22
    ────
    36
    36
    ────
    396

b) 534
    × 54
    ────
    2136
    2670
    ──────
    28,836

c) 4302
    × 107
    ─────
    30114
    0000
    4302
    ───────
    460,314

d) 160
    × 200
    ────
    0
    0
    320
    ─────
    32000

5. a) Jenny pays $275 per month for her car payment. How much does she pay per year?

   b) Frank bought 8 CDs at $15 each. How much did he pay total?

Solution

a) $275
    × 12
    ────
    550
    275
    ────
    $3300

Jenny will pay $3300 per year in car payments.

b)   4
     15
    × 8
    ────
    120

Frank paid $120 total.

1.5 Dividing Whole Numbers

Learning Objectives:

1. Write division problems using other symbols.
2. Divide, if possible.
3. Use short division.
4. Solve application problems.

Solved Examples:

1. Write each division problem using two other symbols.

   a) $6 \div 1$

   b) $7 \overline{)49}$

   c) $9 \div 0$

   d) $\dfrac{56}{8}$

   Solution

   a) $\dfrac{6}{1}$ or $1 \overline{)6}$

   b) $\dfrac{49}{7}$ or $49 \div 7$

   c) $\dfrac{9}{0}$ or $0 \overline{)9}$

   d) $56 \div 8$ or $8 \overline{)56}$

2. Divide. If needed, show remainder with R next to quotient.

   a) $9 \overline{)189}$

   b) $5 \overline{)4255}$

   c) $6 \overline{)51}$

   d) $4 \overline{)1290}$

   Solution

   a)
   $$\begin{array}{r} 21 \\ 9\overline{)189} \\ \underline{18} \\ 09 \\ \underline{9} \\ 0 \end{array}$$

   b)
   $$\begin{array}{r} 851 \\ 5\overline{)4255} \\ \underline{40} \\ 25 \\ \underline{25} \\ 05 \\ \underline{5} \\ 0 \end{array}$$

   c)
   $$\begin{array}{r} 8\,\text{R}\,3 \\ 6\overline{)51} \\ \underline{48} \\ 3 \end{array}$$

   d)
   $$\begin{array}{r} 322\,\text{R}\,2 \\ 4\overline{)1290} \\ \underline{12} \\ 09 \\ \underline{8} \\ 10 \\ \underline{8} \\ 2 \end{array}$$

3. How could you check the answer to this division problem? $6 \overline{)428}$ with quotient $71\,\text{R}2$

   Solution

   $(6 \times 71) + 2 = 426 + 2 = 428$

   The division problem is correct.

4. Which number(s) are divisible by 3?

   a) 6151   b) 7290   c) 32,333

Solution

a) $6 + 1 + 5 + 1 = 13$; 13 is not divisible by 3, so 6151 is not divisible by 3.

b) $7 + 2\ 9 + 0 = 18$; 18 is divisible by 3, so 7290 is divisible by 3.

c) $3 + 2 + 3 + 3 + 3 = 14$; 14 is not divisible by 3, so 32,333 is not divisible by 3.

5. Which number(s) are divisible by 5?

   a) 462   b) 5095   c) 99,990

Solution

a) 462 ends in 2, so it is not divisible by 5.

b) 5095 ends in 5, so it is divisible by 5.

c) 99,990 ends in 0, so it is divisible by 5.

1.6 Long Division

Learning Objectives:

1. Select appropriate answers.
2. Do long division.
3. Solve application problems.

Solved Examples:

1. Divide. If needed, show remainder with R next to quotient.

   a) $61\overline{)488}$

   b) $23\overline{)2272}$

   c) $13\overline{)9360}$

   Solution

   a)
   $$\begin{array}{r} 8 \\ 61\overline{)488} \\ \underline{488} \\ 0 \end{array}$$

   b)
   $$\begin{array}{r} 98\ R\ 18 \\ 23\overline{)2272} \\ \underline{207} \\ 202 \\ \underline{184} \\ 18 \end{array}$$

   c)
   $$\begin{array}{r} 720 \\ 13\overline{)9360} \\ \underline{91} \\ 26 \\ \underline{26} \\ 00 \end{array}$$

2. Divide.

   a) $90 \div 10$

   b) $3700 \div 100$

   c) $613,000 \div 1000$

   Solution

   a) There is one zero in the divisor, so $90 \div 10 = 9$.

   b) There are two zeros in the divisor, so $3700 \div 100 = 37$.

   c) There are three zeros in the divisor, so $613,000 \div 1000 = 613$.

3. Rewrite each of the following by dropping zeros.

   a) $20\overline{)460}$

   b) $30\overline{)19,000}$

   c) $6500\overline{)49,800}$

   Solution

   a) $20\overline{)460} = 2\overline{)46}$

   b) $30\overline{)19,000} = 3\overline{)19,00}$

   c) $6500\overline{)49,800} = 65\overline{)498}$

4. Use multiplication to check the division answers.

a)
$$319\overline{)35,561} \quad 111 \text{ R142}$$

b)
$$760\overline{)47,509} \quad 67 \text{ R207}$$

Solution

a)
$$
\begin{array}{r}
111 \\
\times\ 319 \\
\hline
999 \\
111 \\
333 \\
\hline
35409 \\
+\ \ \ 142 \\
\hline
35,551
\end{array}
$$

The answer is correct.

b)
$$
\begin{array}{r}
760 \\
\times\ \ 67 \\
\hline
5320 \\
4560 \\
\hline
50920 \\
+\ \ \ 207 \\
\hline
51,127
\end{array}
$$

The answer does not check. Rework the original problem.

1.7 Rounding Whole Numbers

Learning Objectives:

1. Round numbers.
2. Estimate by rounding and find exact answers.
3. Solve application problems.

Solved Examples:

1. Round to the nearest ten.

   a) 212   b) 3,487   c) 14

   Solution

   a) To round to the nearest ten, we look at the digit in the ones place. The digit in the ones place is 2, which is less than 5. Therefore, we keep the digit in the tens place and replace all digits to the right of the tens place by 0.
   $212 = 210$ to the nearest ten.

   b) To round to the nearest ten, we look at the digit in the ones place. The digit in the ones place is 7, which is greater than or equal to 5. Therefore, we increase the digit in the tens place by 1 and replace all digits to the right of the tens place by 0.
   $3487 = 3490$ to the nearest ten.

   c) To round to the nearest ten, we look at the digit in the ones place. The digit in the ones place is 4, which is less than 5. Therefore, we keep the digit in the tens place and replace all digits to the right of the tens place by 0.
   $14 = 10$ to the nearest ten.

2. Round to the nearest hundred.

   a) 312   b) 1,267   c) 83

   Solution

   a) To round to the nearest hundred, we look at the digit in the tens place. The digit in the tens place is 1, which is less than 5. Therefore, we keep the digit in the hundreds place and replace all digits to the right of the hundreds place by 0.
   $312 = 300$ to the nearest hundred.

   b) To round to the nearest hundred, we look at the digit in the tens place. The digit in the tens place is 6, which is greater than or equal to 5. Therefore, we increase the digit in the hundreds place by 1 and replace all digits to the right of the hundreds place by 0.
   $1267 = 1300$ to the nearest hundred.

   c) To round to the nearest hundred, we look at the digit in the tens place. The digit in the tens place is 8, which is greater than or equal to 5. Therefore, we increase the digit in the hundreds place by 1 and replace all digits to the right of the hundreds place by 0.
   $83 = 100$ to the nearest hundred.

3. Round to the nearest thousand.

   a) 3,549   b) 677   c) 27,217

Solution

a) To round to the nearest thousand, we look at the digit in the hundreds place. The digit in the hundreds place is 5, which is greater than or equal to 5. Therefore, we increase the digit in the thousands place by 1 and replace all digits to the right of the thousands place by 0. $3549 = 4000$ to the nearest thousand.

b) To round to the nearest thousand, we look at the digit in the hundreds place. The digit in the hundreds place is 6, which is greater than or equal to 5. Therefore, we increase the digit in the thousands place by 1 and replace all digits to the right of the thousands place by 0. $677 = 1000$ to the nearest thousand.

c) To round to the nearest thousand, we look at the digit in the hundreds place. The digit in the hundreds place is 2, which is less than 5. Therefore, we keep the digit in the thousands place and replace all digits to the right of the thousands place by 0. $27,217 = 27,000$ to the nearest thousand.

4. Use front end rounding to find an estimate for each calculation.

   a) $57 + 24 + 88 + 71$ b) 867×72 c) $4,357 \div 213$

Solution

a)
$$
\begin{array}{rcr}
57 & \to & 60 \\
24 & & 20 \\
88 & & 90 \\
+\,71 & & +\,70 \\
\hline
& & 240
\end{array}
$$

We estimate the sum to be 240.

b)
$$
\begin{array}{rcr}
867 & \to & 900 \\
\times\ 72 & & \times\ 70 \\
\hline
& & 63,000
\end{array}
$$

We estimate the product to be 63,000.

c)
$$
213\overline{)4357} \quad \to \quad 200\overline{)4000}^{\,20}
$$

We estimate the quotient to be 20.

1.8 Exponents, Roots, and Order of Operations

Learning Objectives:

1. Identify an exponent and a base.
2. Find the square root of a number.
3. Use the order of operations.

Solved Examples:

1. Write each number in exponential form.

   a) $4 \times 4 \times 4$
   b) $2 \times 2 \times 2 \times 2$
   c) $1 \times 1 \times 1 \times 1 \times 1$
   d) 8

   Solution

   a) $4 \times 4 \times 4 = 4^3$
   b) $2 \times 2 \times 2 \times 2 = 2^4$
   c) $1 \times 1 \times 1 \times 1 \times 1 = 1^5$
   d) $8 = 8^1$

2. Evaluate the expressions.

   a) 3^2
   b) 2^3
   c) 7^1
   d) 9^0

   Solution

   a) $3^2 = 3 \times 3 = 9$
   b) $2^3 = 2 \times 2 \times 2 = 8$
   c) $7^1 = 7$
   d) $9^0 = 1$

3. Evaluate using the correct order of operations.

   a) $5 \times 7 - 6$

   b) $8 \times 3^2 - 6 \div 2$

   c) $5 \times (4 - 3) + 5^2$

   d) $15 \div 3 \times 8 \times 9 \div (12 - 2^3)$

   e) $2^2 + 3^3 + 1^2$

   f) $3 \cdot \sqrt{49} - 7 \cdot 2 + \dfrac{0}{6}$

   Solution

   a)
   $$5 \times 7 - 6 = 35 - 6 \quad \text{Multiply}$$
   $$= 29 \quad \text{Subtract}$$

   b)
   $$8 \times 3^2 - 6 \div 2 = 8 \times 9 - 6 \div 2 \quad \text{Exponents}$$
   $$= 72 - 6 \div 2 \quad \text{Multiplication}$$
   $$= 72 - 3 \quad \text{Division}$$
   $$= 69 \quad \text{Subtraction}$$

   c)
   $$5 \times (4 - 3) + 5^2$$
   $$= 5 \times 1 + 5^2 \quad \text{Inside parentheses}$$
   $$= 5 \times 1 + 25 \quad \text{Exponents}$$
   $$= 5 + 25 \quad \text{Multiplication}$$
   $$= 30 \quad \text{Addition}$$

   d)
   $$15 \div 3 \times 8 \times 9 \div \left(12 - 2^3\right)$$
   $$= 15 \div 3 \times 8 \times 9 \div (12 - 8) \quad \text{Exponent in parentheses}$$
   $$= 15 \div 3 \times 8 \times 9 \div 4 \quad \text{Subtract in parentheses}$$
   $$= 5 \times 8 \times 9 \div 4 \quad \text{Divide}$$
   $$= 40 \times 9 \div 4 \quad \text{Multiply}$$
   $$= 360 \div 4 \quad \text{Multiply}$$
   $$= 90 \quad \text{Divide}$$

e) $2^2 + 3^2 + 1^2$

$= 4 + 9 + 1$ Exponents

$= 13 + 1$ Add

$= 14$ Add

f) $3 \cdot \sqrt{49} - 7 \cdot 2 + \dfrac{0}{6}$

$= 3 \cdot 7 - 7 \cdot 2 + \dfrac{0}{6}$ Square root

$= 21 - 7 \cdot 2 + \dfrac{0}{6}$ Multiplication

$= 21 - 14 + \dfrac{0}{6}$ Multiplication

$= 21 - 14 + 0$ Division

$= 7 + 0$ Subtraction

$= 7$ Addition

1.9 Reading Pictographs, Bar Graphs, and Line Graphs

Learning Objectives:

1. Read and understand a pictograph.
2. Read and understand a bar graph.
3. Read and understand a line graph.

Solved Examples:

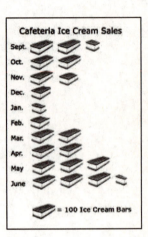

1. The pictograph at the right shows the number of ice cream bar sales during the year 2008. Answer the questions below based on this graph.

   a) Which month had the least sales?

   b) Which month had the most sales?

   c) Approximately how many more ice cream bars were sold in the month of May compared to the month of January?

   d) What was the approximate sales of ice cream bars in September?

   Solution

   a) From the pictograph, January had the least sales.

   b) From the pictograph, June had the most sales.

   c) May has two more symbols than January, so $2 \cdot 100 = 200$ more ice cream bars were sold in May.

   d) There are $2\frac{1}{2}$ symbols, so $2 \cdot 100 + \frac{1}{2} \cdot 100 = 200 + 50 = 250$ ice cream bars were sold in September.

2. The bar graph at the right shows the sales, in dollars, for cookie brands A, B and C.

   a) What was the approximate sales for cookie brand A?

   b) What was the approximate sales for cookie brand B?

   c) By how much did cookie brand A sales exceed sales of cookie brand C?

   Solution

   a) From the bar graph, the sales for cookie brand A were approximately $80,000.

   b) From the bar graph, the sales for cookie brand B were approximately $60,000.

c) According to the bar graph, cookie brand A sales exceeded sales of cookie brand C by
$80,000 - $20,000 = $60,000.

1.10 Solving Application Problems

Learning Objectives:

1. Solve application problems.

Solved Examples:

1. Solve problems involving one type of operation.

   a) Students at a party ordered 8 large pizzas. Each pizza is cut into 6 slices. How many total slices are there?

   b) If a new amusement park covers 54 acres and there are 44,010 square feet in 1 acre, how many square feet of land does the amusement park cover?

Solution

a) (1) Understand the problem.

Mathematics Blueprint for Problem Solving			
Gather the Facts	What Am I Asked to Do?	How Do I Proceed?	Key Points to Remember
8 pizzas were ordered. Each pizza had 6 slices.	Find the total number of slices.	Multiply the amounts.	Write your answer in a complete sentence.

   (2) Solve and state the answer:

$$\begin{array}{r} 8 \\ \times\, 6 \\ \hline 48 \end{array}$$

There are 48 slices of pizza in all.

   (3) **Check**: Estimate to see if the answer is reasonable.

b) (1) Understand the problem.

Mathematics Blueprint for Problem Solving			
Gather the Facts	What Am I Asked to Do?	How Do I Proceed?	Key Points to Remember
The park covers 54 acres. Each acre contains 44,010 square feet.	Find the number of square feet covered by the park.	Multiply the two values.	The final answer will be in square feet.

(2) Solve and state the answer:

$$
\begin{array}{r}
44,010 \\
\times \quad\quad 54 \\
\hline
176040 \\
220050 \\
\hline
2,376,540
\end{array}
$$

The amusement park covers 2,376,540 square feet of land.

(3) **Check**: Estimate to see if the answer is reasonable.

2. Solve problems involving more than one type of operation.

   a) Elizabeth bought six beach towels for $15 each, eight bottles of sunscreen for $5 each, and 2 pairs of sunglasses for $23 each. How much did she spend in all?

   b) Enrico wants to determine the miles-per-gallon rating of he Ford Mustang. He filled the tank when the odometer read 38,861 miles. After ten days, the odometer read 39,437 miles and the tank required 18 gallons to be filled. How many miles per gallon did Enrico's car achieve?

Solution

a) (1) Understand the problem.

Mathematics Blueprint for Problem Solving			
Gather the Facts	What Am I Asked to Do?	How Do I Proceed?	Key Points to Remember
6 towels at $15 each. 8 bottles of sunscreen at $5 each. 2 sunglasses at $23 each.	Find the total amount that Lisette spent.	Multiply each quantity by the selling price, then add the totals. We can do this with an imaginary bill of sale.	The final answer will have units of dollars. Remember order of operations.

(2) Solve and state the answer:

Bill of Sale

Item	Quantity	Price	Amount
Towel	6	$15	$6 \times \$15 = \$90$
Sunscreen	8	$5	$8 \times \$5 = \$40$
Sunglasses	2	$23	$2 \times \$23 = \$46$
		Total	$\$90 + \$40 + \$46 = \$176$

Elizabeth spent a total of $176.

(3) **Check**: Estimate to see if the answer is reasonable.

b) (1) Understand the problem.

Mathematics Blueprint for Problem Solving			
Gather the Facts	What Am I Asked to Do?	How Do I Proceed?	Key Points to Remember
Odometer after 10 days: 39,437 miles Odometer at beginning: 38,861 miles 18 gallons of gas were used.	Find the number of miles per gallon for Egbichi's car.	(a) Subtract the odometer readings to find the number of miles traveled. (b) Divide the result from (a) by the number of gallons of gas.	Final units are in miles per gallon. Remember to watch order of operations if doing multiple steps.

    (2) Solve and state the answer:

$$\text{(a)} \quad \begin{array}{r} 39,437 \\ -\ 38,861 \\ \hline 576 \end{array} \qquad \text{(b)} \quad \begin{array}{r} 32 \\ 18\overline{)576} \\ \underline{54} \\ 36 \\ \underline{36} \\ 0 \end{array}$$

Enrico's car achieved 32 miles per gallon.

(3) **Check**: Estimate to see if the answer is reasonable.

2.1 Basics of Fractions

Learning Objectives:

1. Write fractions to represents diagrams or word problems.
2. Identify the numerator and denominator.
3. Identify proper and improper fractions.
4. Analyze part/whole relationships.

Solved Examples:

1. Use a fraction to represent the shaded part of the object or objects.

   a)

   b)

   Solution

   a) Seven out of fifteen equal parts are shaded. The fraction is $\dfrac{7}{15}$.

   b) Five out of six diamonds are shaded. The fraction is $\dfrac{5}{6}$.

2. Draw a sketch to illustrate the fractional part.

   a) $\dfrac{4}{7}$ of an object

   b) $\dfrac{3}{4}$ of an object

   Solution

   a) We draw a rectangular bar and divide it into 7 equal parts. We then shade 4 parts to show $\dfrac{4}{7}$.

   b) We draw a rectangular bar and divide it into 4 equal parts. We then shade 3 parts to show $\dfrac{3}{4}$.

3. Name the numerator and denominator in each fraction and classify each as proper or improper.

   a) $\dfrac{1}{5}$

   b) $\dfrac{2}{3}$

   c) $\dfrac{17}{15}$

   Solution

   a) The numerator is 1; the denominator is 5. The numerator is smaller than the denominator, so $\dfrac{1}{5}$ is a proper fraction.

   b) The numerator is 2; the denominator is 3. The numerator is smaller than the denominator, so $\dfrac{2}{3}$ is a proper fraction.

   c) The numerator is 17; the denominator is 15. The numerator is larger than the denominator, so $\dfrac{17}{15}$ is an improper fraction.

4. Charlie drove 43 minutes to go to a Celtics game. 17 minutes of his trip were spent in bumper-to-bumper traffic. What fractional part of his time was spent in bumper-to-bumper traffic?

Solution

Charlie spent 17 of 43 minutes in bumper-to-bumper traffic, so he spent $\frac{17}{43}$ of his time in bumper-to-bumper traffic.

2.2 Mixed Numbers

Learning Objectives:

1. Convert between mixed numbers and improper fractions.
2. Identify fractions as proper or improper.

Solved Examples:

1. Change each mixed number to an improper fraction.

a) $2\dfrac{3}{4}$

b) $6\dfrac{2}{9}$

c) $1\dfrac{43}{58}$

d) $103\dfrac{4}{5}$

Solution

a) $2\dfrac{3}{4} = \dfrac{4 \times 2 + 3}{4} = \dfrac{8 + 3}{4} = \dfrac{11}{4}$

b) $6\dfrac{2}{9} = \dfrac{9 \times 6 + 2}{9} = \dfrac{54 + 2}{9} = \dfrac{56}{9}$

c) $1\dfrac{43}{58} = \dfrac{58 \times 1 + 43}{58} = \dfrac{58 + 43}{58} = \dfrac{101}{58}$

d) $103\dfrac{4}{5} = \dfrac{5 \times 103 + 4}{5} = \dfrac{515 + 4}{5} = \dfrac{519}{5}$

2. Change each improper fraction to a mixed number or a whole number.

a) $\dfrac{8}{5}$

b) $\dfrac{81}{9}$

c) $\dfrac{48}{11}$

d) $\dfrac{196}{9}$

Solution

a) Divide 8 by 5.

$$\begin{array}{r} 1 \leftarrow \text{quotient} \\ 5\overline{)8} \\ \underline{5} \\ 3 \leftarrow \text{remainder} \end{array}$$

Thus, $\dfrac{8}{5} = 1\dfrac{3}{5}$.

b) Divide 81 by 9.

$$\begin{array}{r} 9 \leftarrow \text{quotient} \\ 5\overline{)81} \\ \underline{81} \\ 0 \leftarrow \text{remainder} \end{array}$$

Thus, $\dfrac{81}{9} = 9$.

c) Divide 48 by 11.

$$
\begin{array}{r}
4 \quad \leftarrow \text{quotient} \\
11\overline{)48} \\
44 \\
\overline{4} \quad \leftarrow \text{remainder}
\end{array}
$$

Thus, $\dfrac{48}{11} = 4\dfrac{4}{11}$.

d) Divide 196 by 9.

$$
\begin{array}{r}
21 \quad \leftarrow \text{quotient} \\
9\overline{)196} \\
18 \\
\overline{16} \\
9 \\
\overline{7} \quad \leftarrow \text{remainder}
\end{array}
$$

Thus, $\dfrac{196}{9} = 21\dfrac{7}{9}$.

2.3 Factors

Learning Objectives:

1. Find factors of a number.
2. Identify prime numbers.
3. Find prime factorizations.

Solved Examples:

1. Find all whole number factors of each number.

   a) 18

   b) 50

   c) 109

   d) 44

   Solution

   a) $1 \cdot 18 = 18$
      $2 \cdot 9 = 18$
      $3 \cdot 6 = 18$
      The factors of 18 are 1, 2, 3, 6, 9, and 18.

   b) $1 \cdot 50 = 50$
      $2 \cdot 25 = 50$
      The factors of 50 are 1, 2, 25, and 50.

   c) $1 \cdot 109 = 109$
      The factors of 109 are 1 and 109.

   d) $1 \cdot 44 = 44$
      $2 \cdot 22 = 44$
      $4 \cdot 11 = 44$
      The factors of 44 are 1, 2, 11, 22, and 44.

2. Write each number as a product of prime factors.

   a) 6

   b) 20

   c) 30

   d) 84

   Solution

   a) Write 6 as the product of two factors: $6 = 2 \times 3$. We check to see whether each factor is prime. Since both 2 and 3 are prime numbers, we are finished: $6 = 2 \times 3$.

   b) To start, write 20 as the product of any two factors. We will write 20 as 4×5. We check to see whether the factors are prime. If not, we factor them. Since $4 = 2 \times 2$, we have:

   $$20 = \quad 4 \quad \times 5$$
   $$2 \times 2 \times 5$$

   The bottom row of factors are all prime numbers, so the prime factorization of 20 is $20 = 2 \times 2 \times 5$ or $2^2 \times 5$.

c) To start, write 30 as the product of any two factors. We will write 30 as 6×5.
We check to see whether the factors are prime. If not, we factor them. Since $6 = 2 \times 3$, we have:

$$30 = \quad 6 \quad \times 5$$
$$2 \times 3 \times 5$$

The bottom row of factors are all prime numbers, so the prime factorization of 30 is $30 = 2 \times 3 \times 5$.

d) To start, write 84 as the product of any two factors. We will write 84 as 4×21.
We check to see whether the factors are prime. If not, we factor them. Since $4 = 2 \times 2$ and $21 = 3 \times 7$, we have:

$$84 = \quad 4 \quad \times \quad 21$$
$$2 \times 2 \times 3 \times 7$$

The bottom row of factors are all prime numbers, so the prime factorization of 84 is $84 = 2 \times 2 \times 3 \times 7$ or $2^2 \times 3 \times 7$.

3. Determine whether each number is prime or composite. If it is composite, write it as the product of prime factors.

a) 62   b) 13   c) 89   d) 95

Solution

a) 62 is divisible by 2 and 31, so 62 is composite.
$62 = 2 \cdot 31$

b) 13 in only divisible by 1 and itself, so 13 is prime.

c) 89 is only divisible by 1 and itself, so 89 is prime.

d) 95 is divisible by 5 and 19, so 95 is composite.
$95 = 5 \cdot 19$

2.4 Writing a Fraction in Lowest Terms

Learning Objectives:

1. Test divisibility.
2. Write a fraction in lowest terms using common factors.
3. Write a fraction in lowest terms using prime factors.
4. Determine whether two fractions are equivalent.

Solved Examples:

1. Decide whether the given factor is a common factor of both numbers.

   a) 8, 20; 4   b) 21, 35; 5   c) 64, 86; 2   d) 88, 122; 11

   Solution

   a) 4 is a factor of both 8 and 20, so it is a common factor of 8 and 20.

   b) 5 is not a factor of 21, so it is not a common factor of 21 and 35.

   c) 2 is a factor of both 64 and 86, so it is a common factor of 64 and 86.

   d) 11 is not a factor of 122, so it is not a common factor of 88 and 122.

2. Reduce each fraction using common factors.

   a) $\dfrac{5}{10}$ b) $\dfrac{16}{64}$ c) $\dfrac{42}{77}$ d) $\dfrac{88}{90}$

   Solution

   a) $\dfrac{5}{10} = \dfrac{5 \div 5}{10 \div 5} = \dfrac{1}{2}$ b) $\dfrac{16}{64} = \dfrac{16 \div 16}{64 \div 16} = \dfrac{1}{4}$ c) $\dfrac{42}{77} = \dfrac{42 \div 7}{77 \div 7} = \dfrac{6}{11}$ d) $\dfrac{88}{90} = \dfrac{88 \div 2}{90 \div 2} = \dfrac{44}{45}$

3. Reduce each fraction by the method of prime factors.

   a) $\dfrac{4}{16}$ b) $\dfrac{15}{35}$

   c) $\dfrac{27}{72}$ d) $\dfrac{66}{77}$

   Solution

   a) $\dfrac{4}{16} = \dfrac{2 \times 2}{2 \times 2 \times 2 \times 2} = \dfrac{\cancel{2} \times \cancel{2}}{2 \times 2 \times \cancel{2} \times \cancel{2}} = \dfrac{1}{4}$ b) $\dfrac{15}{35} = \dfrac{3 \times 5}{5 \times 7} = \dfrac{3 \times \cancel{5}}{\cancel{5} \times 7} = \dfrac{3}{7}$

   c) $\dfrac{27}{72} = \dfrac{3 \times 3 \times 3}{2 \times 2 \times 2 \times 3 \times 3} = \dfrac{3 \times \cancel{3} \times \cancel{3}}{2 \times 2 \times 2 \times \cancel{3} \times \cancel{3}} = \dfrac{3}{8}$ d) $\dfrac{66}{77} = \dfrac{2 \times 3 \times 11}{7 \times 11} = \dfrac{2 \times 3 \times \cancel{11}}{7 \times \cancel{11}} = \dfrac{6}{7}$

4. Determine whether the fractions are equal.

   a) $\dfrac{1}{3}$ and $\dfrac{2}{6}$

   b) $\dfrac{21}{33}$ and $\dfrac{7}{13}$

   c) $\dfrac{4}{10}$ and $\dfrac{24}{60}$

   d) $\dfrac{18}{54}$ and $\dfrac{24}{48}$

Solution

   a) We determine the cross product:

   $$1 \times 2 \to 3 \times 2 = 6$$
   $$3 \times 6 \to 1 \times 6 = 6$$

   Since $6 = 6$, we know that $\dfrac{1}{3} = \dfrac{2}{6}$.

   b) We determine the cross product:

   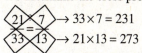
   $$21 \times 7 \to 33 \times 7 = 231$$
   $$33 \times 13 \to 21 \times 13 = 273$$

   Since $231 \neq 273$, we know that $\dfrac{21}{33} \neq \dfrac{7}{13}$.

   c) We determine the cross product:

   $$4 \times 24 \to 10 \times 24 = 240$$
   $$10 \times 60 \to 4 \times 24 = 240$$

   Since $240 = 240$, we know that $\dfrac{4}{10} = \dfrac{24}{60}$.

   d) We determine the cross product:

   $$18 \times 24 \to 54 \times 24 = 1296$$
   $$54 \times 48 \to 18 \times 48 = 864$$

   Since $1296 \neq 864$, we know that $\dfrac{18}{54} \neq \dfrac{24}{48}$.

2.5 Multiplying Fractions

Learning Objectives:

1. Multiply with fractions.
2. Solve area problems.

Solved Examples:

1. Multiply the fractions. Make sure all final answers are simplified.

    a) $\dfrac{1}{2} \cdot \dfrac{3}{4}$

    b) $\dfrac{5}{9} \cdot \dfrac{3}{15}$

    c) $\dfrac{5}{6} \cdot \dfrac{9}{2}$

    d) $\dfrac{5}{24} \cdot \dfrac{36}{25}$

    Solution

    a) $\dfrac{1}{2} \cdot \dfrac{3}{4} = \dfrac{1 \cdot 3}{2 \cdot 4} = \dfrac{3}{8}$

    b) $\dfrac{5}{9} \cdot \dfrac{3}{15} = \dfrac{5 \cdot 3}{9 \cdot 15} = \dfrac{15}{135}$

    Reduce this result:

    $$\dfrac{15}{135} = \dfrac{3 \cdot 5}{3 \cdot 3 \cdot 3 \cdot 5} = \dfrac{\overset{1}{\cancel{3}} \cdot \overset{1}{\cancel{5}}}{3 \cdot 3 \cdot \underset{1}{\cancel{3}} \cdot \underset{1}{\cancel{5}}} = \dfrac{1}{9}.$$

    Thus, $\dfrac{5}{9} \times \dfrac{3}{15} = \dfrac{1}{9}$.

    c) $\dfrac{5}{6} \cdot \dfrac{9}{2} = \dfrac{5 \cdot 9}{6 \cdot 2} = \dfrac{45}{12}$

    Reduce this result:

    $$\dfrac{45}{12} = \dfrac{3 \cdot 3 \cdot 5}{2 \cdot 2 \cdot 3} = \dfrac{\overset{1}{\cancel{3}} \cdot 3 \cdot 5}{2 \cdot 2 \cdot \underset{1}{\cancel{3}}} = \dfrac{15}{4}.$$

    Thus, $\dfrac{5}{6} \times \dfrac{9}{2} = \dfrac{15}{4}$ or $3\dfrac{3}{4}$.

    d) $\dfrac{3}{8} \cdot \dfrac{14}{15} = \dfrac{3}{2 \cdot 2 \cdot 2} \cdot \dfrac{2 \cdot 7}{3 \cdot 5}$

    $$= \dfrac{2 \cdot 3 \cdot 7}{2 \cdot 2 \cdot 2 \cdot 3 \cdot 5}$$

    $$= \dfrac{\overset{1}{\cancel{2}} \cdot \overset{1}{\cancel{3}} \cdot 7}{\underset{1}{\cancel{2}} \cdot 2 \cdot 2 \cdot \underset{1}{\cancel{3}} \cdot 5} = \dfrac{7}{20}$$

2. Multiply the fractions and whole numbers. Make sure all final answers are simplified.

    a) $7 \cdot \dfrac{3}{5}$

    b) $\dfrac{4}{9} \cdot 18$

    c) $1 \cdot \dfrac{18}{36}$

    d) $\dfrac{4}{5} \cdot 25 \cdot \dfrac{35}{16}$

Solution

a) $7 \times \dfrac{3}{5} = \dfrac{7}{1} \times \dfrac{3}{5} = \dfrac{7 \times 3}{1 \times 5} = \dfrac{21}{5}$ or $4\dfrac{1}{5}$

b) $\dfrac{4}{9} \times 18 = \dfrac{4}{9} \times \dfrac{18}{1} = \dfrac{4}{\cancel{9}_{1}} \times \dfrac{\cancel{18}^{2}}{1} = \dfrac{8}{1} = 8$

c) $1 \times \dfrac{18}{36} = \dfrac{1}{1} \times \dfrac{18}{36} = \dfrac{18}{36} = \dfrac{18 \div 18}{36 \div 18} = \dfrac{1}{2}$

d) $\dfrac{4}{5} \times 25 \times \dfrac{35}{16} = \dfrac{\cancel{4}^{1}}{\cancel{5}_{1}} \times \dfrac{\cancel{25}^{5}}{1} \times \dfrac{35}{\cancel{16}_{4}} = \dfrac{175}{4}$ or $43\dfrac{3}{4}$

3. Find the area of each rectangle.

a) Length of $\dfrac{5}{9}$ inch and width of $\dfrac{3}{20}$ inch.

b) Length of $\dfrac{12}{25}$ meter and width of $\dfrac{5}{16}$ meter.

Solution

a) $A = \dfrac{5}{9} \times \dfrac{3}{20} = \dfrac{\cancel{5}^{1}}{\cancel{9}_{3}} \times \dfrac{\cancel{3}^{1}}{\cancel{20}_{4}} = \dfrac{1}{12}$

The area is $\dfrac{1}{12}$ square inch.

b) $A = \dfrac{12}{25} \times \dfrac{5}{16} = \dfrac{\cancel{12}^{3}}{\cancel{25}_{5}} \times \dfrac{\cancel{5}^{1}}{\cancel{16}_{4}} = \dfrac{3}{20}$

The area is $\dfrac{3}{20}$ square meter.

2.6 Applications of Multiplication

Learning Objectives:

1. Solve fraction application problems using multiplication.

Solved Examples:

1. A computer screen measures $\frac{7}{18}$ yard by $\frac{9}{14}$ yard. Find the area of the screen.

   Solution

   $$A = \frac{7}{18} \times \frac{9}{14} = \frac{\overset{1}{\cancel{7}}}{\underset{2}{\cancel{18}}} \times \frac{\overset{1}{\cancel{9}}}{\underset{2}{\cancel{14}}} = \frac{1}{4}$$

   The area is $\frac{1}{4}$ square yard.

2. A parking lot has 560 parking slots of which $\frac{3}{28}$ are classified as *reserved*. How many slots are reserved?

   Solution

   $$560 \times \frac{3}{28} = \frac{\overset{20}{\cancel{560}}}{1} \times \frac{3}{\underset{1}{\cancel{28}}} = 60$$

   There are 60 reserved slots.

3. Jodi bought a car for $15,500. After one year, the car was worth $\frac{4}{5}$ of its original price. What was the car worth after one year?

   Solution

   $$15,500 \times \frac{4}{5} = \frac{\overset{3100}{\cancel{15,500}}}{1} \times \frac{4}{\cancel{5}} = 12,400$$

   The car is worth $12,400 after one year.

4. A family of 4 spends $\frac{3}{20}$ of their total income on taxes. If the family's total income is $98,600, how much is spent on taxes?

Solution

$$98,600 \times \frac{3}{20} = \frac{\overset{4930}{\cancel{98,600}}}{1} \times \frac{3}{\underset{1}{\cancel{20}}} = 14,790$$

$14,790 is spent on taxes.

2.7 Dividing Fractions

Learning Objectives:

1. Find the reciprocal of a fraction.
2. Divide fractions.
3. Solve application problems.

Solved Examples:

1. Find the reciprocal of each number;

   a) 3

   b) $\dfrac{1}{2}$

   c) $\dfrac{9}{5}$

   d) 60

   Solution

   a) $\dfrac{1}{3}$

   b) 2

   c) $\dfrac{5}{9}$

   d) $\dfrac{1}{60}$

2. Divide the fractions. Make sure all final answers are simplified.

   a) $\dfrac{1}{2} \div \dfrac{3}{4}$

   b) $\dfrac{5}{9} \div \dfrac{3}{15}$

   c) $\dfrac{5}{6} \div \dfrac{9}{2}$

   d) $\dfrac{5}{24} \div \dfrac{36}{24}$

   Solution

   a) $\dfrac{1}{2} \div \dfrac{3}{4} = \dfrac{1}{2} \times \dfrac{4}{3} = \dfrac{1}{\cancel{2}} \times \dfrac{\overset{2}{\cancel{4}}}{3} = \dfrac{2}{3}$

   b) $\dfrac{5}{9} \div \dfrac{3}{15} = \dfrac{5}{9} \times \dfrac{15}{3} = \dfrac{5}{\cancel{9}} \times \dfrac{\overset{5}{\cancel{15}}}{3} = \dfrac{25}{9}$ or $2\dfrac{7}{9}$

   c) $\dfrac{5}{6} \div \dfrac{9}{2} = \dfrac{5}{6} \times \dfrac{2}{9} = \dfrac{5}{\cancel{6}} \times \dfrac{\overset{1}{\cancel{2}}}{9} = \dfrac{5}{27}$

   d) $\dfrac{5}{24} \div \dfrac{36}{24} = \dfrac{5}{24} \times \dfrac{24}{36} = \dfrac{5}{\cancel{24}} \times \dfrac{\overset{1}{\cancel{24}}}{36} = \dfrac{5}{36}$

3. Divide the fractions and whole numbers. Make sure all final answers are simplified.

   a) $7 \div \dfrac{3}{5}$

   b) $\dfrac{4}{9} \div 0$

   c) $\dfrac{2}{\frac{18}{36}}$

   d) $\dfrac{\frac{4}{5}}{28}$

Solution

a) $7 \div \dfrac{3}{5} = \dfrac{7}{1} \div \dfrac{3}{5} = \dfrac{7}{1} \times \dfrac{5}{3} = \dfrac{35}{3}$ or $11\dfrac{2}{3}$

b) $\dfrac{4}{9} \div 0$ is undefined.

c) $\dfrac{\dfrac{2}{18}}{36} = 2 \div \dfrac{18}{36} = \dfrac{2}{1} \times \dfrac{36}{18} = \dfrac{2}{1} \times \dfrac{\overset{2}{\cancel{36}}}{\underset{1}{\cancel{18}}} = \dfrac{4}{1} = 4$

d) $\dfrac{\dfrac{4}{5}}{28} = \dfrac{4}{5} \div \dfrac{28}{1} = \dfrac{4}{5} \times \dfrac{1}{28} = \dfrac{\overset{1}{\cancel{4}}}{5} \times \dfrac{1}{\underset{7}{\cancel{28}}} = \dfrac{1}{35}$

4. Irving drove his car to Worcester, a distance of 50 miles, in $\dfrac{5}{6}$ hours. What was his average speed (in miles per hour)?

Solution

We need to divide 50 miles by $\dfrac{5}{6}$ hour to find the average speed.

$50 \div \dfrac{5}{6} = \dfrac{50}{1} \div \dfrac{5}{6} = \dfrac{\overset{10}{\cancel{50}}}{1} \cdot \dfrac{6}{\underset{1}{\cancel{5}}} = \dfrac{60}{1} = 60$

Thus, the average speed is 60 miles per hour.

2.8 Multiplying and Dividing Mixed Numbers

Learning Objectives:

1. Multiply mixed numbers
2. Divide mixed numbers.
3. Solve application problems.

Solved Examples:

1. Multiply the mixed numbers.

   a) $3\frac{1}{2} \cdot 1\frac{1}{4}$

   b) $2\frac{3}{4} \cdot \frac{8}{9}$

   c) $\frac{6}{6} \cdot 8\frac{3}{4}$

   d) $7\frac{1}{2} \cdot 10\frac{2}{15}$

Solution

a) $3\frac{1}{2} \cdot 1\frac{1}{4} = \frac{7}{2} \cdot \frac{5}{4} = \frac{7 \cdot 5}{2 \cdot 4} = \frac{35}{8}$ or $4\frac{3}{8}$

b) $2\frac{3}{4} \cdot \frac{8}{9} = \frac{11}{4} \cdot \frac{8}{9} = \frac{11}{\overset{}{\underset{1}{\cancel{4}}}} \cdot \frac{\overset{2}{\cancel{8}}}{9} = \frac{22}{9}$ or $2\frac{4}{9}$

c) $6 \cdot 8\frac{3}{4} \cdot \frac{1}{6} = \frac{6}{1} \cdot \frac{35}{4} \cdot \frac{1}{6} = \frac{\overset{1}{\cancel{6}}}{1} \cdot \frac{35}{4} \cdot \frac{1}{\underset{1}{\cancel{6}}} = \frac{35}{4}$ or $8\frac{3}{4}$

d) $7\frac{1}{2} \cdot 10\frac{2}{15} = \frac{15}{2} \cdot \frac{152}{15} = \frac{\overset{1}{\cancel{15}}}{\underset{1}{\cancel{2}}} \cdot \frac{\overset{76}{\cancel{152}}}{\underset{1}{\cancel{15}}} = \frac{76}{1} = 76$

2. Divide the mixed numbers.

   a) $3\frac{1}{2} \div 1\frac{1}{6}$

   b) $\frac{5}{3} \div 8\frac{3}{4}$

   c) $1\frac{2}{3} \div 8\frac{3}{4}$

   d) $7\frac{1}{2} \div 1\frac{3}{4}$

Solution

a) $3\frac{1}{2} \div 1\frac{1}{6} = \frac{7}{2} \div \frac{7}{6} = \frac{7}{2} \times \frac{6}{7} = \frac{\overset{1}{\cancel{7}}}{\underset{1}{\cancel{2}}} \times \frac{\overset{3}{\cancel{6}}}{\underset{1}{\cancel{7}}} = \frac{3}{1} = 3$

b) $\frac{5}{3} \div 8\frac{3}{4} = \frac{5}{3} \div \frac{35}{4} = \frac{5}{3} \times \frac{4}{35} = \frac{\overset{1}{\cancel{5}}}{3} \times \frac{4}{\underset{7}{\cancel{35}}} = \frac{4}{21}$

c) $1\frac{2}{3} \div 8\frac{3}{4} = \frac{5}{3} \div \frac{35}{4} = \frac{5}{3} \cdot \frac{4}{35} = \frac{\overset{1}{\cancel{5}}}{3} \cdot \frac{4}{\underset{7}{\cancel{35}}} = \frac{4}{21}$

d) $7\frac{1}{2} \div 1\frac{3}{4} = \frac{15}{2} \div \frac{7}{4} = \frac{15}{\underset{1}{\cancel{2}}} \cdot \frac{\overset{2}{\cancel{4}}}{7} = \frac{30}{7}$ or $4\frac{2}{7}$

3. Jason drove his car to Evansville, a distance of 200 miles, in $4\frac{1}{6}$ hours. What was his average speed (in miles per hour)?

Solution

We need to divide 200 miles by $4\frac{1}{6}$ hours to find the average speed.

$$200 \div 4\frac{1}{6} = \frac{200}{1} \div \frac{25}{6} = \frac{\overset{8}{\cancel{200}}}{1} \cdot \frac{6}{\underset{1}{\cancel{25}}} = \frac{48}{1} = 48$$

Thus, the average speed is 48 miles per hour.

4. Jody is using a recipe that calls for $\frac{1}{4}$ cup of milk per batch. If she has $5\frac{3}{4}$ cups of milk available, how many batches can she make?

Solution

We must divide the amount of milk available by the amount of milk needed per batch. That is, we must find $5\frac{3}{4} \div \frac{1}{4}$.

$$5\frac{3}{4} \div \frac{1}{4} = \frac{23}{4} \div \frac{1}{4} = \frac{23}{4} \times \frac{4}{1} = \frac{23}{\underset{1}{\cancel{4}}} \times \frac{\overset{1}{\cancel{4}}}{1} = 23$$

Therefore, Jody can make 23 batches.

5. What is the area of a rectangular garden that is $11\frac{1}{2}$ yards long and $12\frac{1}{6}$ yards wide?

Solution

The area of a rectangle equals the product of its length and width.

$$\left(11\frac{1}{2}\right)\left(12\frac{1}{6}\right) = \frac{23}{2} \cdot \frac{73}{6} = \frac{1679}{12} = 139\frac{11}{12}$$

Thus, the area of the rectangle is $139\frac{11}{12}$ square yards.

2.9 Adding and Subtracting Like Fractions

Learning Objectives:

1. Add like fractions.
2. Subtract like fractions.
3. Solve application problems.

Solved Examples:

1. Which of the following are like fractions?

   a) $\dfrac{7}{12}, \dfrac{11}{12}$

   b) $\dfrac{8}{18}, \dfrac{1}{18}$

   c) $\dfrac{24}{25}, \dfrac{7}{24}$

   d) $\dfrac{5}{7}, \dfrac{5}{9}, \dfrac{5}{13}$

   Solution

   Like fractions have the same denominator, so $\dfrac{7}{12}$ and $\dfrac{11}{12}$ are like fractions and $\dfrac{8}{18}$ and $\dfrac{1}{18}$ are like fractions.

2. Add. Simplify all answers.

   a) $\dfrac{3}{8}+\dfrac{2}{8}$

   b) $\dfrac{3}{14}+\dfrac{4}{14}$

   c) $\dfrac{1}{10}+\dfrac{3}{10}+\dfrac{4}{10}$

   d) $\dfrac{3}{8}+\dfrac{2}{8}+\dfrac{3}{8}$

   Solution

   a) $\dfrac{3}{8}+\dfrac{2}{8}=\dfrac{3+2}{8}=\dfrac{5}{8}$

   b) $\dfrac{3}{14}+\dfrac{4}{14}=\dfrac{3+4}{14}=\dfrac{7}{14}=\dfrac{7\div 7}{14\div 7}=\dfrac{1}{2}$

   c) $\dfrac{1}{10}+\dfrac{3}{10}+\dfrac{4}{10}=\dfrac{1+3+4}{10}=\dfrac{8}{10}=\dfrac{8\div 2}{10\div 2}=\dfrac{4}{5}$

   d) $\dfrac{3}{8}+\dfrac{2}{8}+\dfrac{3}{8}=\dfrac{8}{8}=1$

3. Subtract. Simplify all answers.

   a) $\dfrac{2}{3}-\dfrac{1}{3}$

   b) $\dfrac{9}{15}-\dfrac{3}{15}$

   c) $\dfrac{7}{9}-\dfrac{4}{9}$

   d) $\dfrac{43}{81}-\dfrac{16}{81}$

   Solution

   a) $\dfrac{2}{3}-\dfrac{1}{3}=\dfrac{2-1}{3}=\dfrac{1}{3}$

   b) $\dfrac{9}{15}-\dfrac{3}{15}=\dfrac{9-3}{15}=\dfrac{6}{15}=\dfrac{6\div 3}{15\div 3}=\dfrac{2}{5}$

c) $\dfrac{7}{9} - \dfrac{4}{9} = \dfrac{3}{9} = \dfrac{3 \div 3}{9 \div 3} = \dfrac{1}{3}$

d) $\dfrac{43}{81} - \dfrac{16}{81} = \dfrac{27}{81} = \dfrac{27 \div 27}{81 \div 27} = \dfrac{1}{3}$

4. Amanda started running on Monday. She ran $\dfrac{3}{16}$ mile. On Wednesday she ran $\dfrac{7}{16}$ mile. How many miles has she run so far this week?

Solution

We must add the two distances: $\dfrac{3}{16} + \dfrac{7}{16} = \dfrac{10}{16} = \dfrac{10 \div 2}{16 \div 2} = \dfrac{5}{8}$

Thus, Amanda has run $\dfrac{5}{8}$ mile so far this week.

5. A highway paving contractor has $\dfrac{11}{12}$ of a highway left to pave. If he paves $\dfrac{5}{12}$ of the highway this week, how much is left to pave?

Solution

We must subtract the two distances: $\dfrac{11}{12} - \dfrac{5}{12} = \dfrac{6}{12} = \dfrac{6 \div 6}{12 \div 6} = \dfrac{1}{2}$

Thus, $\dfrac{1}{2}$ of the highway is left to pave.

2.10 Least Common Multiples

Learning Objectives:

1. Find the least common multiple using multiples of the largest number.
2. Find the least common multiple using any method.
3. Rewrite fractions with the indicated denominator.
4. Find least multiple of the denominators.

Solved Examples:

1. Find the LCM for each pair of numbers using any method.

    a) 4 and 3   b) 6 and 15

    c) 8 and 48   d) 24 and 36

Solution

a) The multiples of 4 are 4, 8, 12, 16, 20, …
The multiples of 3 are 3, 6, 9, 12, 15, …
The first multiple that appears on both lists is the least common multiple. Thus, the number 12 is the LCM of 4 and 3.

b) The multiples of 6 are 6, 12, 18, 24, 30,…
The multiples of 15 are 15, 30, 45, 60, 75,…
The first multiple that appears on both lists is the least common multiple. Thus, the number 30 is the LCM of 6 and 15.

c) Because $8 \times 6 = 48$, we know that 48 is a multiple of 8. Therefore, we can state immediately that the LCM of 8 and 48 is 48.

d) Multiples of 24: 24, 48, 72, 96, 120, …
Multiples of 36: 36, 72, 108, 144, 180, …
The first multiple that appears on both lists is the least common multiple. Thus, the number 72 is the LCM of 24 and 36.

2. Find the LCM of the denominators of each pair of fractions using any method.

    a) $\dfrac{7}{9}$, $\dfrac{5}{6}$ b) $\dfrac{3}{15}$, $\dfrac{13}{20}$ c) $\dfrac{7}{12}$, $\dfrac{5}{18}$

    d) $\dfrac{1}{9}$, $\dfrac{5}{7}$, $\dfrac{14}{63}$ e) $\dfrac{7}{8}$, $\dfrac{9}{14}$, $\dfrac{11}{16}$ f) $\dfrac{1}{20}$, $\dfrac{13}{16}$, $\dfrac{3}{4}$

Solution

a) Write each denominator as the product of primes:
$9 = 3 \times 3$ and $6 = 2 \times 3$.
The LCM will contain the factor 2 once and the factor 3 twice. Therefore,
$LCM = 2 \times 3 \times 3 = 18$.

b) Write each denominator as the product of primes:
$15 = 3 \times 5$ and $20 = 2 \times 2 \times 5$.
The LCM will contain the factor 2 twice, the factor 3 once, and the factor 5 once. Therefore,
$LCM = 2 \times 2 \times 3 \times 5 = 60$.

c) Write each denominator as the product of primes:
$12 = 2 \times 2 \times 3$ and $18 = 2 \times 3 \times 3$.
The LCM will contain the factor 2 twice and the factor 3 twice. Therefore,
$LCM = 2 \times 2 \times 3 \times 3 = 36$.

d) Write each denominator as the product of primes: $9 = 3 \times 3$, 7 is prime, and $63 = 3 \times 3 \times 7$. The LCM will contain the factor 3 twice and the factor 7 once. Therefore, $LCM = 3 \times 3 \times 7 = 63$.

e) Write each denominator as the product of primes: $8 = 2 \times 2 \times 2$, $14 = 2 \times 7$, and $16 = 2 \times 2 \times 2 \times 2$. The LCM will contain the factor 2 four times and the factor 7 once. Thus, $LCM = 2 \times 2 \times 2 \times 2 \times 7 = 112$

f) Write each denominator as the product of primes: $20 = 2 \times 2 \times 5$, $16 = 2 \times 2 \times 2 \times 2$, and $4 = 2 \times 2$. The LCM will contain the factor 2 four times and the factor 5 once. Thus, $LCM = 2 \times 2 \times 2 \times 2 \times 5 = 80$.

3. In the following problems, the divisions have already been worked out. Find the least common multiple.

a)
$$
\begin{array}{r|cc}
2 & 12 & 16 \\
\hline
2 & 6 & 8 \\
\hline
2 & \not{3} & 4 \\
\hline
2 & \not{3} & 2 \\
\hline
3 & 3 & \not{1} \\
\hline
& 1 & 1
\end{array}
$$

b)
$$
\begin{array}{r|ccc}
3 & 9 & 21 & \not{35} \\
\hline
3 & 3 & \not{7} & 35 \\
\hline
5 & \not{1} & \not{7} & 35 \\
\hline
7 & \not{1} & 7 & 7 \\
\hline
& 1 & 1 & 1
\end{array}
$$

Solution

a) LCM = $2 \cdot 2 \cdot 2 \cdot 2 \cdot 3 = 48$

b) LCM = $3 \cdot 3 \cdot 5 \cdot 7 = 315$

4. Rewrite each fraction with the indicated denominator.

a) $\dfrac{1}{4} = \dfrac{?}{12}$

b) $\dfrac{5}{7} = \dfrac{?}{49}$

c) $\dfrac{3}{8} = \dfrac{?}{32}$

Solution

a) $\dfrac{1}{4} \times \dfrac{c}{c} = \dfrac{?}{12}$
We know that $4 \times 3 = 12$, so the value c that we multiply the numerator and denominator by is 3. Therefore,
$\dfrac{1}{4} = \dfrac{1}{4} \times \dfrac{3}{3} = \dfrac{3}{12}$

b) $\dfrac{5}{7} \times \dfrac{c}{c} = \dfrac{?}{49}$
We know that $7 \times 7 = 49$, so the value c that we multiply the numerator and denominator by is 7. Therefore,
$\dfrac{5}{7} = \dfrac{5}{7} \times \dfrac{7}{7} = \dfrac{35}{49}$

c) $\dfrac{3}{8} = \dfrac{?}{32}$
We know that $8 \times 4 = 32$, so we multiply the numerator and denominator by 4. So,
$\dfrac{3}{8} = \dfrac{3}{8} \times \dfrac{4}{4} = \dfrac{12}{32}$

2.11 Adding and Subtracting Unlike Fractions

Learning Objectives:

1. Add unlike fractions.
2. Subtract unlike fractions.
3. Solve application problems.

Solved Examples:

1. Add. Simplify all answers.

a) $\dfrac{2}{4}+\dfrac{1}{8}$

b) $\dfrac{5}{6}+\dfrac{7}{8}$

c) $\dfrac{7}{12}+\dfrac{9}{30}$

d) $\dfrac{2}{3}+\dfrac{2}{24}+\dfrac{1}{6}$

Solution

a) $\dfrac{2}{4}+\dfrac{1}{8}=\dfrac{2}{4}\cdot\dfrac{2}{2}+\dfrac{1}{8}=\dfrac{4}{8}+\dfrac{1}{8}=\dfrac{5}{8}$

b) LCD = 24

$\dfrac{5}{6}=\dfrac{5}{6}\cdot\dfrac{4}{4}=\dfrac{20}{24}$; $\dfrac{7}{8}=\dfrac{7}{8}\cdot\dfrac{3}{3}=\dfrac{21}{24}$

$\dfrac{5}{6}+\dfrac{7}{8}=\dfrac{20}{24}+\dfrac{21}{24}=\dfrac{41}{24}$ or $1\dfrac{17}{24}$

c) LCD = 60

$\dfrac{7}{12}=\dfrac{7}{12}\cdot\dfrac{5}{5}=\dfrac{35}{60}$; $\dfrac{9}{30}=\dfrac{9}{30}\cdot\dfrac{2}{2}=\dfrac{18}{60}$

$\dfrac{7}{12}+\dfrac{9}{30}=\dfrac{35}{60}+\dfrac{18}{60}=\dfrac{53}{60}$

d) LCD = 24

$\dfrac{2}{3}=\dfrac{2}{3}\cdot\dfrac{8}{8}=\dfrac{16}{24}$; $\dfrac{2}{24}=\dfrac{2}{24}$; $\dfrac{1}{6}=\dfrac{1}{6}\cdot\dfrac{4}{4}=\dfrac{4}{24}$

$\dfrac{2}{3}+\dfrac{2}{24}+\dfrac{1}{6}=\dfrac{16}{24}+\dfrac{2}{24}+\dfrac{4}{24}=\dfrac{22}{24}=\dfrac{11}{12}$

2. Subtract. Simplify all answers.

a) $\dfrac{3}{4}-\dfrac{5}{8}$

b) $\dfrac{2}{3}-\dfrac{3}{16}$

c) $\dfrac{5}{6}-\dfrac{8}{12}$

d) $\dfrac{4}{5}-\dfrac{8}{10}$

Solution

a) $\dfrac{3}{4}-\dfrac{5}{8}=\dfrac{3}{4}\cdot\dfrac{2}{2}-\dfrac{5}{8}=\dfrac{6}{8}-\dfrac{5}{8}=\dfrac{1}{8}$

b) LCD = 48

$\dfrac{2}{3}=\dfrac{2}{3}\cdot\dfrac{16}{16}=\dfrac{32}{48}$; $\dfrac{3}{16}=\dfrac{3}{16}\cdot\dfrac{3}{3}=\dfrac{9}{48}$

$\dfrac{2}{3}-\dfrac{3}{16}=\dfrac{32}{48}-\dfrac{9}{48}=\dfrac{23}{48}$

c) $\dfrac{5}{6} - \dfrac{8}{12} = \dfrac{5}{6} \cdot \dfrac{2}{2} - \dfrac{8}{12} = \dfrac{10}{12} - \dfrac{8}{12} = \dfrac{2}{12} = \dfrac{1}{6}$

d) $\dfrac{4}{5} - \dfrac{8}{10} = \dfrac{4}{5} \cdot \dfrac{2}{2} - \dfrac{8}{10} = \dfrac{8}{10} - \dfrac{8}{10} = \dfrac{0}{10} = 0$

3. At *Big-M University*, $\dfrac{1}{4}$ of the students major in engineering and $\dfrac{3}{20}$ major in accounting. If the two disciplines are combined, what is the result?

Solution

We must add the two disciplines. $\dfrac{1}{4} + \dfrac{3}{20} = \dfrac{1}{4} \cdot \dfrac{5}{5} + \dfrac{3}{20} = \dfrac{5}{20} + \dfrac{3}{20} = \dfrac{8}{20} = \dfrac{2}{5}$

The combined result is $\dfrac{2}{5}$.

4. $\dfrac{1}{2}$ of a farm is devoted to raising corn, and $\dfrac{1}{5}$ of the farm is devoted to raising soybeans. The farmer decides to reduce the amount of land producing corn by the amount used to produce soybeans. How much will be devoted to corn production?

Solution

We must subtract $\dfrac{1}{2} - \dfrac{1}{5}$ to find the amount of land that will be devoted to corn production

The LCD of $\dfrac{1}{2}$ and $\dfrac{1}{5}$ is 10.

$\dfrac{1}{2} \cdot \dfrac{5}{5} = \dfrac{5}{10}; \qquad \dfrac{1}{5} \cdot \dfrac{2}{2} = \dfrac{2}{10}$

$\dfrac{1}{2} - \dfrac{1}{5} = \dfrac{5}{10} - \dfrac{2}{10} = \dfrac{3}{10}$

Thus, $\dfrac{3}{10}$ of the farm will be devoted to corn production.

2.12 Adding and Subtracting Mixed Numbers

Learning Objectives:

1. Add mixed numbers.
2. Subtract mixed numbers.
3. Add or subtract by changing mixed numbers to improper fractions.
4. Solve application problems.

Solved Examples:

1. Add or subtract. Express the answer as a mixed or whole number. Simplify all answers.

a) $8\frac{3}{10}+1\frac{1}{10}$

b) $8\frac{7}{15}+3\frac{3}{10}$

c) $12-4\frac{3}{8}$

d) $16\frac{2}{9}-10\frac{11}{12}$

Solution

a)
$$8\frac{3}{10}$$
$$+1\frac{1}{10}$$

Add the whole numbers $\rightarrow 9\frac{4}{10} \leftarrow$ Add the fractions

$$= 9\frac{2}{5} \leftarrow \text{Reduce the fraction}$$

b) The LCD of $\frac{7}{15}$ and $\frac{3}{10}$ is 30.

$$\frac{7}{15}=\frac{7}{15}\times\frac{2}{2}=\frac{14}{30}; \quad \frac{3}{10}=\frac{3}{10}\times\frac{3}{3}=\frac{9}{30}$$

$$8\frac{7}{15} = 8\frac{14}{30}$$
$$+ 3\frac{3}{10} = + 3\frac{9}{30}$$

Add the whole numbers $\rightarrow 11\frac{23}{30} \leftarrow$ Add the fractions

c) The LCD is 8. We must borrow 1 from 12 to obtain $11\frac{8}{8}$.

$$12 = 11\frac{8}{8}$$
$$- 4\frac{3}{8} = - 4\frac{3}{8}$$

Subtract the whole numbers $\rightarrow 7\frac{5}{8} \leftarrow$ Subtract the fractions

d) The LCD of $\frac{2}{9}$ and $\frac{11}{12}$ is 36.

$$\frac{2}{9}=\frac{2}{9}\times\frac{4}{4}=\frac{8}{36}; \quad \frac{11}{12}=\frac{11}{12}\times\frac{3}{3}=\frac{33}{36}$$

We cannot subtract $\frac{8}{36}-\frac{33}{36}$, so we will need to borrow 1 from 16 to obtain

$$16\frac{2}{9}=16\frac{8}{36}=15+1\frac{8}{36}=15+\frac{44}{36}=15\frac{44}{36}$$

$$16\frac{2}{9} = 16\frac{8}{36} = 15\frac{44}{36}$$
$$- 10\frac{11}{12} = -10\frac{33}{36} = - 10\frac{33}{36}$$

Subtract the whole numbers $\rightarrow 5\frac{11}{36} \leftarrow$ Subtract the fractions

2.   Jan walked $49\frac{1}{5}$ yards and Mark walked $21\frac{3}{10}$ yards. How much farther did Jan walk?

Solution

We need to subtract $49\frac{1}{5} - 21\frac{3}{10}$ to find how much further Jan walked.

The LCD of $\frac{1}{5}$ and $\frac{3}{10}$ is 10.

$$\frac{1}{5} = \frac{1}{5} \times \frac{2}{2} = \frac{2}{10}$$

We cannot subtract $\frac{2}{10} - \frac{3}{10}$, so we will need to borrow 1 from 49 to obtain

$$49\frac{1}{5} = 49\frac{2}{10} = 48 + 1\frac{2}{10} = 48 + \frac{12}{10} = 48\frac{12}{10}$$

$$
\begin{array}{ccccc}
49\frac{1}{5} & = & 49\frac{2}{10} & = & 48\frac{12}{10} \\
-\ 21\frac{3}{10} & = & -\ 21\frac{3}{10} & = & -\ 21\frac{3}{10} \\
\end{array}
$$

Subtract the whole numbers $\rightarrow 27\frac{9}{10} \leftarrow$ Subtract the fractions

Jan walked $27\frac{9}{10}$ yards farther than Mark.

2.13 Order Relations and the Order of Operations

Learning Objectives:

1. Locate fractions on the number line.
2. Identify the greater of two fractions.
3. Use exponents with fractions.
4. Use the order of operations.
5. Solve application problems.
6. Simplify.

Solved Examples:

1. Write the following by using the symbols < and >.

   a) $\dfrac{1}{2}$ _____ $\dfrac{4}{5}$

   b) $\dfrac{5}{4}$ _____ $\dfrac{13}{9}$

   c) $\dfrac{9}{13}$ _____ $\dfrac{5}{12}$

   Solution

   a) The LCD is 10.

   $\dfrac{1}{2} = \dfrac{5}{10}$ and $\dfrac{4}{5} = \dfrac{8}{10}$

   $\dfrac{5}{10} < \dfrac{8}{10}$, so $\dfrac{1}{2} < \dfrac{4}{5}$.

   b) The LCD is 36.

   $\dfrac{5}{4} = \dfrac{45}{36}$ and $\dfrac{13}{9} = \dfrac{52}{36}$

   $\dfrac{45}{36} < \dfrac{52}{36}$, so $\dfrac{5}{4} < \dfrac{13}{9}$.

   c) The LCD is 156.

   $\dfrac{9}{13} = \dfrac{108}{156}$ and $\dfrac{5}{12} = \dfrac{65}{156}$

   $\dfrac{108}{156} > \dfrac{65}{156}$, so $\dfrac{9}{13} > \dfrac{5}{12}$.

2. Evaluate using the correct order of operations.

   a) $\left(\dfrac{2}{3}\right)^2$

   b) $\left(\dfrac{1}{2}\right)^2$

   c) $\dfrac{4}{5} - \dfrac{1}{2} \cdot \dfrac{6}{5}$

   d) $\dfrac{1}{4} \cdot \dfrac{3}{4} + \dfrac{3}{8} \div \dfrac{3}{4}$

   e) $\left(\dfrac{5}{6} - \dfrac{7}{12}\right) \cdot \dfrac{7}{9}$

   f) $\dfrac{4}{3} \div \left(\dfrac{3}{5} - \dfrac{3}{10}\right)$

   g) $\left(\dfrac{1}{3}\right)^2 \cdot \dfrac{27}{30}$

   h) $\left(\dfrac{2}{3}\right)^2 \div \dfrac{5}{9}$

   i) $\left(\dfrac{7}{15} - \dfrac{3}{10}\right)^2$

   Solution

   a) $\left(\dfrac{2}{3}\right)^2 = \dfrac{2}{3} \cdot \dfrac{2}{3} = \dfrac{4}{9}$

   b) $\left(\dfrac{1}{2}\right)^2 = \dfrac{1}{2} \cdot \dfrac{1}{2} = \dfrac{1}{4}$

   c) $\dfrac{4}{5} - \dfrac{1}{2} \cdot \dfrac{6}{5} = \dfrac{4}{5} - \dfrac{1}{\underset{1}{\cancel{2}}} \cdot \dfrac{\overset{3}{\cancel{6}}}{5}$

   $= \dfrac{4}{5} - \dfrac{3}{5} = \dfrac{1}{5}$

d) $\dfrac{1}{4} \cdot \dfrac{3}{4} + \dfrac{3}{8} \div \dfrac{3}{4}$

$= \dfrac{3}{16} + \dfrac{3}{8} \div \dfrac{3}{4}$

$= \dfrac{3}{16} + \dfrac{\overset{1}{\cancel{3}}}{\underset{2}{\cancel{8}}} \cdot \dfrac{\overset{1}{\cancel{4}}}{\underset{1}{\cancel{3}}}$

$= \dfrac{3}{16} + \dfrac{1}{2} \quad \text{LCD is 16}$

$= \dfrac{3}{16} + \dfrac{8}{16}$

$= \dfrac{11}{16}$

e) $\left(\dfrac{5}{6} - \dfrac{7}{12}\right) \cdot \dfrac{7}{9} \quad \text{LCD is 12}$

$= \left(\dfrac{10}{12} - \dfrac{7}{12}\right) \cdot \dfrac{7}{9}$

$= \dfrac{\overset{1}{\cancel{3}}}{\underset{4}{\cancel{12}}} \cdot \dfrac{7}{9}$

$= \dfrac{7}{36}$

f) $\dfrac{4}{3} \div \left(\dfrac{3}{5} - \dfrac{3}{10}\right) \quad \text{LCD is 10}$

$= \dfrac{4}{3} \div \left(\dfrac{6}{10} - \dfrac{3}{10}\right)$

$= \dfrac{4}{3} \div \dfrac{3}{10}$

$= \dfrac{4}{3} \div \dfrac{10}{3}$

$= \dfrac{40}{9} = 4\dfrac{4}{9}$

g) $\left(\dfrac{1}{3}\right)^2 \cdot \dfrac{27}{30}$

$= \dfrac{1}{3} \cdot \dfrac{1}{3} \cdot \dfrac{27}{30}$

$= \dfrac{1}{\cancel{9}} \cdot \dfrac{\overset{\cancel{3}}{\cancel{27}}}{\underset{10}{\cancel{30}}}$

$= \dfrac{1}{10}$

h) $\left(\dfrac{2}{3}\right)^2 \div \dfrac{5}{9}$

$= \left(\dfrac{2}{3}\right)^2 \div \dfrac{5}{9}$

$= \dfrac{2}{3} \cdot \dfrac{2}{3} \div \dfrac{5}{9}$

$= \dfrac{4}{9} \div \dfrac{5}{9}$

$= \dfrac{4}{\cancel{9}} \cdot \dfrac{\overset{1}{\cancel{9}}}{5}$

i) $\left(\dfrac{7}{15} - \dfrac{3}{10}\right)^2$

$= \left(\dfrac{14}{30} - \dfrac{9}{30}\right)^2$

$= \left(\dfrac{5}{30}\right)^2$

$= \dfrac{25}{900} = \dfrac{1}{36}$

3.1 Reading and Writing Decimals

Learning Objectives:

1. Identify the digit with a given place value.
2. Write a decimal number given specified place values.
3. Write decimals as fractions or mixed numbers.
4. Write decimals in words.
5. Write decimals in numbers.

Solved Examples:

1. Write a word name for each decimal.

    a) 0.35   b) 0.007   c) 4.12

    d) 32.1   e) $144.25   f) $34,256.09

Solution

a) 0.35   thirty-five *hundredths*   b) 0.007   seven *thousandths*   c) 4.12   four and twelve *hundredths*

d) 32.1   thirty-two and one *tenth*   e) $144.25 one hundred forty-four and $\frac{25}{100}$ dollars   f) $34,256.09 thirty-four thousand, two hundred fifty-six and $\frac{9}{100}$ dollars

2. Write in decimal notation.

    a) five tenths   b) thirty three hundredths

    c) seven hundred sixty four thousandths   d) six and eleven ten-thousandths

Solution

a) five *tenths* $\rightarrow \dfrac{5}{10}$

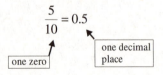

b) thirty-three *hundredths* $\rightarrow \dfrac{33}{100}$

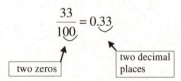

c) seven hundred sixty-four *thousandths*

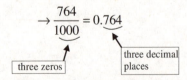

d) six and eleven *ten-thousandths*

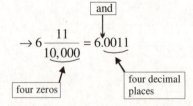

3. Change each decimal into a fraction. Reduce whenever possible.

    a) 0.3   b) 0.8

    c) 3.25   d) 122.004

Solution

    a) $0.3 = \dfrac{3}{10}$ b) $0.8 = \dfrac{8}{10} = \dfrac{4}{5}$

    c) $3.25 = 3\dfrac{25}{100} = 3\dfrac{1}{4}$ d) $122.004 = 122\dfrac{4}{1000} = 122\dfrac{1}{250}$

3.2 Rounding Decimals

Learning Objectives:

1. Round decimals to a given place.
2. Round decimals in application problems.

Solved Examples:

1   Round each number to the place indicated.

a)   5.38 to the nearest tenth

b)   0.753 to the nearest tenth

c)   103.843 to the nearest hundredth

d)   7.385 to the nearest one

e)   19.1299 to the nearest thousandth

f) $5.247 to the nearest cent

g) $819.983 to the nearest cent

h) $5.247 to the nearest dollar

i) $819.983 to the nearest dollar

Solution

a)   5.38

⌐ We locate the hundredths place.
Since the digit to the right of tenths is greater than 5, we round up (increase the value of the tenths place by one) and drop all digits to the right. The answer is 5.4 .

b)   0.753

⌐ We locate the hundredths place.
Since the digit to the right of tenths is 5, we round up (increase the value of the tenths place by one) and drop all digits to the right. The answer is 0.8 .

c)   103.843

⌐ We locate the thousandths place.
Since the digit to the right of hundredths is less than 5, we drop all digits to the right. The answer is 103.84 .

d)   7.385

⌐ We locate the tenths place.
Since the digit to the right of ones is less than 5, we drop all digits to the right. The answer is 7.

e)   19.1299

⌐ We locate the ten-thousandths place.
Since the digit to the right of thousandths is greater than 5, we round up (increase the value of the thousandths place by one) and drop all digits to the right. Since the digit in the thousandths place was 9 and we increase by 1, the digit becomes 0 and we add 1 to the digit in the hundredths place. The answer is 19.130 .

f) $5.247

⌐ We locate the thousandths place.
Since the digit to the right of hundredths is greater than 5, we round up (increase the value of the hundredths place by one) and drop all digits to the right. The answer is $5.25 .

g) $819.983

   ⌐ We locate the thousandths place.

   Since the digit to the right of hundredths is less than 5, we drop all digits to the right. The answer is $819.98 .

h) $5.247

   ⌐ We locate the tenths place.

   Since the digit to the right of ones is less than 5, we drop all digits to the right. The answer is $5.

i)   Since the digit to the right of hundredths is greater than 5, we round up (increase the value of the hundredths place by one) and drop all digits to the right. The answer is $5.25 .

   Since the digit to the right of ones is greater than 5, we round up (increase the value of the ones place by one) and drop all digits to the right. Since the digit in the ones place was 9 and we increase by 1, the digit becomes 0 and we add 1 to the digit in the tens place. The answer is $820.

2. *Pep-Me-Up* energy drink sells for $6.89 for a six-pack. So one bottle is $1.148\overline{3}. How much does one bottle cost rounded to the nearest cent?

Solution

$1.148\overline{3}

   ⌐ We locate the thousandths place.

Since the digit to the right of hundredths is greater than 5, we round up (increase the value of the hundredths place by one) and drop all digits to the right. One bottle costs $1.15 rounded to the nearest cent.

3. Emily's state income tax is $7564.23. Round this amount to the nearest dollar.

Solution

$7564.233

   ⌐ We locate the tenths place.

Since the digit to the right of ones is less than 5, we drop all digits to the right. Emily's state income tax rounded to the nearest dollar is $7564.

3.3 Adding and Subtracting Decimals

Learning Objectives:

1. Add or subtract decimals.
2. Solve application problems.
3. Add or subtract decimals with rounding.
4. Use estimation skills.

Solved Examples:

1. Add the following numbers.

   a) $56.3 + 12.2$ b) $56.3 + 19.8$ c) $1.665 + 9.888$

   d) $1.84 + 20.749$ e) $10 + 1.24$ f) $1.2 + 0.337 + 6$

Solution

a)
$$\begin{array}{r} 56.3 \\ +12.2 \\ \hline 68.5 \end{array}$$

b)
$$\begin{array}{r} {\scriptstyle 1\,1} \\ 56.3 \\ +19.8 \\ \hline 76.1 \end{array}$$

c)
$$\begin{array}{r} {\scriptstyle 1\,1\,1} \\ 1.665 \\ +9.888 \\ \hline 11.553 \end{array}$$

d)
$$\begin{array}{r} {\scriptstyle 1} \\ 1.840 \\ +20.749 \\ \hline 22.589 \end{array}$$

e)
$$\begin{array}{r} 10.00 \\ +1.24 \\ \hline 11.24 \end{array}$$

f)
$$\begin{array}{r} 1.200 \\ 0.337 \\ +6.000 \\ \hline 7.537 \end{array}$$

2. Subtract the following numbers.

   a) $48.7 - 42.3$ b) $48.7 - 2.9$ c) $30.44 - 16.3$

   d) $7 - 4.1$ e) $5.00725 - 1.06921$ f) $7238.24 - 6125.08$

Solution

a)
$$\begin{array}{r} 48.7 \\ -42.3 \\ \hline 6.4 \end{array}$$

b)
$$\begin{array}{r} {\scriptstyle 7\ 17} \\ 4\cancel{8}.\cancel{7} \\ -2.9 \\ \hline 45.8 \end{array}$$

c)
$$\begin{array}{r} {\scriptstyle 2\,10} \\ 3\cancel{0}.44 \\ -16.30 \\ \hline 14.14 \end{array}$$

d)
$$\begin{array}{r} {\scriptstyle 6\ 10} \\ \cancel{7}.\cancel{0} \\ -4.1 \\ \hline 2.9 \end{array}$$

e)
$$\begin{array}{r} {\scriptstyle 9\ \ 9} \\ {\scriptstyle 4\ 10\ 10\ 17} \\ \cancel{5}.\cancel{0}\,\cancel{0}\,\cancel{7}\,25 \\ -1.06921 \\ \hline 3.93804 \end{array}$$

f)
$$\begin{array}{r} {\scriptstyle 1\ 14} \\ 7238.2\cancel{4} \\ -6125.08 \\ \hline 1113.16 \end{array}$$

3. Add or subtract as needed to solve the following application problems.

   a) Jasmine drove on a summer trip. When she began, the odometer read 32,046.22 miles. When she was done, the odometer read 32,731.19 miles. How many miles did she travel during her trip?

   b) Keith purchased some clothing at the mall. He bought a shirt for $28.99, a pair of jeans for $39.99, and a watch for $24.50. How much did he spend?

Solution

a) Subtract the initial odometer reading from the final reading.

$$\begin{array}{r} \overset{6\ \ 2\ 10\ 11}{32,\cancel{7}\cancel{3}\cancel{1}.\cancel{1}9} \\ -\ 32,0\ 4\ 6.22 \\ \hline 6\ 8\ 4.97 \end{array}$$

Jasmine traveled 684.97 miles during her trip.

b) Add the amounts spent on the shirt, jeans, and watch.

$$\begin{array}{r} \overset{2\ 2\ 1}{\$28.99} \\ 39.99 \\ +\ 24.50 \\ \hline \$93.48 \end{array}$$

Keith spent a total of $93.48.

3.4 Multiplying Decimals

Learning Objectives:

1. Multiply decimals.
2. Solve application problems.

Solved Examples:

1. Multiply the following numbers.

    a) 2.2×6 b) 0.22×0.6

    c) 22×0.6 d) $\begin{array}{r} 0.581 \\ \times\ \ 2.9 \end{array}$

    e) $\begin{array}{r} 73.12 \\ \times\ 22.34 \end{array}$ f) $\begin{array}{r} 0.6288 \\ \times\ \ 5003 \end{array}$

Solution

    a) $\begin{array}{l} 2.2 \quad \text{1 decimal place} \\ \underline{\times\ 6} \quad \text{0 decimal places} \\ 13.2 \quad \text{1 decimal place } (1 + 0 = 1) \end{array}$

    b) $\begin{array}{l} 0.22 \quad \text{2 decimal places} \\ \underline{\times\ 0.6} \quad \text{1 decimal place} \\ 0.132 \quad \text{3 decimal places } (2 + 1 = 3) \end{array}$

    c) $\begin{array}{l} 22 \quad \text{0 decimal place} \\ \underline{\times\ 0.6} \quad \text{1 decimal place} \\ 13.2 \quad \text{1 decimal place } (0 + 1 = 1) \end{array}$

    d) $\begin{array}{l} 0.581 \quad \text{3 decimal places} \\ \underline{\times\ \ 2.9} \quad \text{1 decimal place} \\ 5229 \\ \underline{1162} \\ 1.6849 \quad \text{4 decimal places} \\ \qquad\quad (3 + 1 = 4) \end{array}$

    e) $\begin{array}{l} 73.12 \quad \text{2 decimal places} \\ \underline{\times\ \ \ 22.34} \quad \text{2 decimal places} \\ 29248 \\ 21936 \\ 14624 \\ \underline{14624} \\ 1633.5008 \quad \text{4 decimal places} \\ \qquad\qquad (2 + 2 = 4) \end{array}$

    f) $\begin{array}{l} 0.6288 \quad \text{4 decimal places} \\ \underline{\times\ \ \ 5003} \quad \text{0 decimal places} \\ 18864 \\ \underline{3144000} \\ 3145.8864 \quad \text{4 decimal places} \\ \qquad\qquad (4 + 0 = 4) \end{array}$

2. Use front-end rounding to round each problem and estimate the answer. Then find the exact answer.

    a) $\begin{array}{r} 48.3 \\ \times\ 7.9 \end{array}$ b) $\begin{array}{r} 32.05 \\ \times\ \ 96 \end{array}$

Solution

a) *Estimate:*   *Exact:*   b) *Estimate:*   *Exact:*

50	48.3	1 decimal place	30	32.05	1 decimal place
× 8	× 7.9	1 decimal place	× 100	× 96	1 decimal place
400	4347		3000	19230	
	3381			28845	
	381.57	2 decimal places		3076.80	2 decimal places
		(1 + 1 = 2)			(1 + 1 = 2)

3. a) A retail store purchases 100 sweaters at $35.65 each. How much did the store pay for the order?

     b) A college is purchasing carpeting for a new student lounge. What is the price of a carpet that is 21.3 yards wide and 180.4 yards long if the cost is $11.25 per square yard?

Solution

a) Multiply the number of sweaters by the price.
$$\$35.65 \times 100 = \$3565$$

The power of ten has two zeros, so the decimal point moves two places to the right. The retail store paid $3565 for the order.

b) First find the area of the lounge by finding the product of the length and width.

```
    180.4   1 decimal place
×    21.3   1 decimal place
    5412
   1804
   3608
  3842.52   2 decimal places
             (1 + 1 = 2)
```

Determine the price of the carpet by finding the product of the area and the price per square yard.

```
     3842.52   2 decimal places
×      11.25   2 decimal places
    1921260
     768504
     384252
     384252
 $43,228.3500   4 decimal places
                (2 + 2 = 4)
```

It would cost the college $43,228.35 to carpet the new lounge.

3.5 Dividing Decimals

Learning Objectives:

1. Divide decimals.
2. Solve application problems.
3. Use the order of operations with decimals.

Solved Examples:

1. Divide until there is a remainder of zero.

   a) $8\overline{)50.4}$ b) $0.8112 \div 0.06$ c) $\dfrac{1.62}{6}$

Solution

a)
$$
\begin{array}{r}
6.3 \\
8\overline{)50.4} \\
\underline{48} \\
24 \\
\underline{24} \\
0
\end{array}
$$

b)
$$
\begin{array}{r}
13.52 \\
0.06_\wedge\overline{)0.81_\wedge12} \\
\underline{6} \\
21 \\
\underline{18} \\
31 \\
\underline{30} \\
12 \\
\underline{12} \\
0
\end{array}
$$

c)
$$
\begin{array}{r}
0.27 \\
6\overline{)1.62} \\
\underline{1\,2} \\
42 \\
\underline{42} \\
0
\end{array}
$$

2. Divide and round to the nearest hundredth.

   a) $0.6\overline{)0.557}$ b) $4.399 \div 0.13$

Solution

a)
$$
\begin{array}{r}
0.928 = 0.93 \text{ to nearest hundredth} \\
0.6_\wedge\overline{)0.5_\wedge570} \\
\underline{5\,4} \\
17 \\
\underline{12} \\
50 \\
\underline{48} \\
2
\end{array}
$$

Note that we added an extra zero at the end of the dividend to carry out the division.

b)
$$
\begin{array}{r}
33.838 = 33.84 \text{ to nearest hundredth} \\
0.13_\wedge\overline{)4.39_\wedge900} \\
\underline{3\,9} \\
49 \\
\underline{39} \\
10\,9 \\
\underline{10\,4} \\
50 \\
\underline{39} \\
110 \\
\underline{104} \\
6
\end{array}
$$

Note that we added two extra zeros at the end of the dividend to carry out the division.

3. Evaluate using the correct order of operations.

   a) $6.2 + (4.3)^2 - 9.72$

   b) $2.25 + 1.06 \times 4.85$

   c) $2.25 - 1.06 \times (4.85 - 3.95)$

   d) $25.1 + 11.4 \div 7.5 \times 3.75$

   e) $(0.2)^3 + (7 - 2.4) \times 5.5$

   f) $4.9 \times 3.6 \times 2.1 - 0.1 \times 0.2 \times 0.3$

Solution

a) $6.2 + (4.3)^2 - 9.72$
   $= 6.2 + 18.49 - 9.72$ Exponents
   $= 24.69 - 9.72$ Addition
   $= 14.97$ Subtraction

b) $2.25 + 1.06 \times 4.85$
   $= 2.25 + 5.141$ Multiplication
   $= 7.391$ Addition

c) $2.25 - 1.06 \times (4.85 - 3.95)$
   $= 2.25 - 1.06 \times 0.9$ Parentheses
   $= 2.25 - 0.954$ Mutliplication
   $= 1.296$ Subtraction

d) $25.1 + 11.4 \div 7.5 \times 3.75$
   $= 25.1 + 1.52 \times 3.75$ Division
   $= 25.1 + 5.7$ Multiplication
   $= 30.8$ Addition

e) $(0.2)^3 + (7 - 2.4) \times 5.5$
   $= (0.2)^3 + 4.6 \times 5.5$ Parentheses
   $= 0.008 + 4.6 \times 5.5$ Exponents
   $= 0.008 + 25.3$ Multiplication
   $= 25.308$ Addition

f) We first work through all the multiplications moving from left to right. Our last operation will be the subtraction.
$4.9 \times 3.6 \times 2.1 - 0.1 \times 0.2 \times 0.3$
$= 17.64 \times 2.1 - 0.1 \times 0.2 \times 0.3$
$= 37.044 - 0.1 \times 0.2 \times 0.3$
$= 37.044 - 0.02 \times 0.3$
$= 37.044 - 0.006$
$= 37.038$

4. Mark owns a Ford Escort that travels 360 miles on 15.5 gallons of gas. How many miles per gallon does it achieve? (Round your answer to the nearest tenth.)

Solution

Divide the number of miles by the number of gallons used.

$$
\begin{array}{r}
23.22 \\
15.5_\wedge \overline{)360.0_\wedge 00} \\
310 \\
\overline{500} \\
465 \\
\overline{350} \\
310 \\
\overline{400} \\
310 \\
\overline{90}
\end{array}
$$

Mark's Ford Escort gets 23.2 miles per gallon.
(Note that we added two extra zeros at the end of the dividend to carry out the division.)

3.6 Writing Fractions and Decimals

Learning Objectives:

1. Write fractions as equivalent decimals.
2. Compare the size of fractions and decimals.

Solved Examples:

1. Write as an equivalent decimal. Divide until there is a remainder of zero or a repeating decimal.

    a) $\dfrac{1}{4}$ b) $4\dfrac{5}{6}$

Solution

a) To write $\frac{1}{4}$ as a decimal, we divide the denominator into the numerator until the remainder becomes zero or we obtain a repeating remainder.

$$\begin{array}{r} 0.25 \\ 4\overline{)1.00} \\ \underline{8} \\ 20 \\ \underline{20} \\ 0 \end{array}$$

Thus, $\dfrac{1}{4} = 0.25$.

b) $4\dfrac{5}{6}$ means $4 + \dfrac{5}{6}$.

To write $\frac{5}{6}$ as a decimal, we divide the denominator into the numerator until the remainder becomes zero or we obtain a repeating remainder.

$$\begin{array}{r} 0.833 \\ 6\overline{)5.000} \\ \underline{4\,8} \\ 20 \\ \underline{18} \\ 20 \\ \underline{18} \\ 2 \end{array}$$

repeating remainders

Thus, $\dfrac{5}{6} = 0.8\overline{3}$ and $4\dfrac{5}{6} = 4.8\overline{3}$.

2. Write as an equivalent decimal or decimal approximation. Round your answer to the nearest thousandth if needed.

    a) $\dfrac{39}{23}$ b) $\dfrac{11}{19}$

Solution

a)
$$\begin{array}{r} 1.6956 \\ 23\overline{)39.0000} \\ \underline{23} \\ 160 \\ 138 \\ \overline{220} \\ 207 \\ \overline{130} \\ 115 \\ \overline{150} \\ 138 \\ \overline{12} \end{array}$$

Rounding to the nearest thousandth, we round 1.6956 to 1.696. Thus, $\dfrac{39}{23} \approx 1.696$.

b)
$$\begin{array}{r} 0.5789 \\ 19\overline{)11.0000} \\ \underline{95} \\ 150 \\ 133 \\ \overline{170} \\ 152 \\ \overline{180} \\ 171 \\ \overline{9} \end{array}$$

Rounding to the nearest thousandth, we round 0.5789 to 0.579. Thus, $\dfrac{11}{19} \approx 0.579$.

3. Fill in the blank with one of the symbols <, =, or >.

    a) 0.33 ___ 0.31

    c) 22.001 ___ 21.001

    b) 1.56 ___ 1.560

    d) 0.006 ___ $\dfrac{6}{100}$

Solution

a) $0.3\underline{3}$ ___ $0.3\underline{1}$
The numbers in the tenths place are the same, but the numbers in the hundredths place are different. Since $3 > 1$, we can write $0.33 > 0.31$. We also note that 0.33 is further to the right on a number line than 0.31, so 0.33 is larger than 0.31.

c) d) $2\underline{2}.001$ ___ $2\underline{1}.001$
The numbers in the tens place are the same, but the numbers in the ones place are different. Since $2 > 1$, we can write $22.001 > 21.001$. We also note that 22.001 is further to the right on a number line than 21.0001, so 22.001 is larger.

b) Start by adding an additional zero to the number on the left.
1.56 ___ $1.560 \rightarrow 1.560$ ___ 1.560
Since the two numbers are identical, we use the = sign. $1.56 = 1.560$

d) Rewrite the fraction in decimal form. $\dfrac{6}{100}$ is six *hundredths*, which can also be written as 0.06. Add an additional zero at the end to obtain the same number of decimal places.

0.006 ___ $\dfrac{6}{100} \rightarrow 0.00\underline{6}$ ___ $0.06\underline{0}$
The numbers in the tenths place are the same, but the numbers in the hundredths place differ. Since $0 < 6$, we can write $0.006 < 0.060$, or

$0.006 < \dfrac{6}{100}$.

4. Arrange each set of decimals from smallest to largest.

    a) 0.415, 0.42, 0.409, 0.4102

    b) 23.082, 23.02, 23.088, 23.079

Solution

a) It is helpful to add additional zeros and place the decimals that begin with 0.41 together.
0.4150, 0.4102, 0.4200, 0.4090
In order, from smallest to largest, we have 0.4090, 0.4102, 0.4150, 0.4200.

b) It is helpful to add additional zeros and place the decimals that begin with 23.08 together.
23.020, 23.082, 23.088, 23.079
In order, from smallest to largest, we have 23.020, 23.079, 23.082, 23.088.

4.1 Ratios

Learning Objectives:

1. Write ratios as fractions.
2. Write ratios in application problems.

Solved Examples:

1. Write the ratio in lowest terms.

   a) $9 : 17$

   b) 72 to 76

   c) 6 to $9\frac{1}{2}$

   d) 3.6 to 12.6

   Solution

   a) $9 : 17 = \dfrac{9}{17}$

   b) 72 to 76
   $$= \frac{72}{76} = \frac{72 \div 4}{76 \div 4}$$
   $$= \frac{18}{19}$$

   c) $\dfrac{6\frac{1}{3}}{9\frac{1}{2}} = 6\frac{1}{3} \div 9\frac{1}{2}$
   $$= \frac{19}{3} \div \frac{19}{2} = \frac{19}{3} \cdot \frac{2}{19}$$
   $$= \frac{2}{3}$$

   d) 3.6 to 12.6
   $$= \frac{3.6}{12.6} = \frac{3.6 \cdot 10}{12.6 \cdot 10}$$
   $$= \frac{36 \div 18}{126 \div 18} = \frac{2}{7}$$

2. Write each ratio in lowest terms.

   a) 6 pounds of sunflower seed to 8 pounds of sunflower seed

   b) 12 quarts of green beans to 18 quarts of green beans

   c) 10 minutes to 1 hour

   d) 24 ounces of candy to 4 pounds of candy

   Solution

   a) $\dfrac{6 \text{ pounds}}{8 \text{ pounds}} = \dfrac{6}{8} = \dfrac{6 \div 2}{8 \div 2} = \dfrac{3}{4}$

   b) $\dfrac{12 \text{ quarts}}{18 \text{ quarts}} = \dfrac{12}{18} = \dfrac{12 \div 6}{18 \div 6} = \dfrac{2}{3}$

   c) 1 hour = 60 minutes

   $\dfrac{10 \text{ minutes}}{1 \text{ hour}} = \dfrac{10 \text{ minutes}}{60 \text{ minutes}} = \dfrac{10 \div 10}{60 \div 10} = \dfrac{1}{6}$

   d) $\dfrac{24 \text{ ounces}}{4 \text{ ounces}} = \dfrac{24}{4} = \dfrac{24 \div 4}{4 \div 4} = \dfrac{6}{1}$

3. A couple went out for the evening and spent $38 on dinner and $24 at the movies.
   What is the ratio of dollars spent on dinner to the total amount spent for the evening?

   Solution

   a) The couple spent a total of $24 + $38 = 62. Thus, the ratio of dollars spent on dinner to total amount
   spent is $\dfrac{38}{62} = \dfrac{19}{31}$.

4. A fitness club raises their monthly dues from $24 to $30. Find the ratio of the increase in dues to the
   original amount of dues.

   Solution

   The dues increased $30 - $24 = 6, so the ratio of the increase in dues to the original amount of dues is
   $\dfrac{\$6}{\$24} = \dfrac{6}{24} = \dfrac{6 \div 6}{24 \div 6} = \dfrac{1}{4}$.

4.2 Rates

Learning Objectives:

1. Write rates as fractions.
2. Find unit rates.
3. Find the best buy.
4. Solve application problems.

Solved Examples:

1. Write as a rate in simplest form.

   a) 5 cars for 20 people   b) 186 miles in 8 hours   c) 82 hours for 18 projects

   Solution

   a) $\dfrac{5 \text{ cars}}{20 \text{ people}} = \dfrac{5 \div 5 \text{ cars}}{20 \div 5 \text{ people}}$

       $= \dfrac{1 \text{ car}}{4 \text{ people}}$

   b) $\dfrac{186 \text{ miles}}{8 \text{ hours}} = \dfrac{186 \div 2 \text{ miles}}{8 \div 2 \text{ hours}}$

       $= \dfrac{93 \text{ miles}}{2 \text{ hours}}$

   c) $\dfrac{82 \text{ hours}}{18 \text{ projects}} = \dfrac{82 \div 2 \text{ hours}}{18 \div 2 \text{ projects}}$

       $= \dfrac{41 \text{ hours}}{9 \text{ projects}}$

2. Write as a unit rate.

   a) 132 miles on 3 gallons of gas   b) 1200 cars in 400 households

   c) 243 miles in 9 hours   d) $950 earned in 5 weeks

   Solution

   a) $\dfrac{132 \text{ miles}}{3 \text{ gallons}} = \dfrac{132 \div 3 \text{ miles}}{3 \div 3 \text{ gallons}} = \dfrac{44 \text{ miles}}{1 \text{ gallon}}$

   The unit rate is 44 miles/hour.

   b) $\dfrac{1200 \text{ cars}}{400 \text{ households}} = \dfrac{1200 \div 400 \text{ cars}}{400 \div 400 \text{ households}}$

       $= \dfrac{3 \text{ cars}}{1 \text{ household}}$

   The unit rate is 3 cars/household.

   c) $\dfrac{243 \text{ miles}}{9 \text{ hours}} = \dfrac{243 \div 9 \text{ miles}}{9 \div 9 \text{ hours}} = \dfrac{27 \text{ miles}}{1 \text{ hour}}$

   The unit rate is 27 miles/hour.

   d) $\dfrac{\$950}{5 \text{ weeks}} = \dfrac{\$950 \div 5}{5 \div 5 \text{ weeks}} = \dfrac{\$190}{1 \text{ week}}$

   The unit rate is $190/week.

3. Find the actual cost of a gallon of gas if 16 gallons cost $63.36.

   Solution

   Find the unit rate.

   $\dfrac{16 \text{ gallons}}{\$63.36} = \dfrac{16 \div 16 \text{ gallons}}{\$63.36 \div 16} = \dfrac{1 \text{ gallon}}{\$3.96}$

   One gallon of gas costs $3.96.

4. One jar of jelly costs $2.32 for 16 ounces. Another jar costs $2.03 for 13 ounces.
   Find which is the better buy. Round unit prices to two decimal places.

Solution

We find the unit rate for each size jar.

16 ounce jar: $\dfrac{\$2.32}{16 \text{ ounces}} = \$0.145/\text{ounce}$

13 ounce jar: $\dfrac{\$2.03}{13 \text{ ounces}} \approx \$0.156/\text{ounce}$

The 16 ounce jar of jelly for $2.32 is the better buy because its unit rate in dollars per ounce is lower.

4.3 Proportions

Learning Objectives:

1. Write proportions.
2. Write ratios in lowest terms to determine whether proportions are true or false.
3. Use cross products to determine whether proportions are true or false.

Solved Examples:

1. Write a proportion.

   a) 3 is to 5 as 6 is to 10

   b) 6.5 is to 5 as 52 is to 40

   c) $3\frac{1}{3}$ is to 4 as $4\frac{1}{6}$ is to 5

   Solution

   a) $\dfrac{3}{5} = \dfrac{6}{10}$

   b) $\dfrac{6.5}{5} = \dfrac{52}{40}$

   c) $\dfrac{3\frac{1}{3}}{4} = \dfrac{4\frac{1}{6}}{5}$

2. Determine whether the equation is a proportion.

   a) $\dfrac{1}{2} = \dfrac{3}{6}$

   b) $\dfrac{4}{10} = \dfrac{16}{39}$

   c) $\dfrac{1}{2} = \dfrac{4.8}{9.6}$

   d) $\dfrac{40}{39.2} = \dfrac{5}{5.3}$

   e) $\dfrac{2\frac{5}{9}}{5} = \dfrac{5\frac{1}{9}}{10}$

   f) $\dfrac{318 \text{ feet}}{4 \text{ rolls}} = \dfrac{954 \text{ feet}}{12 \text{ rolls}}$

   Solution

   a) We check to see if the cross products are equal.

   $$\dfrac{1}{2} \diagdown\mkern-18mu\diagup \dfrac{3}{6} \quad \begin{array}{l} \to\ 2 \times 3 = 6 \\ \to\ 1 \times 6 = 6 \end{array}$$

   The cross products are equal.

   Thus, $\dfrac{1}{2} = \dfrac{3}{6}$. This is a proportion.

   b) We check to see if the cross products are equal.

   $$\dfrac{4}{10} \diagdown\mkern-18mu\diagup \dfrac{16}{39}$$

   $\to\ 10 \times 16 = 160$

   $\to\ 4 \times 39 = 156$

   The cross products are not equal. Thus, $\dfrac{4}{10} \neq \dfrac{16}{39}$. This is not a proportion.

   c) We check to see if the cross products are equal.

   $$\dfrac{1}{2} \diagdown\mkern-18mu\diagup \dfrac{4.8}{9.6} \quad \begin{array}{l} \to\ 2 \times 4.8 = 9.6 \\ \to\ 1 \times 9.6 = 9.6 \end{array}$$

   The cross products are equal.

   Thus, $\dfrac{1}{2} = \dfrac{4.8}{9.6}$. This is a proportion.

d) We check to see if the cross products are equal.

$$\frac{40}{39.2} \diagdown \frac{5}{5.3}$$

$\rightarrow 39.2 \times 5 = 196$

$\rightarrow 40 \times 5.3 = 212$

The cross products are not equal. Thus, $\dfrac{40}{39.2} \neq \dfrac{5}{5.3}$. This is not a proportion.

e) We check to see if the cross products are equal.

$$\frac{2\frac{5}{9}}{5} \diagdown \frac{5\frac{1}{9}}{10}$$

$\rightarrow 5 \times 5\dfrac{1}{9} = \dfrac{5}{1} \times \dfrac{46}{9} = \dfrac{230}{9}$

$\rightarrow 2\dfrac{5}{9} \times 10 = \dfrac{23}{9} \times \dfrac{10}{1} = \dfrac{230}{9}$

The cross products are equal.

Thus, $\dfrac{2\frac{5}{9}}{5} = \dfrac{5\frac{1}{9}}{10}$. This is a proportion.

f) We check to see if the cross products are equal.

$$\frac{318}{4} \diagdown \frac{954}{12}$$

$\rightarrow 4 \times 954 = 3816$

$\rightarrow 318 \times 12 = 3816$

The cross products are equal. Thus, $\dfrac{318 \text{ feet}}{4 \text{ rolls}} = \dfrac{954 \text{ feet}}{12 \text{ rolls}}$. This is a proportion.

3. Solve the following application problems. Answer yes or no, and provide a reason for your answer.

a) A car traveled 578 miles in 8.5 hours. A truck traveled 272 miles in 4 hours. Did they travel at the same speed?

b) Sharon earned gross pay of $793.80 working 42 hours each week in a web design agency. Jesse's gross weekly pay was $737.10 for a 39-hour work week with a different agency. Was Sharon's pay per hour the same as Jesse's?

Solution

a) We must check $\dfrac{578 \text{ miles}}{8.5 \text{ hours}} \overset{?}{=} \dfrac{272 \text{ miles}}{4 \text{ hours}}$.

To do so, we check to see if the cross products are equal.

$$\frac{578}{8.5} \diagdown \frac{272}{4} \qquad \begin{array}{l} \rightarrow 8.5 \times 272 = 2312 \\ \rightarrow 578 \times 4 = 2312 \end{array}$$

The cross products are equal. Thus, $\dfrac{578 \text{ miles}}{8.5 \text{ hours}} = \dfrac{272 \text{ miles}}{4 \text{ hours}}$. This is a proportion, which means that two rates are equal. In other words, the car and truck did travel at the same speed.

b) We must check $\dfrac{\$793.80}{42 \text{ hours}} \overset{?}{=} \dfrac{\$737.10}{39 \text{ hours}}$.

To do so, we check to see if the cross products are equal.

$$\frac{793.80}{42} \diagdown \frac{737.10}{39} \qquad \begin{array}{l} \rightarrow 42 \times 737.10 = 30,958.2 \\ \rightarrow 793.80 \times 39 = 30,958.2 \end{array}$$

The cross products are equal. Thus, $\dfrac{\$793.80}{42 \text{ hours}} = \dfrac{\$737.10}{39 \text{ hours}}$. This is a proportion, which means that two rates are equal. In other words, Sharon's pay per hour is the same as Jesse's pay per hour.

4.4 Solving Proportions

Learning Objectives:

1. Find the unknown number in a proportion.

Solved Examples:

1. Find the missing number in a proportion. Round to the nearest tenth if needed.

a) $\dfrac{x}{10} = \dfrac{8}{20}$

b) $\dfrac{3}{x} = \dfrac{9}{15}$

c) $\dfrac{1}{2} = \dfrac{x}{17}$

d) $\dfrac{2}{7} = \dfrac{3}{x}$

e) $\dfrac{2}{x} = \dfrac{0.6}{1.2}$

f) $\dfrac{x}{6.8} = \dfrac{0.08}{5}$

g) $\dfrac{5}{\frac{2}{9}} = \dfrac{45}{x}$

h) $\dfrac{1}{5\frac{1}{2}} = \dfrac{x}{11}$

Solution

a) $\dfrac{x}{10} = \dfrac{8}{20}$

$20 \times x = 10 \times 8$ Find the cross product.

$20 \times x = 80$ Simplify.

$\dfrac{20 \times x}{20} = \dfrac{80}{20}$ Divide each side by 20.

$x = 4$

b) $\dfrac{3}{x} = \dfrac{9}{15}$

$9 \times x = 3 \times 15$ Find the cross product.

$9 \times x = 45$ Simplify.

$\dfrac{9 \times x}{9} = \dfrac{45}{9}$ Divide each side by 9.

$x = 5$

c) $\dfrac{1}{2} = \dfrac{x}{17}$

$17 \times 1 = 2 \times x$ Find the cross product.

$17 = 2 \times x$ Simplify.

$\dfrac{17}{2} = \dfrac{2 \times x}{2}$ Divide each side by 2.

$8.5 = x$

d) $\dfrac{2}{7} = \dfrac{3}{x}$

$3 \times 7 = 2 \times x$ Find the cross product.

$21 = 2 \times x$ Simplify.

$\dfrac{21}{2} = \dfrac{2 \times x}{2}$ Divide each side by 2.

$10.5 = x$

e) $\dfrac{2}{x} = \dfrac{0.6}{1.2}$

$0.6 \times x = 2 \times 1.2$ Find the cross product.

$0.6 \times x = 2.4$ Simplify.

$\dfrac{0.6 \times x}{0.6} = \dfrac{2.4}{0.6}$ Divide each side by 0.6.

$x = 4$

f) $\dfrac{x}{6.8} = \dfrac{0.08}{5}$

$5 \times x = 6.8 \times 0.08$ Find the cross product.

$5 \times x = 0.544$ Simplify.

$\dfrac{5 \times x}{5} = \dfrac{0.544}{5}$ Divide each side by 5.

$x \approx 0.1$

g) $\dfrac{5}{\dfrac{2}{9}} = \dfrac{45}{x}$

$5 \times x = \dfrac{2}{9} \times 45$ Find the cross product.

$5 \times x = 10$ Simplify.

$\dfrac{5 \times x}{5} = \dfrac{10}{5}$ Divide each side by 5.

$x = 2$

h) $\dfrac{1}{5\dfrac{1}{2}} = \dfrac{x}{11}$

$5\dfrac{1}{2} \times x = 11 \times 1$ Find the cross product.

$5\dfrac{1}{2} \times x = 11$ Simplify.

$\dfrac{5\dfrac{1}{2} \times x}{5\dfrac{1}{2}} = \dfrac{11}{5\dfrac{1}{2}}$ Divide each side by $5\dfrac{1}{2}$.

$x = 2$

2. Find the value of n. Round to the nearest hundredth when necessary.

a) $\dfrac{n \text{ ounces}}{23 \text{ quarts}} = \dfrac{39.6 \text{ ounces}}{9 \text{ quarts}}$

b) $\dfrac{27 \text{ liters}}{n \text{ grams}} = \dfrac{4 \text{ liters}}{18.8 \text{ grams}}$

c) $\dfrac{3 \text{ kilometers}}{1.86 \text{ miles}} = \dfrac{n \text{ kilometers}}{5 \text{ miles}}$

d) $\dfrac{2\dfrac{1}{5} \text{ feet}}{6 \text{ pounds}} = \dfrac{n \text{ feet}}{10 \text{ pounds}}$

Solution

a) $\dfrac{n \text{ ounces}}{23 \text{ quarts}} = \dfrac{39.6 \text{ ounces}}{9 \text{ quarts}}$

$9 \times n = 23 \times 39.6$ Find the cross product.

$9 \times n = 910.8$ Simplify.

$\dfrac{9 \times n}{9} = \dfrac{910.8}{9}$ Divide each side by 9.

$n = 101.2$

b) $\dfrac{27 \text{ liters}}{n \text{ grams}} = \dfrac{4 \text{ liters}}{18.8 \text{ grams}}$

$4 \times n = 27 \times 18.8$ Find the cross product.

$4 \times n = 507.6$ Simplify.

$\dfrac{4 \times n}{4} = \dfrac{507.6}{4}$ Divide each side by 4.

$n = 126.9$

c) $\dfrac{3 \text{ kilometers}}{1.86 \text{ miles}} = \dfrac{n \text{ kilometers}}{5 \text{ miles}}$

$1.86 \times n = 5 \times 3$ Find the cross product.

$1.86 \times n = 15$ Simplify.

$\dfrac{1.86 \times n}{1.86} = \dfrac{15}{1.86}$ Divide each side by 1.86.

$n \approx 8.06$

d) $\dfrac{2\dfrac{1}{5} \text{ feet}}{6 \text{ pounds}} = \dfrac{n \text{ feet}}{10 \text{ pounds}}$

$6 \times n = 2\dfrac{1}{5} \times 10$ Find the cross product.

$6 \times n = 22$ Simplify.

$\dfrac{6 \times n}{6} = \dfrac{22}{6}$ Divide each side by 6.

$n \approx 3.67$

4.5 Solving Application Problems with Proportions

Learning Objectives:

1. Solve application problems.

Solved Examples:

1. It takes Kim 22 minutes to type and spell check 14 pages of a manuscript. Find how long it takes her to type and spell check 77 pages. Round your answer to the nearest whole number if needed.

   Solution

   Set up a proportion comparing time to number of pages.

   Let n = the unknown number of minutes.

   $$\frac{22 \text{ minutes}}{14 \text{ pages}} = \frac{n \text{ minutes}}{77 \text{ pages}}$$

   $22 \times 77 = 14 \times n$ Find the cross product.

   $1694 = 14 \times n$ Simplify.

   $\dfrac{1694}{14} = \dfrac{14 \times n}{14}$ Divide each side by 14.

   $121 = n$

   Kim will need 121 minutes to type and spell check 77 pages of manuscript.

   Check: Estimate to see if the answer is reasonable.

2. On an architect's blueprint, 1 inch corresponds to 12 feet. If an exterior wall is 44 feet long, find how long the blueprint measurement should be. Write your answer as a mixed number, if needed.

   Solution

   Set up a proportion comparing the blueprint length to the real length.

   Let n = the unknown measurement.

   $$\frac{12 \text{ feet}}{1 \text{ inch}} = \frac{44 \text{ feet}}{n \text{ inches}}$$

   $12 \times n = 44 \times 1$ Find the cross product.

   $12 \times n = 44$ Simplify.

   $\dfrac{12 \times n}{12} = \dfrac{44}{12}$ Divide each side by 12.

   $n = \dfrac{11}{3} = 2\dfrac{2}{3}$

   An exterior wall that is 44 feet long is represented by a blueprint measurement of $2\dfrac{2}{3}$ inches.

   Check: Estimate to see if the answer is reasonable.

3. It is recommended that there be at least 11.2 square feet of ground space in a garden for every newly planted shrub. A garden is 25.6 feet by 21 feet. Find the maximum number of shrubs the garden can accommodate.

Solution

Determine the area of the garden. Then, set up a proportion comparing the number of shrubs to area.

Let n = the maximum number of shrubs.

$$\frac{11.2 \text{ sq. ft.}}{1 \text{ shrub}} = \frac{537.6 \text{ sq. ft.}}{n \text{ shrubs}}$$

$11.2 \times n = 537.6 \times 1$ Find the cross product.

$11.2 \times n = 537.6$ Simplify.

$\dfrac{11.2 \times n}{11.2} = \dfrac{537.6}{11.2}$ Divide each side by 11.2.

$n = 48$

The garden can accommodate a maximum of 48 shrubs.

Check: Estimate to see if the answer is reasonable.

4. At a college in eastern Minnesota, 7 out of every 10 students worked either a full-time or part-time job in addition to their studies. If 4900 students were enrolled at the college, how many did not have a full-time or part-time job?

Solution

Set up a proportion comparing students who do not have a full-time or part-time job to the total number of students.

Let n = the number of students who did not have a full-time or part-time job.

Since 7 out of every 10 students have a full-time or part-time job, this means that 3 out of every 10 students do not have a full-time or part-time job.

$$\frac{3 \text{ without job}}{10 \text{ students}} = \frac{n \text{ without job}}{4900 \text{ students}}$$

$4900 \times 3 = 10 \times n$

$14,700 = 10 \times n$

$\dfrac{14,700}{10} = \dfrac{10 \times n}{10}$

$1470 = n$

Thus, 1470 students did not have a full-time or part-time job.

Check: Estimate to see if the answer is reasonable.

4.6 Basics of Percents

Learning Objectives:

1. Write percents as decimals.
2. Write decimals as percents.
3. Convert between decimals and percents.
4. Solve application problems.

Solved Examples:

1. Write as a percent.

   a) $\dfrac{4}{100}$

   b) $\dfrac{70}{100}$

   c) $\dfrac{357}{100}$

   d) $\dfrac{0.3}{100}$

   e) $\dfrac{6.2}{100}$

   f) $\dfrac{0.039}{100}$

   Solution

   a) $\dfrac{4}{100} = 4\%$

   b) $\dfrac{70}{100} = 70\%$

   c) $\dfrac{357}{100} = 357\%$

   d) $\dfrac{0.3}{100} = 0.3\%$

   e) $\dfrac{6.2}{100} = 6.2\%$

   f) $\dfrac{0.039}{100} = 0.039\%$

2. Write each percent as a decimal.

   a) 30%

   b) 0.22%

   c) 623%

   d) 0.4%

   e) 97.61%

   f) 100%

   Solution

   a) Recall that % means "parts out of 100".
   $$30\% = \frac{30}{100} = 0.3$$

   b) Drop the % and move the decimal point two places to the left.
   $$0.22\% = 0.0022$$

   c) Drop the % and move the decimal point two places to the left.
   $$623\% = 6.23$$

   d) Drop the % and move the decimal point two places to the left.
   $$0.4\% = 0.004$$

   e) Drop the % and move the decimal point two places to the left.
   $$97.61\% = 0.9761$$

   f) Drop the % and move the decimal point two places to the left.
   $$100\% = 1.00$$

3. Write each decimal as a percent.

   a) 0.25

   b) 0.8

   c) 0.031

   d) 1.0

   e) 3.33

   f) 0.00037

Solution

a) Move the decimal point two places to the right, then write the % symbol at the end of the number.
$0.25 = 25\%$

b) Move the decimal point two places to the right, then write the % symbol at the end of the number.
$0.8 = 80\%$

c) Move the decimal point two places to the right, then write the % symbol at the end of the number.
$0.031 = 3.1\%$

d) Move the decimal point two places to the right, then write the % symbol at the end of the number.
$1.0 = 100\%$

e) Move the decimal point two places to the right, then write the % symbol at the end of the number.
$3.33 = 333\%$

f) Move the decimal point two places to the right, then write the % symbol at the end of the number.
$0.00037 = 0.037\%$

4. Find the following percents.

   a) 100% of 50 states

   b) 300% of $50

   c) 50% of 16 athletes

   d) 1% of $1500

Solution

a) 100% is all of the number. So 100% of 50 states is 50 states.

b) 300% is three times the amount. So, 300% of $50 is $3 \cdot 50 = \$150$.

c) 50% is half of the athletes.
So, 50% of 16 athletes is $\frac{1}{2} \cdot 16 = 8$ athletes.

d) 1% is found by moving the decimal point two places to the left.
So, 1% of $1500 is $15.

5. In a survey of 100 people, 41 preferred onions on their hot dogs. What percent preferred onions?

Solution

41 out of 100 people preferred onions on their hot dogs.

$$\frac{41}{100} = 0.41 = 41\%$$

4.7 Percents and Fractions

Learning Objectives:

1. Write percents as fractions or mixed numbers.
2. Write fractions as percents.
3. Convert between fractions, decimals, and percents.
4. Solve application problems.

Solved Examples:

1. Write each percent as a fraction or as a mixed number.

   a) 92%

   b) 15.5%

   c) 9.76%

   d) 320%

   e) $\dfrac{1}{2}\%$

   f) $2\dfrac{5}{6}\%$

   Solution

   a) $92\% = \dfrac{92}{100} = \dfrac{23}{25}$

   b) $15.5\% = 0.155$
   $= \dfrac{155}{1000} = \dfrac{31}{200}$

   c) $9.76\% = 0.0976$
   $= \dfrac{976}{10,000} = \dfrac{61}{625}$

   d) $320\% = 3.20$
   $= 3\dfrac{20}{100} = 3\dfrac{1}{5}$

   e) $\dfrac{1}{2}\% = 0.5\%$
   $= 0.005$
   $= \dfrac{5}{1000} = \dfrac{1}{200}$

   f) $2\dfrac{5}{6}\% = \dfrac{2\frac{5}{6}}{100} = 2\dfrac{5}{6} \div \dfrac{100}{1}$
   $= \dfrac{17}{6} \div \dfrac{100}{1} = \dfrac{17}{6} \cdot \dfrac{1}{100}$
   $= \dfrac{17}{600}$

2. Write as a percent. Round to the nearest hundredth of a percent when necessary.

   a) $\dfrac{1}{2}$

   b) $\dfrac{3}{4}$

   c) $\dfrac{9}{20}$

   d) $\dfrac{7}{12}$

   e) $2\dfrac{1}{5}$

   f) $4\dfrac{5}{8}$

   Solution

   a) $\dfrac{1}{2} = 0.5$
   $= 50\%$

   b) $\dfrac{3}{4} = 0.75$
   $= 75\%$

   c)
   $$\begin{array}{r} 0.45 \\ 20\overline{)9.000} \\ \underline{80} \\ 100 \\ \underline{100} \\ 0 \end{array}$$

   Thus, $\dfrac{9}{20} = 0.45 = 45\%$.

d) Divide to find a decimal approximation.

$$\begin{array}{r} 0.58333 \\ 12\overline{)7.00000} \end{array}$$

$$\underline{60}$$
$$100$$
$$\underline{96}$$
$$40$$
$$\underline{36}$$
$$40$$
$$\underline{36}$$
$$40$$
$$\underline{36}$$
$$4$$

Thus,

$$\frac{7}{12} \approx 0.5833 = 58.33\% .$$

e) $2\frac{1}{5} = 2.2$

$$= 220\%$$

f) Obtain a decimal equivalent for $\frac{5}{8}$.

$$\begin{array}{r} 0.625 \\ 8\overline{)5.000} \end{array}$$

$$\underline{48}$$
$$20$$
$$\underline{16}$$
$$40$$
$$\underline{40}$$
$$0$$

Thus,

$$4\frac{5}{8} = 4.625 = 462.5\% .$$

3. Write as a percent containing a fraction.

a) $\dfrac{3}{7}$

b) $\dfrac{9}{14}$

c) $\dfrac{13}{90}$

d) $\dfrac{7}{40}$

Solution

a) Do the division, but stop after two decimal places and write the remainder in fraction form.

$$\begin{array}{r} 0.42 \\ 7\overline{)3.00} \end{array}$$

$$\underline{28}$$
$$20$$
$$\underline{14}$$
$$6$$

Thus, $\dfrac{3}{7} = 0.42\dfrac{6}{7} = 42\dfrac{6}{7}\% .$

b) Do the division, but stop after two decimal places and write the remainder in fraction form.

$$\begin{array}{r} 0.64 \\ 14\overline{)9.00} \end{array}$$

$$\underline{84}$$
$$60$$
$$\underline{56}$$
$$4$$

Thus, $\dfrac{9}{14} = 0.64\dfrac{4}{14} = 0.64\dfrac{2}{7} = 64\dfrac{2}{7}\% .$

c) Do the division, but stop after two decimal places and write the remainder in fraction form.

$$\begin{array}{r} 0.14 \\ 90\overline{)13.00} \end{array}$$

$$\underline{90}$$
$$400$$
$$\underline{360}$$
$$40$$

Thus, $\dfrac{13}{90} = 0.14\dfrac{40}{90} = 0.14\dfrac{4}{9} = 14\dfrac{4}{9}\% .$

d) Do the division, but stop after two decimal places and write the remainder in fraction form.

$$\begin{array}{r} 0.17 \\ 40\overline{)7.00} \end{array}$$

$$\underline{40}$$
$$300$$
$$\underline{280}$$
$$20$$

Thus, $\dfrac{7}{40} = 0.17\dfrac{20}{40} = 0.17\dfrac{1}{2} = 17\dfrac{1}{2}\% .$

4. Fill in the chart. Round decimals to ten-thousandths and percents to hundredths, as needed.

	Fraction	Decimal	Percent
a)	$\frac{2}{7}$		
b)		0.05	
c)			5.75%

Solution

a) In the first row, the number is written as a fraction. We change the fraction to a decimal and a percent.

$$\frac{2}{7} \rightarrow \quad 7\overline{)2.00000}^{\,0.28571} \quad \rightarrow 28.57\%$$

b) In the second row, the number is written as a decimal. We first write this as a percent, then as a fraction.

$$0.05 = 5\% \qquad 0.05 = \frac{5}{100} = \frac{1}{20}$$

c) In the third row, the number is written as a percent. We first write the number as a decimal, then as a fraction.

$$5.75\% = \frac{5.75}{100} = 0.0575 \qquad 0.0575 = \frac{575}{10,000} = \frac{23}{400}$$

The completed table would be

Fraction	Decimal	Percent
$\frac{2}{7}$	0.2857	28.57%
$\frac{1}{20}$	0.05	5%
$\frac{23}{400}$	0.0575	5.75%

4.8 Using the Percent Proportion and Identifying the Components in a Percent Problem

Learning Objectives:

1. Solve for an unknown value in a percent proportion.
2. Solve problems using a percent proportion.
3. Set up percent proportions.
4. Set up percent proportions for application problems.

Solved Examples:

1. Identify the part, whole, and percent. Do not solve for any unknowns.

    a) 75% of 720 is 434.   b) What is 66% of 39?   c) 56% of what is 2443?

Solution

a) percent = 75

    The whole is the entire quantity. It follows the word *of.* Here the whole is 720.

    The part is the portion being compared to the whole. Here the part is 434.

b) percent = 66

    The part is unknown. We represent the part by the variable *x.*

    The whole is the entire quantity. It follows the word *of.* Here the whole is 39.

c) percent = 56

    The whole is unknown. We represent the whole by the variable *x.*

    The part is the portion being compared to the whole. Here the part is 2443.

2. Solve using the percent proportion. Find the missing part.

    a) The percent is 25 and the whole is 80.   b) The percent is 8.1 and the whole is 300.

Solution

a) percent = 25, whole = 80

$$\frac{\text{part}}{\text{whole}} = \frac{\text{percent}}{100}$$

$$\frac{x}{80} = \frac{25}{100}$$

$$\frac{x}{80} = \frac{1}{4}$$

$$4x = 80(1)$$

$$4x = 80$$

$$\frac{4x}{4} = \frac{80}{4}$$

$$x = 20$$

Thus, 25% of 80 is 20.

b) percent = 8.1, whole = 300

$$\frac{\text{part}}{\text{whole}} = \frac{\text{percent}}{100}$$

$$\frac{x}{300} = \frac{8.1}{100}$$

$$100x = 300(8.1)$$

$$100x = 2430$$

$$\frac{100x}{100} = \frac{2430}{100}$$

$$x = 24.3$$

Thus, 8.1% of 300 is 24.3.

3. Solve using the percent proportion. Find the missing whole.

   a) The percent is 50 and the part is 90.

   b) The percent is 4 and the part is 15.

   Solution

   a) percent = 50, part = 90

   $$\frac{part}{whole} = \frac{percent}{100}$$

   $$\frac{90}{x} = \frac{50}{100}$$

   $$\frac{90}{x} = \frac{1}{2}$$

   $$2(90) = x$$

   $$180 = x$$

   Thus, 50% of 180 is 90.

   b) percent = 4, part = 15

   $$\frac{part}{whole} = \frac{percent}{100}$$

   $$\frac{15}{x} = \frac{4}{100}$$

   $$\frac{15}{x} = \frac{1}{25}$$

   $$25(15) = x$$

   $$375 = x$$

   Thus, 4% of 375 is 15.

4. Solve using the percent proportion. Find the missing percent.

   a) The part is 30 and the whole is 150.

   b) The part is 0.4 and the whole is 20.

   Solution

   a) whole = 150, part = 30

   $$\frac{part}{whole} = \frac{percent}{100}$$

   $$\frac{30}{150} = \frac{x}{100}$$

   $$\frac{1}{5} = \frac{x}{100}$$

   $$100(1) = 5x$$

   $$100 = 5x$$

   $$\frac{100}{5} = \frac{5x}{5}$$

   $$20 = x$$

   Thus, 30 is 20% of 150.

   b) whole = 20, part = 0.4

   $$\frac{part}{whole} = \frac{percent}{100}$$

   $$\frac{0.4}{20} = \frac{p}{100}$$

   $$100(0.4) = 20p$$

   $$40 = 20p$$

   $$\frac{40}{20} = \frac{20p}{20}$$

   $$2 = p$$

   Thus, 0.4 is 2% of 20.

5. Set up the percent proportion and write *unknown* for the value that is not given. Do not solve for the unknown.

   a) 25% of doctors in a hospital are female. If there are 840 doctors altogether, how many doctors are female?

   b) An inspector found 60 defective watches during an inspection. If this is 0.005% of the total number of watches inspected, how many watches were inspected?

   c) In a recent survey of 180 people, 36 said that their favorite color of car was white. What percent of the people surveyed liked white cars?

Solution

a) The basic situation here is that 25% of 840 doctors is some number.

percent = 25, whole = 840

$$\frac{\text{part}}{\text{whole}} = \frac{\text{percent}}{100}$$

$$\frac{x}{840} = \frac{25}{100}$$

b) The basic situation here is that 60 is 0.005% of some number of watches. percent = 0.005

$p = 0.005$, part = 60

$$\frac{\text{part}}{\text{whole}} = \frac{\text{percent}}{100}$$

$$\frac{60}{b} = \frac{0.005}{100}$$

c) We want to determine what percent 36 people are of 180 people.

whole = 180, part = 36

$$\frac{\text{part}}{\text{whole}} = \frac{\text{percent}}{100}$$

$$\frac{36}{180} = \frac{p}{100}$$

4.9 Using Proportions to Solve Percent Problems

Learning Objectives:

1. Find the part using the multiplication shortcut.
2. Find the whole using the percent proportion.
3. Find the percent using the percent proportion.
4. Solve application problems.

Solved Examples:

1. Translate into a mathematical equation.

   a) What is 50% of 80?   b) 70% of what number is 32?   c) What % of 30 is 24?

   Solution

   a) What is 50% of 80?
   $$\downarrow \quad \downarrow \downarrow \quad \downarrow \downarrow$$
   $$x \; = 50\% \; \bullet \; 80$$

   b) 70% of $\underline{\text{what number}}$ is 32?
   $$\downarrow \quad \downarrow \qquad \downarrow \qquad \downarrow \downarrow$$
   $$70\% \bullet \qquad x \qquad = 32$$

   c) $\underline{\text{What \%}}$ of 30 is 24?
   $$\downarrow \quad \downarrow \downarrow \downarrow$$
   $$x \quad \bullet \; 30 = 24$$

2. Solve. Find the missing part.

   a) What is 10% of 70?   b) What is 32% of 224?   c) Find 190% of 375.

   Solution

   a) What is 10% of 70?
   $$\downarrow \quad \downarrow \downarrow \quad \downarrow \downarrow$$
   $$x \; = 10\% \; \bullet \; 70$$
   $$x \; = (0.10)(70)$$
   $$x \; = 7$$

   b) What is 32% of 224?
   $$\downarrow \quad \downarrow \downarrow \quad \downarrow \downarrow$$
   $$x \; = 32\% \; \bullet \; 224$$
   $$x \; = (0.32)(224)$$
   $$x \; = 71.68$$

   c) What is 190% of 375?
   $$\downarrow \quad \downarrow \quad \downarrow \downarrow$$
   $$x \; = 190\% \; \bullet \; 375$$
   $$x \; = (1.90)(375)$$
   $$x \; = 712.5$$

3. Solve. Find the missing whole.

   a) 50% of what number is 30?   b) 65% of what is 91?   c) 6.6 is 33% of what?

   Solution

   a) 50% of $\underline{\text{what number}}$ is 30?
   $$\downarrow \quad \downarrow \qquad \downarrow \qquad \downarrow \downarrow$$
   $$50\% \bullet \qquad x \qquad = 30$$
   $$(0.50)x = 30$$
   $$x = \frac{30}{0.5}$$
   $$x = 60$$

   b) 65% of $\underline{\text{what}}$ is 91?
   $$\downarrow \quad \downarrow \quad \downarrow \quad \downarrow \downarrow$$
   $$65\% \bullet \quad x \quad = 91$$
   $$(0.65)x = 91$$
   $$x = \frac{91}{0.65}$$
   $$x = 140$$

   c) 6.6 is 33% of $\underline{\text{what}}$?
   $$\downarrow \downarrow \downarrow \quad \downarrow \downarrow$$
   $$6.6 = 33\% \; \bullet \quad x$$
   $$6.6 = (0.33)x$$
   $$\frac{6.6}{0.33} = x$$
   $$20 = x$$

4. Solve. Find the missing percent.

    a) 25 is what % of 125?   b) What percent of 80 is 0.8?   c) 126 is what % of 28?

Solution

a) 25 is <u>what %</u> of 125?

$$\downarrow\downarrow \quad \downarrow \quad \downarrow \quad \downarrow$$
$$25 = \quad x \quad \bullet \quad 125$$
$$25 = 125x$$
$$\frac{25}{125} = x$$
$$x = 0.2 = 20\%$$

b) <u>What %</u> of 80 is 0.8?

$$\downarrow \quad \downarrow\downarrow\downarrow$$
$$x \quad \bullet\ 80 = 0.8$$
$$80x = 0.8$$
$$x = \frac{0.8}{80}$$
$$x = 0.01 = 1\%$$

c) 126 is <u>what %</u> of 28?

$$\downarrow\downarrow \quad \downarrow \quad \downarrow \quad \downarrow$$
$$126 = \quad x \quad \bullet \quad 28$$
$$126 = 28x$$
$$\frac{126}{28} = x$$
$$x = 4.5 = 450\%$$

5. a) The Smith family paid 22% of the purchase price of a $231,000 home as a down payment. Determine the amount of the down payment.

    b) One day 144 office workers were sick with colds. If this was 72% of the total number of office workers, how many office workers were there?

    c) Lisa bought a share of stock for $48.24. She was paid a dividend of $12.06. Determine what percent of the stock price is the dividend.

Solution

a) The question is asking What is 22% of $231,000?

$$\downarrow \quad \downarrow\downarrow \quad \downarrow\downarrow$$
$$x \quad = 22\% \ \bullet \ 231,000$$
$$x \quad = (0.22)(231,000)$$
$$x \quad = 50,820$$

The Smith family paid $50,820 as a down payment.

b) The question is asking 72% of <u>what</u> is 144?

$$\downarrow \quad \downarrow \quad \downarrow \quad \downarrow\downarrow$$
$$72\% \ \bullet \quad x \quad = 144$$
$$(0.72)x = 144$$
$$x = \frac{144}{0.72}$$
$$x = 200$$

There were 200 office workers.

c) The question is asking 12.06 is <u>what %</u> of 48.24?

$$\downarrow \quad \downarrow \quad \downarrow \quad \downarrow \quad \downarrow$$
$$12.06 = \quad x \quad \bullet \ 48.24$$
$$12.06 = 48.24x$$
$$\frac{12.06}{48.24} = x$$
$$x = 0.25 = 25\%$$

The dividend was 25% of the stock price.

4.10 Using the Percent Equation

Learning Objectives:

1. Find the part using the percent equation.
2. Find the whole using the percent equation.
3. Find the percent using the percent equation.
4. Solve application problems.

Solved Examples:

1. a) Rita now earns $10.50 per hour. This is 20% more than what she earned last year. What did she earn per hour last year?

    b) Jeff puts aside $72.50 per week for his monthly car payment. He earns $362.50 per week. What percent of his income is set aside for car payments?

    c) Every day Elise orders either cappuccino or espresso from the coffee bar. She had 73 espressos and 292 cappuccinos this year. What percent of the coffees were espressos?

Solution

a) The percent is 120. Use x for the unknown whole. The part is 10.50.

$$\frac{10.50}{x} = \frac{120}{100}$$

$$\frac{10.50}{x} = \frac{6}{5}$$

$$5(10.50) = 6x$$

$$52.5 = 6x$$

$$\frac{52.5}{6} = \frac{6x}{6}$$

$$8.75 = x$$

Rita earned $8.75 per hour last year.

b) Use x for the unknown percent. The whole is 362.50. The part is 72.50.

$$\frac{\text{part}}{\text{whole}} = \frac{\text{percent}}{100}$$

$$\frac{72.50}{362.50} = \frac{x}{100}$$

$$\frac{1}{5} = \frac{x}{100}$$

$$100 = 5x$$

$$\frac{100}{5} = x$$

$$20 = x$$

Thus, Jeff puts aside 20% of his income for car payments.

c) Use x for the unknown percent. The whole is $73 + 262 = 365$. The part is 73.

$$\frac{\text{part}}{\text{whole}} = \frac{\text{percent}}{100}$$

$$\frac{73}{365} = \frac{x}{100}$$

$$\frac{1}{5} = \frac{x}{100}$$

$$100 = 5x$$

$$\frac{100}{5} = x$$

$$20 = x$$

Thus, 20% of the coffees were espressos.

2. a) Lisa has $36.00 to spend on dinner. She wants to tip the waitress 20% of the cost of her meal. How much money can she spend on the meal itself?

b) Nolan is building a new house. When the house is finished it will cost $322,400. The price of building the house this year is 4% higher than it would have been last year. What would the price of the house have been last year?

c) In total a medical research facility spent $21,000,000 to develop a drug. 40% of this cost was for staff, and 35% was for equipment. How much was spent to cover both the staff and the equipment?

Solution

a) Let n = the meal cost.

Total Cost = Meal cost + 20% tip

$$36.00 = n + 20\% \text{ of } n$$

$$36.00 = n + 0.2n$$

$$36.00 = 1.2n$$

$$\frac{36.00}{1.2} = \frac{1.2n}{1.2}$$

$$30 = n$$

Lisa can spend $30.00 on the meal itself.

b) Let x = the house price last year.

Current price = price last year + increase

$$322,400 = x + 4\% \text{ of } x$$

$$322,400 = x + 0.04x$$

$$322,400 = 1.04x$$

$$\frac{322,400}{1.04} = \frac{1.04x}{1.04}$$

$$310,000 = x$$

The price of the house last year would have been $310,000.

c) Let n = amount spent to cover both the staff and the equipment.

$$\binom{\text{amount}}{\text{for both}} = \binom{\text{amount}}{\text{for staff}} + \binom{\text{amount}}{\text{for equipment}}$$

$$n = 40\% \text{ of } 21,000,000 + 35\% \text{ of } 21,000,000$$

$$n = (0.4)(21,000,000) + (0.35)(21,000,000)$$

$$n = 8,400,000 + 7,350,000$$

$$n = 15,750,000$$

The medical research facility spent a total of $15,750,000 on both the staff and equipment.

4.11 Solving Application Problems with Percent

Learning Objectives:

1. Solve application problems about sales tax and commission.
2. Solve discount application problems.
3. Solve mixed application problems.

Solved Examples:

1. The sales tax rate in a certain state is 8%. The selling price of a digital television at *Big-Discount Electronics* is $640.

   a) What is the sales tax?

   b) What is the total cost of the television?

   Solution

   a) sales tax = rate of tax • cost of item
      $$= (8\%)(640)$$
      $$= (0.08)(640)$$
      $$= \$51.20$$
      The amount of sales tax is $51.20.

   b) The total cost is $640 + $51.20 = $691.20.

2. a) A sales representative is paid a commission rate of 3.3%. Find her commission if she sold $52,020 worth of goods last month.

   b) A salesperson earned a commission of $5316 for selling $44,300 worth of batteries to various stores. Find the commission rate.

   c) A sales representative for a medical supply company was paid $175,500 in commissions last year. If his commission rate was 5%, what was the sales total for the medical supplies he sold last year?

   Solution

   a) $\text{commission} = \left(\begin{array}{c}\text{commission} \\ \text{rate}\end{array}\right) \times \left(\begin{array}{c}\text{value} \\ \text{of sales}\end{array}\right)$
      $$= 3.3\% \times \$52,020$$
      $$= 0.033 \times 52,020$$
      $$= 1716.66$$
      Her commission is $1716.66.

b) $\text{commission} = \left(\begin{array}{c}\text{commission}\\\text{rate}\end{array}\right) \times \left(\begin{array}{c}\text{value}\\\text{of sales}\end{array}\right)$

$\$5316 = \left(\begin{array}{c}\text{commission}\\\text{rate}\end{array}\right) \times \$44,300$

$5316 = \left(\begin{array}{c}\text{commission}\\\text{rate}\end{array}\right) \times 44,300$

$\dfrac{5316}{44,300} = \text{commission rate}$

$0.12 = \text{commission rate}$

The salesperson's commission rate is 12%.

c) $\text{commission} = \left(\begin{array}{c}\text{commission}\\\text{rate}\end{array}\right) \times \left(\begin{array}{c}\text{value}\\\text{of sales}\end{array}\right)$

$\$175,500 = 5\% \times \left(\begin{array}{c}\text{value}\\\text{of sales}\end{array}\right)$

$175,500 = 0.05 \times \left(\begin{array}{c}\text{value}\\\text{of sales}\end{array}\right)$

$\dfrac{175,500}{0.05} = \text{value of sales}$

$3,510,000 = \text{value of sales}$

The salesperson had total sales of \$3,510,000.

3. a) Find the amount of discount. The original price is \$121.40, the discount rate is 60%.

b) Find the sale price. The original price is \$17,700.00, the discount rate is 22%.

c) A \$2500 table is on sale at 5% off. Find the discount.

Solution

a) Let x = the discount.
    discount = discount rate $\times$ orig. amount
    $d = 60\% \times \$121.40$
    $d = (0.60)(121.4)$
    $d = 72.84$
    The discount was \$72.84.

b) Let x = the sale price.
    Sale price = original price $-$ discount
    $x = 17,700.00 - 22\%$ of $17,700.00$
    $x = 17,700.00 - (0.22)(17,700.00)$
    $x = 17,700.00 - 3,894.00$
    $x = 13,806.00$
    The sale price is \$13,806.00.

c) Let x = the discount.
    discount = discount rate $\times$ orig. amount
    $d = (5\%)(\$2500)$
    $d = (0.05)(2500)$
    $d = 125$
    The discount was \$125.

4. a) Find the percent of increase when the original amount is 20 and the new amount is 28.

b) Find the percent of decrease when the original amount is 170 and the new amount is 136.

c) One share of stock which originally sold for $120 now sells for $108. What is the percent of decrease?

Solution

a) Amount of increase $= 28 - 20 = 8$

$$\text{percent of increase} = \frac{\text{amt. of increase}}{\text{original amount}}$$

$$\text{percent of increase} = \frac{8}{20}$$

$$= 0.4$$

The percent increase is 40%.

b) Amount of decrease $= 170 - 136 = 34$

$$\text{percent of decrease} = \frac{\text{amt. of decrease}}{\text{original amount}}$$

$$\text{percent of decrease} = \frac{34}{170}$$

$$= 0.2$$

The percent decrease is 20%.

c) Amount of decrease $= 120 - 108 = 12$

$$\text{percent of decrease} = \frac{\text{amt. of decrease}}{\text{original amount}}$$

$$\text{percent of decrease} = \frac{12}{120}$$

$$= 0.10$$

The percent decrease is 10%.

4.12 Simple Interest

Learning Objectives:

1. Find the simple interest.
2. Find the total amount due.
3. Solve application problems.

Solved Examples:

1. Find the interest. Round to the nearest cent, if necessary.

   a) Principal $300, rate 10%, time 1 year

   b) Principal $2000, rate 8%, time 5 years

   c) Principal $14,000, rate $8\frac{1}{2}$%, time $4\frac{1}{2}$ years

   Solution

   a) $I = p \cdot r \cdot t$
   $= (300)(0.10)(1)$
   $= 30$

   The interest is $30.

   b) $I = p \cdot r \cdot t$
   $= (2000)(0.08)(5)$
   $= 800$

   The interest is $800.

   c) $I = p \cdot r \cdot t$
   $= (14,000)(0.085)(5)$
   $= 5950$

   The interest is $5950.

2. Find the amount due. Round to the nearest cent if necessary.

   a) Principal $580, rate 4%, time 3 years

   b) Principal $25,000, rate 6%, time 6 months

   Solution

   a) $I = p \cdot r \cdot t$
   $= (580)(0.04)(3)$
   $= 69.6$

   The interest is $69.60.

   Amount due = principal + interest
   $= \$580 + \$69.60 = \$649.60$

   b) $I = p \cdot r \cdot t$
   $= (3000)(0.0475) \cdot \frac{9}{12}$
   $= 106.88$ (rounded)

   The interest is $106.88.

   Amount due = principal + interest
   $= \$3000 + \$106.88 = \$3106.88$

3. Rachel placed $2000 in a one-year CD paying simple interest of 7.5% for one year. How much will Rachel earn in one year? Round to the nearest cent, if necessary.

   Solution

   $I = p \cdot r \cdot t$
   $= (2000)(0.075)(1)$
   $= 150$

   Rachel will earn $150.

4. Tony borrowed $4000 to finish his education at an interest rate of 4% per year at simple interest. How much will Tony have to pay back in 3 years? Round to the nearest cent if necessary.

Solution

$I = p \cdot r \cdot t$

$\quad = (4000)(0.04)(3)$

$\quad = 480$

Tony will have to pay back $4000 + $480 = $4480.

4.13 Compound Interest

Learning Objectives:

1. Find the compound interest and compound amount.
2. Solve application problems.

Solved Examples:

1. Find the compound amount on each of the following using multiplication.

   a) $5000, paying 8% interest, compounded annually, for 6 years.

   b) $2000, paying 5% interest, compounded annually, for 8 years.

   c) $8,000, paying 6% interest, compounded annually, for 4 years.

   Solution

   a) $100\% + 8\% = 108\% = 1.08$
      $5000(1.08)(1.08)(1.08)(1.08)(1.08)(1.08) \approx 7934.37$
      The compound amount is $7934.37.

   b) $100\% + 5\% = 105\% = 1.05$
      $2000(1.05)(1.05)(1.05)(1.05)(1.05)(1.05)(1.05)(1.05) \approx 2954.91$
      The compound amount is $2954.91.

   c) $100\% + 6\% = 106\% + 1.06\%$
      $8000(1.06)(1.06)(1.06)(1.06) \approx 10,099.82$
      The compound amount is $10,099.82.

2. Use the compound interest table below to find the compound amount for each of the following:

Time Periods	3.00%	3.5%	4.00%	5.00%	5.50%	6.00%
1	1.0300	1.0350	1.0400	1.0500	1.0550	1.0600
2	1.0609	1.0712	1.08016	1.1025	1.1130	1.1236
3	1.0927	1.1087	1.1249	1.1576	1.1742	1.1910
4	1.1255	1.1475	1.1699	1.2155	1.2388	1.3605
5	1.1593	1.1877	1.2167	1.2763	1.3070	1.3382
6	1.1941	1.2293	1.2653	1.3401	1.3788	1.4185
7	1.2299	1.2723	1.3159	1.4071	1.4547	1.5036
8	1.2668	1.3168	1.3686	1.4775	1.5347	1.5938
9	1.3048	1.3629	1.4233	1.5513	1.6191	1.6895
10	1.3439	1.4106	1.4820	1.6289	1.7081	1.7908
11	1.3842	1.4600	1.5395	1.7103	1.8021	1.8983

   a) $1 deposited at 4% interest rate for 7 years.

   b) $1 deposited at 6% interest rate for 10 years.

   c) $16,000 deposited at 3.5% interest rate for 5 years.

Solution

a) 4% column, row 7 of the table gives 1.3159.
   compound amount = ($1)(1.3159) ≈ $1.14

b) 6% column, row 10 of the table gives 1.7908.
   compound amount = ($1)(1.7908) ≈ $1.79

c) 3.5% column, row 5 of the table gives 1.1877.
   compound amount = ($16,000)(1.1877) = $19,003.20

3. West Central Savings Bank loaned the Brown Flying School $30,000 to upgrade their buildings. The loan is to be repaid in a lump sum after 5 years at 6% interest, compounded annually.

   a) What will be the total amount due at the end of 5 years?

   b) How much interest will be due on the loan?

Solution

a) 6% column, row 5 from the table gives 1.3382.
   compound amount = ($30,000)(1.3382) = $40,146
   The total amount due at the end of 5 years is $40,146.

b) The amount of interest due is $40,146 - $30,000 = $10,146$.

5.1 Lines and Angles

Learning Objectives:

1. Use vocabulary to describe lines and angles.
2. Identify and calculate measures of complementary and supplementary angles.
3. Identify and calculate measures of congruent angles.
4. Calculate measures of angles related to parallel lines.

Solved Examples:

1. Identify each figure as a line, line segment, or ray and name it using the appropriate symbol.

a)

b)

c)

Solution

a) This is a line named $\overleftrightarrow{MN}$ or $\overleftrightarrow{NM}$. A line goes on forever in both directions.

b) This is a ray named $\overrightarrow{MN}$ or $\overrightarrow{NM}$. A ray is part of a line that has only one endpoint and goes on forever in one direction.

c) This is a line segment named $\overline{MN}$ or $\overline{NM}$. A line segment has two endpoints.

2. Identify each pair of lines as appearing to be parallel, perpendicular, or intersecting.

a)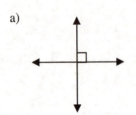

b)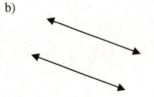

c)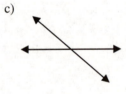

Solution

a) The lines intersect at right angles, so they are perpendicular.

b) The lines appear to be parallel. Parallel lines in the same plane never intersect.

c) The lines intersect, so they are *not* parallel. The do *not* form a right angle at their intersection, so they are *not* perpendicular. Thus, the lines are intersecting.

3. Identify which angles, if any, are the type listed.

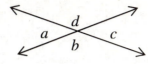

a) obtuse

b) acute

c) supplementary

d) complementary

e) congruent

f) right

Solution

a) Angle b and angle d are both obtuse angles

b) Angle a and angle c are both acute angles.

c) Angle a and angle b are supplementary; angle b and angle c are supplementary; angle c and angle d are supplementary; angle d and angle a are supplementary.

d) No pair of angles shown in the figure are complementary.

e) Angle a and angle c are vertical angles, so they have the same measure. Likewise, angle b and angle d are vertical angles, so they have the same measure.

f) No right angles are shown.

4. Find the measure of the angle(s).

a)

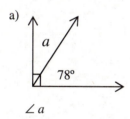

$\angle a$

b)

a c 37° b

$\angle a$, $\angle b$, and $\angle c$

c)

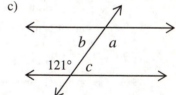

$\angle a$, $\angle b$, and $\angle c$
(Assume horizontal lines are parallel.

Solution

a) The angles shown are complementary. Since complementary angle have a sum of $90°$, we have $\angle a = 90° - 78° = 12°$.

b) Angle a and the angle measuring $37°$ are vertical angles. So, $\angle a = 37°$. Angle b and angle a are supplementary. Since supplementary angles have a sum of $180°$, we have $\angle b = 180° - 37° = 143°$. Angle c and the angle b are vertical angles. So, $\angle c = 143°$.

c) Angle a and the angle measuring $121°$ are alternate interior angles. So, $\angle a = 121°$. Angle b and angle a are supplementary. Since supplementary angles have a sum of $180°$, we have $\angle b = 180° - 121° = 59°$. Angle c and the angle b are alternate interior angles. So, $\angle c = 59°$.

5.2 Rectangles and Squares

Learning Objectives:

1. Find the perimeter and area of rectangles and squares.
2. Find the perimeter and area of composite shapes.
3. Solve application problems.

Solved Examples:

1. Find the perimeters and areas of the rectangle and square.

a)
8 cm
1.2 cm 1.2 cm
8 cm

b)
5 ft
5 ft 5 ft
5 ft

Solution

a) $P = 2 \cdot \text{length} + 2 \cdot \text{width}$

$= 2 \cdot 8 \text{ cm} + 2 \cdot 1.2 \text{ cm}$

$= 16 \text{ cm} + 2.4 \text{ cm}$

$= 18.4 \text{ cm}$

$A = \text{length} \cdot \text{width}$

$= 8 \text{ cm} \cdot 1.2 \text{ cm}$

$= 9.6 \text{ cm}^2$

b) $P = 4s$

$= 4 \cdot 5 \text{ ft}$

$= 20 \text{ ft}$

$A = s^2$

$= 5 \text{ ft} \cdot 5 \text{ ft}$

$= 25 \text{ ft}^2$

2. Find the perimeters and areas of the shapes.

a)
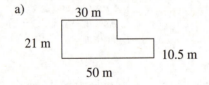
30 m
21 m
10.5 m
50 m

b)
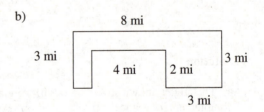
8 mi
3 mi
4 mi 2 mi
3 mi
3 mi

Solution

a) $P = 30 \text{ m} + 10.5 \text{ m} + 20 \text{ m}$

$+ 10.5 \text{ m} + 50 \text{ m} + 21 \text{ m}$

$= 142 \text{ m}$

Draw a vertical line to break up the figure into two rectangles.

Rectangle with length 30 m and width 21 m:

b) $P = 3 \text{ mi} + 8 \text{ mi} + 3 \text{ mi} + 3 \text{ mi}$

$+ 2 \text{ mi} + 4 \text{ mi} + 2 \text{ mi} + 1 \text{ mi}$

$= 26 \text{ mi}$

Draw two vertical lines to break up the figure into two rectangles and a square.

Rectangle with length 3 mi and width 1 mi:

$A = 30 \text{ m} \cdot 21 \text{ m}$

$= 630 \text{ m}^2$

Rectangle with length 20 m and width 10.5 m:

$A = 20 \text{ m} \cdot 10.5 \text{ m}$

$= 210 \text{ m}^2$

Total area $= 630 \text{ m}^2 + 210 \text{ m}^2$

$= 840 \text{ m}^2$

$A = 3 \text{ mi} \cdot 1 \text{ mi}$

$= 3 \text{ mi}^2$

Rectangle with length 4 mi and width 1 mi:

$A = 4 \text{ mi} \cdot 1 \text{ mi}$

$= 4 \text{ mi}^2$

Square with $s = 3$ mi:

$A = 4 \text{ mi} \cdot 4 \text{ mi}$

$= 16 \text{ mi}^2$

Total area $= 3 \text{ mi}^2 + 4 \text{ mi}^2 + 9 \text{ mi}^2$

$= 16 \text{ mi}^2$

3. A badminton area will feature four badminton courts, covering a total area of 400 ft by 80 ft. If sand costs $0.37 per square foot, how much will it cost to sand the court area?

Solution

Find the area of the courts.

$A = 400 \text{ ft} \cdot 80 \text{ ft}$

$= 32,000 \text{ ft}^2$

Multiply the area by the cost per square foot.

$\text{Cost} = 32,0000 \text{ ft}^2 \cdot \$0.37$

$= \$11,840$

The total cost to sand the court area is $11,840.

5.3 Parallelograms and Trapezoids

Learning Objectives:

1. Find the perimeter of parallelograms and trapezoids.
2. Find the area of parallelograms and trapezoids.
3. Solve application problems.
4. Find the area of composite shapes.

Solved Examples:

1. Draw an example of the shape.

   a) parallelogram   b) rhombus   c) trapezoid

   Solution

   a) 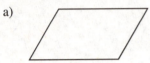 b)   c)

2. Find the perimeter and area of each shape.

   a)

   13.4 in.

   8.2 in.   8.2 in. 6.5 in.

   13.4 in.

   b)

   4 m

   4 m   4 m 3 m

   4 m

   c)

   25 ft

   23 ft   23 ft 20 ft

   18 ft

   Solution

   a) $P = 2(8.2 \text{ in}) + 2(13.4 \text{ in})$ $A = bh$
     $= 16.4 \text{ in} + 26.8 \text{ in}$ $= (13.4 \text{ in})(6.5 \text{ in})$
     $= 43.2 \text{ in}$ $= 87.1 \text{ in}^2$

   b) $P = 4(4 \text{ m})$ $A = bh$
     $= 16 \text{ m}$ $= (4 \text{ m})(3 \text{ m})$
              $= 12 \text{ m}^2$

c) $P = 25 \text{ ft} + 23 \text{ ft}$
  $+ 18 \text{ ft} + 23 \text{ ft}$
  $= 89 \text{ ft}$

$$A = \frac{h(b + B)}{2}$$

$$= \frac{(20 \text{ ft})(25 \text{ ft} + 18 \text{ ft})}{2}$$

$$= \frac{(20 \text{ ft})(43 \text{ ft})}{2}$$

$$= \frac{860}{2} \text{ ft}^2 = 430 \text{ ft}^2$$

3. Find the perimeter and area of the shape made up of a parallelogram and trapezoid.

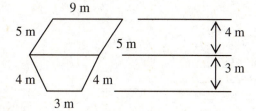

Solution

To find the perimeter, we add up the lengths of all edges.

$P = 9 \text{ m} + 5 \text{ m} + 4 \text{ m} + 3 \text{ m} + 4 \text{ m} + 5 \text{ m} = 30 \text{ m}$

To find the area, we find the area of the parallelogram and the area of the trapezoid separately, and add the result together.

Area of parallelogram:

$A = bh = (9 \text{ m})(4 \text{ m}) = 36 \text{ m}^2$

Area of trapezoid:

$$A = \frac{h(b + B)}{2}$$

$$= \frac{(3 \text{ m})(9 \text{ m} + 3 \text{ m})}{2}$$

$$= \frac{(3 \text{ m})(12 \text{ m})}{2}$$

$$= \frac{36}{2} \text{ m}^2 = 18 \text{ m}^2$$

Total Area: $36 \text{ m}^2 + 18 \text{ m}^2 = 54 \text{ m}^2$

5.4 Triangles

Learning Objectives:

1. Find the perimeter and area of a triangle.
2. Find the measure of angles in a triangle.
3. Solve application problems.

Solved Examples:

1. Fill in the missing information.

   a) A triangle is a three-sided figure with _____ angles.
   b) The sum of the measures of the angles in a triangle is _____.
   c) A triangle with two equal sides is called an _____.
   d) A triangle with three equal sides is called an _____.
   e) A triangle with one 90° angle is called a _____.

   Solution

   a) three
   b) 180°
   c) isosceles triangle
   d) equilateral triangle
   e) right triangle

2. Find the missing angle in the triangle.

   a) Two angles are 28° and 72°.   b) Two angles are 139° and 20°.

   Solution

   a) We will use the fact that the sum of the measures of the angles of a triangle is 180°. Let x represent the missing angle.
   $$28 + 72 + x = 180$$
   $$100 + x = 180$$
   What number x when added to 100 equals 180? Since $100 + 80 = 180$, x must equal 80. So, the measure of the missing angle is 80°.

   b) We will use the fact that the sum of the measures of the angles of a triangle is 180°. Let x represent the missing angle.
   $$139 + 20 + x = 180$$
   $$159 + x = 180$$
   What number x when added to 159 equals 180? Since $159 + 21 = 180$, x must equal 21. So, the measure of the missing angle is 21°.

3. Find the perimeter and area of the triangle.

   a)
   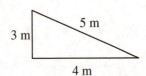

   b)
   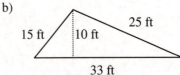

   c) A triangle with all three sides of 6.8 yd and height of 5.4 yd.

Solution

a) To find the perimeter, we add up the lengths of the three sides.

$P = 3 \text{ m} + 4 \text{ m} + 5 \text{ m} = 12 \text{ m}$

The triangle appears to be a right triangle. So, the height is 3 m. The base is 4 m.

$A = \dfrac{bh}{2} = \dfrac{(4 \text{ m})(3 \text{ m})}{2} = \dfrac{12}{2} \text{ m}^2 = 6 \text{ m}^2$

b) To find the perimeter, we add up the lengths of the three sides.

$P = 33 \text{ ft} + 25 \text{ ft} + 15 \text{ ft} = 73 \text{ ft}$

The base of the triangle is 33 ft. The height is 10 ft.

$A = \dfrac{bh}{2} = \dfrac{(33 \text{ ft})(10 \text{ ft})}{2} = \dfrac{330}{2} \text{ ft}^2 = 165 \text{ ft}^2$

c) To find the perimeter, we add up the lengths of the three sides.

$P = 6.8 \text{ yd} + 6.8 \text{ yd} + 6.8 \text{ yd} = 20.4 \text{ yd}$

The base of the triangle is 6.8 yd. The height is 5.4 yd.

$$A = \dfrac{bh}{2}$$
$$= \dfrac{(6.8 \text{ yd})(5.4 \text{ yd})}{2}$$
$$= \dfrac{36.72}{2} \text{ yd}^2 = 18.36 \text{ yd}^2$$

5.5 Circles

Learning Objectives:

1. Find the radius and diameter of a circle.
2. Find circumference and area of a circle.
3. Find the area.
4. Solve application problems.

Examples:

1. Find the diameter or radius of each circle. Round to the nearest hundredth.

   a) radius r = 10 ft

   b) radius r = 6.2 cm

   c) radius r = 34.9 m

   d) diameter d = 49 yd

   e) diameter d = 0.02 cm

   f) diameter d = 13.4 in.

   Solution

   a) $d = 2r$
   $= 2(10 \text{ ft}) = 20 \text{ ft}$

   b) $d = 2r$
   $= 2(6.2 \text{ cm}) = 12.4 \text{ cm}$

   c) $d = 2r$
   $= 2(34.9 \text{ m}) = 69.8 \text{ m}$

   d) $r = \dfrac{d}{2}$
   $= \dfrac{49 \text{ yd}}{2} = 24.5 \text{ yd}$

   e) $r = \dfrac{d}{2}$
   $= \dfrac{0.02 \text{ cm}}{2} = 0.01 \text{ cm}$

   f) $r = \dfrac{d}{2}$
   $= \dfrac{13.4 \text{ in}}{2} = 6.7 \text{ in}$

2. Find the circumference and area of each circle. Round to the nearest tenth.

   a) radius = 12 cm

   b) radius = 24.06 ft

   c) radius = 0.3 in.

   d) diameter = 44.4 m

   e) diameter = 1.004 yd

   f) diameter = 2300 mi

   Solution

   a) $d = 2r$
   $= 2(12 \text{ cm}) = 24 \text{ cm}$
   $C = \pi d$
   $\approx (3.14)(24 \text{ cm})$
   $= 75.36 \text{ cm}$
   $\approx 75.4 \text{ cm}$
   $A = \pi r^2$
   $\approx (3.14)(12 \text{ cm})^2$
   $= (3.14)(12 \text{ cm})(12 \text{ cm})$
   $= (3.14)(144 \text{ cm}^2)$
   $= 452.16 \text{ cm}^2$
   $\approx 452.2 \text{ cm}^2$

   b) $d = 2r$
   $= 2(24.06 \text{ ft}) = 48.12 \text{ ft}$
   $C = \pi d$
   $\approx (3.14)(48.12 \text{ ft})$
   $= 151.0968 \text{ ft}$
   $\approx 151.1 \text{ ft}$
   $A = \pi r^2$
   $\approx (3.14)(24.06 \text{ ft})^2$
   $= (3.14)(24.06 \text{ ft})(24.06 \text{ ft})$
   $= (3.14)(578.8836 \text{ ft}^2)$
   $= 1817.694504 \text{ ft}^2$
   $\approx 1817.7 \text{ ft}^2$

   c) $d = 2r$
   $= 2(0.3 \text{ in}) = 0.6 \text{ in}$
   $C = \pi d$
   $\approx (3.14)(0.6 \text{ in})$
   $= 1.884 \text{ in}$
   $\approx 1.9 \text{ in}$
   $A = \pi r^2$
   $\approx (3.14)(0.3 \text{ in})^2$
   $= (3.14)(0.3 \text{ in})(0.3 \text{ in})$
   $= (3.14)(0.09 \text{ in}^2)$
   $= 0.2826 \text{ in}^2$
   $\approx 0.3 \text{ in}^2$

d) $C = \pi d$

$\approx (3.14)(44.4 \text{ m})$

$= 139.416 \text{ m}$

$\approx 139.4 \text{ m}$

$r = \dfrac{d}{2} = \dfrac{44.4 \text{ m}}{2} = 22.2 \text{ m}$

$A = \pi r^2$

$\approx (3.14)(22.2 \text{ m})^2$

$= (3.14)(22.2 \text{ m})(22.2 \text{ m})$

$= (3.14)(492.84 \text{ m}^2)$

$= 1547.5176 \text{ m}^2$

$\approx 1547.5 \text{ m}^2$

e) $C = \pi d$

$\approx (3.14)(1.004 \text{ yd})$

$= 3.15256 \text{ yd}$

$\approx 3.2 \text{ yd}$

$r = \dfrac{d}{2} = \dfrac{1.004 \text{ yd}}{2} = 0.502 \text{ yd}$

$A = \pi r^2$

$\approx (3.14)(0.502 \text{ yd})^2$

$= (3.14)(0.502 \text{ yd})(0.502 \text{ yd})$

$= (3.14)(0.252004 \text{ yd}^2)$

$= 0.79129256 \text{ yd}^2$

$\approx 0.8 \text{ yd}^2$

f) $C = \pi d$

$\approx (3.14)(2300 \text{ mi}) = 7222 \text{ mi}$

$r = \dfrac{d}{2} = \dfrac{2300 \text{ mi}}{2} = 1150 \text{ mi}$

$A = \pi r^2$

$\approx (3.14)(1150 \text{ mi})^2$

$= (3.14)(1150 \text{ mi})(1150 \text{ mi})$

$= (3.14)(1,322,500 \text{ mi}^2)$

$= 4,152,650 \text{ mi}^2$

3. Solve the following application problems. Round to the nearest hundredth.

a) A bicycle wheel makes six revolutions. Determine how far the bicycle travels in feet if the diameter of the wheel is 24 in.

b) A radio station sends out radio waves in all directions from a tower at the center of the circle of broadcast range. Determine how large an area is reached if the diameter of the broadcast range is 135 mi.

c) Find the area of the shaded region.

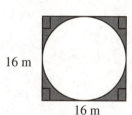

16 m

16 m

Solution

a) *Understand the problem.* The distance the wheel travels when it makes 1 revolution is the circumference of the tire. Since our tire makes 6 revolutions. We will multiply the circumference by 6.

Solve and state the answer. We are given that the diameter is 24 in, which is equivalent to 2 ft.

$C = \pi d \approx (3.14)(2 \text{ ft}) = 6.28 \text{ ft}$

So, the distance traveled when the wheel makes six revolutions is $(6)(6.28 \text{ ft}) = 37.68 \text{ ft} \approx 37.7 \text{ ft}$

Check: Estimate to see if the answer is reasonable.

b) *Understand the problem.* We need to find the area of a circle with diameter $d = 135$ mi.

Solve and state the answer.

$$r = \frac{d}{2} = \frac{135 \text{ mi}}{2} = 67.5 \text{ mi}$$

$$A = \pi r^2$$

$$\approx (3.14)(67.5 \text{ mi})^2$$

$$= (3.14)(67.5 \text{ mi})(67.5 \text{ mi})$$

$$= (3.14)(4556.25 \text{ mi}^2)$$

$$= 14,306.625 \text{ mi}^2$$

$$\approx 14,306.6 \text{ mi}^2$$

Check: Estimate to see if the answer is reasonable.

c) *Understand the problem.* We need to find the shaded area. To do this, we will subtract the area of the circle with diameter $d = 16$ m from the area of the square with side $s = 16$ m.

Solve and state the answer.

Area of the square: $A = s^2 = (16 \text{ m})^2 = 256 \text{ m}^2$

Area of the circle:

$$r = \frac{d}{2} = \frac{16 \text{ m}}{2} = 8 \text{ m}$$

$$A = \pi r^2$$

$$\approx (3.14)(8 \text{ m})^2$$

$$= (3.14)(64 \text{ m}^2)$$

$$= 200.96 \text{ m}^2$$

So, the area of the shaded region is $256 \text{ m}^2 - 200.96 \text{ m}^2$

$$= 55.04 \text{ m}^2$$

$$\approx 55.0 \text{ m}^2$$

Check: Estimate to see if the answer is reasonable.

5.6 Volume and Surface Area

Learning Objectives:

1. Name solids and find the volume.
2. Solve application problems involving volume.
3. Find the volume and surface area of a cylinder or rectangular solid.

Solved Examples:

1. Find the volume of each rectangular solid. Round to the nearest tenth.

   a) width = 16 mm, length = 22 mm, height = 21 mm

   b) width = 1.3 ft, length = 4.5 ft, height = 3.1 ft

   Solution

   a) $V = (22 \text{ mm})(16 \text{ mm})(21 \text{ mm})$

   $= (352)(21) \text{ mm}^3$

   $= 7392 \text{ mm}^3$

   b) $V = lwh$

   $= (4.5 \text{ ft})(1.3 \text{ ft})(3.1 \text{ ft})$

   $= (5.85)(3.1) \text{ ft}^3$

   $= 18.135 \text{ ft}^3$

   $\approx 18.1 \text{ ft}^3$

2. Find the volume and surface area of each cylinder. Round to the nearest tenth.

   a) diameter = 12 cm, height = 34 cm

   b) diameter = 55.4 mm, height = 3.3 mm

   Solution

   a) $r = \dfrac{d}{2} = \dfrac{12 \text{ cm}}{2} = 6 \text{ cm}$

   $V = \pi r^2 h$

   $= (3.14)(6 \text{ cm})^2 (34 \text{ cm})$

   $= (3.14)(36 \text{ cm}^2)(34 \text{ cm})$

   $= (113.04 \text{ cm}^2)(34 \text{ cm})$

   $= 3843.36 \text{ cm}^3$

   $\approx 3843.4 \text{ cm}^3$

   $S = 2\pi rh + 2\pi r^2$

   $= 2(3.14)(6 \text{ cm})(34 \text{ cm}) + 2(3.14)(6 \text{ cm})^2$

   $= 1281.12 \text{ cm}^2 + 226.08 \text{ cm}^2$

   $= 1507.2 \text{ cm}^2$

   b) $r = \dfrac{d}{2} = \dfrac{55.4 \text{ mm}}{2} = 27.7 \text{ mm}$

   $V = \pi r^2 h$

   $= (3.14)(27.7 \text{ mm})^2 (3.3 \text{ mm})$

   $= (3.14)(767.29 \text{ mm}^2)(3.3 \text{ mm})$

   $= (2409.2906 \text{ mm}^2)(3.3 \text{ mm})$

   $= 7950.65898 \text{ mm}^3$

   $\approx 7950.7 \text{ mm}^3$

   $S = 2\pi rh + 2\pi r^2$

   $= 2(3.14)(27.7 \text{ mm})(3.3 \text{ mm}) + 2(3.14)(27.7 \text{ mm})^2$

   $= 574.0548 \text{ mm}^2 + 4778.68212 \text{ mm}^2$

   $= 5352.72792 \text{ mm}^2$

   $\approx 5352.7 \text{ mm}^2$

3. Find the volume of each sphere. Round to the nearest tenth.

   a) radius = 4 in.
                                      b) hemisphere, radius = 2 cm

Solution

a) $V = \dfrac{4\pi r^3}{3}$

$= \dfrac{(4)(3.14)(4 \text{ in})^3}{3}$

$= \dfrac{(4)(3.14)(4)(4)(4) \text{ in}^3}{3}$

$= \dfrac{803.84 \text{ in}^3}{3}$

$\approx 267.9 \text{ in}^3$

b) The volume of a hemisphere is $\frac{1}{2}$ of the volume of a sphere.

a) $V = \dfrac{1}{2} \cdot \dfrac{4\pi r^3}{3}$

$= \dfrac{1}{2} \cdot \dfrac{(4)(3.14)(2 \text{ cm})^3}{3}$

$= \dfrac{1}{\cancel{2}} \cdot \dfrac{\cancel{(4)}^2 (3.14)(2)(2)(2) \text{ cm}^3}{3}$

$= \dfrac{50.92 \text{ cm}^3}{3}$

$\approx 134.0 \text{ cm}^3$

4. Find the volume of each cone. Round to the nearest tenth.

   a) height = 14 in., radius = 6 in.
                         b) height = 22 cm, radius = 13 cm

Solution

a) $V = \dfrac{\pi r^2 h}{3}$

$= \dfrac{(3.14)(6 \text{ in})^2 (14 \text{ in})}{3}$

$= \dfrac{(3.14)(6 \text{ in})(6 \text{ in})(14 \text{ in})}{3}$

$= (3.14)(2)(6)(14) \text{ in}^3$

$= (6.28)(84) \text{ in}^3$

$\approx 527.5 \text{ in}^3$

b) $V = \dfrac{\pi r^2 h}{3}$

$= \dfrac{(3.14)(13 \text{ cm})^2 (22 \text{ cm})}{3}$

$= \dfrac{(3.14)(169 \text{ cm}^2)(22 \text{ cm})}{3}$

$= \dfrac{(530.66 \text{ cm}^2)(22 \text{ cm})}{3}$

$= \dfrac{11{,}674.52 \text{ cm}^3}{3}$

$\approx 3891.5 \text{ cm}^3$

5. Find the volume of each pyramid. Round to the nearest thousandth.

   a) height 12 m, rectangular base 6 m by 10 m

   b) height 0.5 cm, rectangular base 0.2 cm by 0.2 cm

Solution

a) Area of base $= B = (6 \text{ m})(10 \text{ m}) = 60 \text{ m}^2$

$$V = \frac{Bh}{3} = \frac{(60 \text{ m}^2)(12 \text{ m})}{3}$$

$$= (60)(4) \text{ m}^3$$

$$= 240 \text{ m}^3$$

b) Area of base $= B = (0.2 \text{ cm})(0.2 \text{ cm})$

$$= 0.04 \text{ cm}^2$$

$$V = \frac{Bh}{3}$$

$$= \frac{(0.04 \text{ cm}^2)(0.5 \text{ cm})}{3}$$

$$= \frac{0.02 \text{ cm}^3}{3}$$

$$\approx 0.007 \text{ cm}^3$$

5.7 Pythagorean Theorem

Learning Objectives:

1. Find square roots.
2. Find unknown lengths using the Pythagorean Theorem.
3. Solve application problems.

Solved Examples:

1. Evaluate the square root.

   a) $\sqrt{4}$　　　　　　　b) $\sqrt{25}$　　　　　　　c) $\sqrt{16}$

   d) $\sqrt{121}$　　　　　　e) $\sqrt{144}$　　　　　　f) $\sqrt{169}$

   Solution

   a) $\sqrt{4} = 2$ because (2)(2) = 4.

   b) $\sqrt{25} = 5$ because (5)(5) = 25.

   c) $\sqrt{16} = 4$ because (4)(4) – 16.

   d) $\sqrt{121} = 11$ because (11)(11) = 121.

   e) $\sqrt{144} = 12$ because (12)(12) = 144.

   f) $\sqrt{169} = 13$ because (13)(13) = 169.

2. Approximate the square root to the nearest thousandth.

   a) $\sqrt{8}$　　　b) $\sqrt{33}$　　　c) $\sqrt{105}$　　　d) $\sqrt{136}$　　　e) $\sqrt{50}$

   Solution

   We use a calculator to approximate each root.

   a) $\sqrt{8} \approx 2.828$　　b) $\sqrt{33} \approx 5.745$　　c) $\sqrt{105} \approx 10.247$　　d) $\sqrt{136} \approx 11.662$　　e) $\sqrt{50} \approx 7.701$

3. Draw a right triangle and label the sides as "leg", "leg", and "hypotenuse". Then write the Pythagorean Theorem.

   Solution

   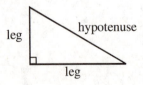

   $(\text{hypotenuse})^2 = (\text{leg})^2 + (\text{leg})^2$

4. Find the missing length in the right triangle. Round to the nearest thousandth.

a)

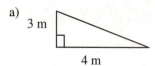

3 m

4 m

b)

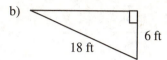

6 ft

18 ft

c) leg = 12 m, leg = 4 m

d) hypotenuse = 22 yd, leg = 14 yd

e) leg = 6 ft, leg = 6 ft

f) leg = 53 cm, hypotenuse = 75 cm

Solution

a) $\text{hypotenuse} = \sqrt{(\text{leg})^2 + (\text{leg})^2}$

$= \sqrt{(3 \text{ m})^2 + (4 \text{ m})^2}$

$= \sqrt{9 \text{ m}^2 + 16 \text{ m}^2}$

$= \sqrt{25 \text{ m}^2}$

$= 5 \text{ m}$

b) $\text{leg} = \sqrt{(\text{hypotenuse})^2 - (\text{leg})^2}$

$= \sqrt{(18 \text{ ft})^2 - (6 \text{ ft})^2}$

$= \sqrt{324 \text{ ft}^2 - 36 \text{ ft}^2}$

$= \sqrt{288 \text{ ft}^2}$

$\approx 16.971 \text{ ft}$

c) $\text{hypotenuse} = \sqrt{(\text{leg})^2 + (\text{leg})^2}$

$= \sqrt{(12 \text{ m})^2 + (4 \text{ m})^2}$

$= \sqrt{144 \text{ m}^2 + 16 \text{ m}^2}$

$= \sqrt{160 \text{ m}^2}$

$\approx 12.649 \text{ m}$

d) $\text{leg} = \sqrt{(\text{hypotenuse})^2 - (\text{leg})^2}$

$= \sqrt{(22 \text{ yd})^2 - (14 \text{ yd})^2}$

$= \sqrt{484 \text{ yd}^2 - 196 \text{ yd}^2}$

$= \sqrt{288 \text{ yd}^2}$

$\approx 16.971 \text{ yd}$

e) $\text{hypotenuse} = \sqrt{(\text{leg})^2 + (\text{leg})^2}$

$= \sqrt{(6 \text{ ft})^2 + (6 \text{ ft})^2}$

$= \sqrt{36 \text{ ft}^2 + 36 \text{ ft}^2}$

$= \sqrt{72 \text{ ft}^2}$

$\approx 8.485 \text{ ft}$

f) $\text{leg} = \sqrt{(\text{hypotenuse})^2 - (\text{leg})^2}$

$= \sqrt{(75 \text{ cm})^2 - (53 \text{ cm})^2}$

$= \sqrt{5625 \text{ cm}^2 - 2809 \text{ cm}^2}$

$= \sqrt{2816 \text{ cm}^2}$

$\approx 53.066 \text{ cm}$

5. a) A 30-ft ladder is placed against a college classroom building at a point 22 ft above the ground. What is the distance from the base of the ladder to the building?

b) Betty's kite is flying 45 yd directly above a rock. The rock is 23 yd from where she is standing. Find the length of the string holding the kite.

Solution

a) In a $30° \text{-} 60° \text{-} 90°$ triangle, the leg opposite of the $30°$ angle is $\frac{1}{2}$ of the hypotenuse.

$$\text{leg} = \frac{1}{2} \times 10 \text{ in.} = 5 \text{ in.}$$

We now know the hypotenuse and one leg of a right triangle, so we can find the other leg by using the Pythagorean Theorem.

$$\text{leg} = \sqrt{(\text{hypotenuse})^2 - (\text{leg})^2}$$
$$= \sqrt{(10 \text{ in.})^2 - (5 \text{ in.})^2}$$
$$= \sqrt{100 \text{ in.}^2 - 25 \text{ in.}^2}$$
$$= \sqrt{75 \text{ in.}^2}$$
$$\approx 8.7 \text{ in.}$$

b) $45° \text{-} 45° \text{-} 90°$ triangles are isosceles, so the other leg must also equal 8 m.

In a $45° \text{-} 45° \text{-} 90°$, the hypotenuse equals the length of one leg multiplied by $\sqrt{2}$.

$$\text{hypotenuse} = \sqrt{2} \times \text{leg}$$
$$\approx 1.414 \times (8 \text{ m})$$
$$\approx 11.3 \text{ m}$$

5.8 Congruent and Similar Triangles

Learning Objectives:

1. Identify similar triangles.
2. Write ratios for corresponding sides.
3. Find unknown lengths in similar triangles.
4. Solve application problems.

Solved Examples:

1. Identify which pairs of triangles shown appear to be *congruent*, *similar*, or *neither*.

a)   b)   c)

Solution

a) The triangles have the same shape, but different size, so they are similar.

b) The triangles have the same shape and the same size, so they are congruent.

c) The triangles do not have the same shape or size, so they are neither similar nor congruent.

2. Determine which of these methods can be used to prove that each pair of triangles is congruent: Angle-Side-Angle (ASA), Side-Side-Side (SSS), or Side-Angle-Side (SAS).

a)   b) 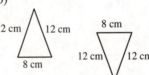   c)

Solution

a) On both triangles, two corresponding sides and the angle between them measure the same, so the Side-Angle-Side (SAS) method can be used to prove that the triangles are congruent.

b) Each pair of corresponding sides has the same length, so the Side-Side-Side (SSS) method can be used to prove the triangles are congruent.

c) On both triangles, two corresponding angles and the side that connect them measure the same, so the Angle-Side-Angle (ASA) method can be used to prove that the triangles are congruent.

3. Find the perimeter of each triangle. Assume the triangles are similar.

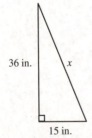

Solution

Write a proportion to find x.

$$\frac{x}{13} = \frac{36}{12} \text{ or } \frac{x}{13} = \frac{3}{1}$$

$$x \cdot 1 = 3 \cdot 13$$

$$x = 39 \text{ in.}$$

The perimeter of the triangle is 36 in. + 15 in. + 39 in. = 90 in.

Write a proportion to find y.

$$\frac{15}{y} = \frac{36}{12} \text{ or } \frac{15}{y} = \frac{3}{1}$$

$$3 \cdot y = 15 \cdot 1$$

$$\frac{3 \cdot y}{3} = \frac{15}{3}$$

$$y = 5 \text{ in.}$$

The perimeter of the triangle is 12 in. + 5 in. + 13 in. = 30 in.

4. A 5-foot-tall person casts a shadow that is 3 feet long. At the same time, another person casts a shadow that is 3.6 feet long. Find the height of the second person.

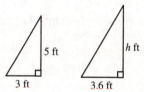

Solution

Write a proportion to find h.

$$\frac{3}{3.6} = \frac{5}{h}$$

$$3 \cdot h = 3.6 \cdot 5$$

$$\frac{3 \cdot h}{3} = \frac{18}{3}$$

$$h = 6 \text{ ft}$$

The second person is 6 feet tall.

6.1 Circle Graphs

Learning Objectives:

1. Interpret circle graphs.
2. Construct circle graphs.
3. Solve application problems.

Solved Examples:

1. An Olympic softball pitcher has thrown 680 pitches during the first part of the softball season. The following circle graph shows the results of her pitches. Use the circle graph to answer the accompanying questions.

Results of 680 Pitches

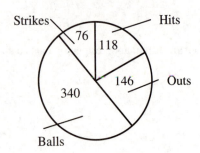

a) What category had the fewest pitches?

b) How many pitches were balls?

c) How many pitches were either hits or balls?

d) What is the ratio of the number of strikes to the total number of pitches?

e) What is the ratio of the number of balls to the number of strikes?

Solution

a) The category with fewest pitches is "strikes" with 76 pitches. Note that this category has the smallest pie-shaped section of the circle.

b) From the graph, 340 pitches were balls.

c) 118 pitches were hits and 340 pitches were balls. If the add these numbers, we have $118 + 340 = 458$. Thus, 458 pitches were either hits or balls.

d) 76 pitches were strikes and 680 pitches were thrown in total.

$$\frac{76}{680} = \frac{76 \div 4}{680 \div 4} = \frac{19}{170}$$

So, the ratio of the number of strikes to the total number of pitches is $\frac{19}{170}$.

e) 340 pitches were balls and 76 pitches were strikes.

$$\frac{340}{76} = \frac{340 \div 4}{76 \div 4} = \frac{85}{19}$$

So, the ratio of the number of balls to the number of strikes is $\frac{85}{19}$.

2. The following circle graph indicates the percent of the total game points that were scored in each quarter of a Celtics game. Use the circle graph to answer the accompanying questions.

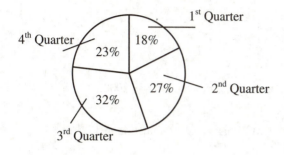

Celtics Scoring per Quarter

a) In which quarter did the Celtics score the most points?

b) What percent of the points were scored during the first two quarters?

c) If the total number of points scored during the game was 90, how many points were score during the fourth quarter?

d) How many points were scored in the last two quarters?

Solution

a) The largest percent corresponds to the quarter with the most points, which is the 3rd quarter. Thus, the Celtics scored the most points in the 3rd quarter.

b) We add 18% for the 1st quarter and 27% for the 2nd quarter: $18\% + 27\% = 45\%$. Thus, 45% of the points were scored during the first two quarters.

c) Since 23% of the points were scored in the 4th quarter, we must find 23% of 90:
    $(0.23)(90) = 20.7$. Rounding to the nearest whole number, approximately 21 points were scored during the 4th quarter.

d) We add 32% for the 3rd quarter and 23% for the 4th quarter: $32\% + 23\% = 55\%$. Thus, 55% of the points were scored in the last two quarters. We must find 55% of 90: $(0.55)(90) = 49.5$. Rounding, we conclude that approximately 50 points were scored in the last two quarters.

3. At *Big-M University*, the total enrollment is 12,000 students. If the number of accounting majors is 2760, how many degrees should be used for the accounting sector when constructing a circle graph?

Solution

Find the percent of the total enrollment that accounting majors represent.
$$\frac{2760}{12,000} = 0.23 = 23\%$$

Multiply the percent by 360 to find the size of the sector.
$(0.23)(360°) = 82.8°$

$82.8°$ should be used for the accounting sector.

6.2 Bar Graphs and Line Graphs

Learning Objectives:

1. Interpret bar graphs.
2. Interpret line graphs.

Solved Examples:

1. The following bar graph shows the population of Greencastle, IN, from 1940 to 2000.

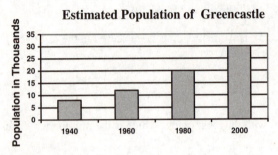

a) What was the population in 1980?

b) What was the population in 1940?

c) Between what years did the population of Greencastle increase the smallest amount? What is the amount?

Solution

a) The bar that represents 1980 rises to 20,000. Thus, the population in 1980 was 20,000.

b) The bar that represents 1940 rises to 8,000. Thus, the population in 1940 was 8,000.

c) The smallest difference in the heights of the bars occurs between 1940 and 1960. The population in 1940 was 8000; the population in 1960 was 12,000. The difference is $12,000 - 8000 = 4000$.

2. The following double bar graph shows the points scored per quarter in a basketball game.

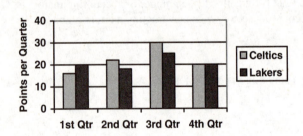

a) How many points did the Celtics score in the 3rd quarter?

b) How many points did the Lakers score in the 4th quarter?

c) In what quarter was there the biggest difference in scoring? What was the difference?

Solution

a) The bar representing the Celtics rises to 30 for the 3rd quarter. Thus, 30 points were scored by the Celtics in the 3rd quarter.

b) The bar representing the Lakers rises to 20 for the 4th quarter. Thus, 20 points were scored by the Lakers in the 4th quarter.

c) Looking at the double-bar graph, the biggest difference in the heights of the bars representing the Celtics and Lakers (by quarter) occurs in the 3rd quarter.

In the 3rd quarter, the bar the represents the Celtics rises to 30 and the bar that represents the Lakers rises to 25. Now, $30 - 25 = 5$. Thus, the Celtics scored about 5 points more than the Lakers in the 3rd quarter.

3.   The following line graph shows the average monthly rent in Bloomington, IN.

Monthly Rent in Bloomington

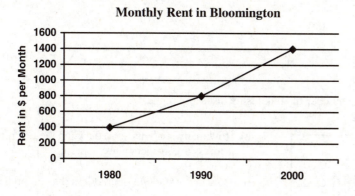

a) What was the rent in 1990?

b) Between which 10 years did the rent increase the most?

c) What was the increase in rent between 1980 and 2000?

Solution

a) The point for 1990 lies at 800. Thus, rent was $800 in 1990.

b) The line is steepest between 1990 and 2000. Thus, rent increased the most between 1990 and 2000.

c) The rent in 1980 was $400; the rent in 2000 was $1400. The difference is $1400 - \$400 = \$1000$.

6.3 Frequency Distributions and Histograms

Learning Objectives:

1. Interpret histograms.
2. Complete frequency tables.

Solved Examples:

1. The following histogram shows the number of students by test score range.

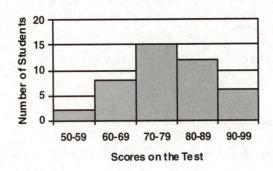

a) How many students scored a C on the test if the Professor considers a test score of 70-79 a C?

b) How many students scored less than 70 on the test?

c) How many students scored between 80 and 99 on the test?

Solution

a) Since the 70–79 bar rises to a height of 15, fifteen students scored a C on the test.

b) From the histogram, we see that there are two different bar heights to be included. Two tests were scored 50–59 and eight tests were scored 60–69. We add to find $2 + 8 = 10$. Thus, 10 students scored less than 70 on the test.

c) From the histogram, we see that there are two different bar heights to be included. Twelve tests were scored 80–89 and six tests were scored 90–99. We add to find $12 + 6 = 18$. Thus, 18 students scored between 80 and 99 on the test.

2. Each italicized/bold number below is the number of miles the Rose family has driven in a week.

a) Determine the frequency of the class intervals.

				Miles Driven (Class Interval)	Tally	Frequency
152	*236*	*563*	*363*	100 – 199	____	____
162	*333*	*257*	*380*	200 – 299	____	____
415	*498*	*140*	*299*	300 – 399	____	____
385	*213*	*545*	*367*	400 – 499	____	____
315	*293*	*415*	*312*	500 – 599	____	____

b) Construct a histogram using the frequency data.

Solution

a)   Miles Driven

(Class Interval)	Tally	Frequency
100 – 199	|||	3
200 – 299	LHN	5
300 – 399	LHN ||	7
400 – 499	|||	3
500 – 599	||	2

b)

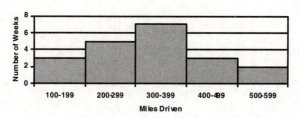

6.4 Mean, Median, and Mode

Learning Objectives:

1. Find the mean and weighted mean.
2. Find the median.
3. Find the mode.
4. Solve application problems.

Solved Examples:

1. Find the mean of each list of numbers. Round to the nearest thousandth when necessary.

   a) The numbers of points Joe scored playing basketball over the last four games were as follows: 18, 7, 25, 18, 12, 15, 22, 16

   b) The numbers of miles Jackie ran over the last four days were as follows: 5, 9, 10, 7, 8

   c) The captain of the college baseball team achieved the following results:

	Hits	Times at Bat
Game 1	1	4
Game 2	3	6
Game 3	2	5

   Find his batting average by dividing his total number of hits by the total times at bat.

   Solution

   a) To find the mean, we divide the sum of the values by the number of values.
   $$\frac{18+7+25+18+12+15+22+16}{8} = \frac{133}{8} = 16.625$$
   Thus, Joe scored a mean of 16.625 points per game.

   b) To find the mean, we divide the sum of the values by the number of values.
   $$\frac{5+9+10+7+8}{5} = \frac{39}{5} = 7.8$$
   Thus, Jackie ran a mean of 7.8 miles per day.

   c) To find his batting average, we divide his total number of hit by his total times at bat.
   $$\frac{1+3+2}{4+6+5} = \frac{6}{15} = 0.4$$
   Extending to the thousandths place, his batting average was 0.400 (hits per at-bat).

2. On Mr. Vetere's first math exam, 5 students received a B, 4 students received a C, 3 students received an A, and 3 students received a D. Letter grades are valued 4.0, 3.0, 2.0, 1.0, and 0.0 for A, B, C, D, and F, respectively. Compute the weighted average of the scores on this exam.

Solution

Number of Students	Grade	Numbers of Students • Grade
5	B (= 3.0)	5•3 = 15
4	C (= 2.0)	4•2 = 8
3	A (= 4.0)	3•4 = 12
3	D (= 1.0)	3•1 = 3

$$\text{weighted mean} = \frac{\text{sum of products}}{\text{total number of students}}$$
$$= \frac{15+8+12+3}{5+4+3+3}$$
$$= \frac{38}{15}$$
$$\approx 2.53$$

The weighted average of the scores on this exam is 2.53.

3. Find the median of each list of numbers. Round to the nearest tenth, if necessary.

a) Number of hits in 5 softball games: 5, 11, 15, 7, 2

b) Number of pages in 6 books: 323, 250, 346, 311, 358, 409

c) Annual salaries of local cable television employees: $28,500, $34,340, $30,885, $45,429, $33,479

Solution

a) To find the median, we must first arrange the numbers in order from smallest to largest. Since the number of values is odd, the median is the middle value.

$$\underbrace{2,\ 5}_{\substack{\text{two}\\\text{numbers}}}\ \underbrace{7}_{\substack{\text{middle}\\\text{number}}}\ \underbrace{11,\ 15}_{\substack{\text{two}\\\text{numbers}}}$$

Thus, the median is 7 hits.

b) To find the median, we must first arrange the numbers in order from smallest to largest. Since the number of values is even, the median is the mean of the middle two values.

$$\underbrace{250,\ 311}_{\substack{\text{two}\\\text{numbers}}}\ \underbrace{323,\ 346}_{\substack{\text{two middle}\\\text{number}}}\ \underbrace{358,\ 409}_{\substack{\text{two}\\\text{numbers}}}$$

Therefore, the median is $\dfrac{323+346}{2} = \dfrac{669}{2} = 334.5$ pages.

c) To find the median, we must first arrange the numbers in order from smallest to largest. Since the number of values is odd, the median is the middle value.

$$\underbrace{\$28,500,\ \$30,855}_{\substack{\text{two}\\\text{numbers}}}\ \underbrace{\$33,479}_{\substack{\text{middle}\\\text{number}}}\ \underbrace{\$34,340,\ \$45,429}_{\substack{\text{two}\\\text{numbers}}}$$

Thus, the median is $33,479.

4. Find the mode(s) of each list of numbers.

    a) Number of touchdowns scored in the first eight football games: 3, 2, 2, 1, 0, 3, 2, 4

    b) Daily high temperatures $(^\circ F)$ for one week: 83°, 85°, 88°, 90°, 88°, 83°, 81°

    c) Flight time (in minutes) for five flights from St. Louis to Chicago: 65, 67, 73, 68, 78

Solution

a) The value 2 occurs three times, whereas each of the other values occur few than three times. Therefore, the mode is 2 touchdowns.

b) The value 83 occurs twice, as does the value 88. All of the other values occur only once. Thus, we have two modes: $83^\circ F$ and $88^\circ F$.

c) Each value occurs only once. Therefore, the set of data has no mode.

7.1 Exponents, Order of Operations, and Inequality

Learning Objectives:

1. Evaluate exponential expressions.
2. Use order of operations.
3. Use inequality symbols.
4. Translate between word statements and symbols.
5. Reverse inequality symbols.
6. Solve applications.

Solved Examples:

1. Find the value of each exponential expression.

   a) 6^2

   b) 3^4

   c) $\left(\dfrac{3}{4}\right)^4$

   Solution

   a) $6^2 = 6 \cdot 6 = 36$

   b) $3^4 = 3 \cdot 3 \cdot 3 \cdot 3 = 81$

   c) $\left(\dfrac{3}{4}\right)^4 = \dfrac{3}{4} \cdot \dfrac{3}{4} \cdot \dfrac{3}{4} \cdot \dfrac{3}{4} = \dfrac{81}{256}$

2. Find the value of each expression.

   a) $4 - 8 \div 2$

   b) $2^2 \cdot 3 - 3$

   c) $10 - 3^2 + 2$

   d) $16 - 32 \div 2^3$

   e) $6 + \dfrac{16 - 4}{2 + 2^2} - 2$

   f) $24 \div \dfrac{3^2}{9 - 6} + 5$

   g) $2(3+1) + \left[3(4-2) + 4\right]$

   Solution

   a) $4 - 8 \div 2 = 4 - (8 \div 2)$ Divide first.
   $= 4 - 4$ Subtract.
   $= 0$

   b) $2^2 \cdot 3 - 3 = 4 \cdot 3 - 3$ Apply the exponent.
   $= 12 - 3$ Multiply.
   $= 9$ Subtract.

   c) $10 - 3^2 + 2 = 10 - 9 + 2$ Apply the exponent.
   $= 3$ Add and subtract, working from left to right.

   d) $16 - 32 \div 2^3 = 16 - 32 \div 8$
          Apply the exponent.
   $= 16 - 4$ Divide.
   $= 12$ Subtract.

e) $6 + \dfrac{16-4}{2+2^2} - 2 = 6 + \dfrac{16-4}{2+4} - 2$

$\qquad\qquad$ Apply the exponent.

$\qquad = 6 + \dfrac{12}{6} - 2$

$\qquad\qquad$ Simplify the
$\qquad\qquad$ numerator and
$\qquad\qquad$ denominator.

$\qquad = 6 + 2 - 2$ $\quad$ Divide.

$\qquad = 6$ $\qquad\qquad$ Add and subtract,
$\qquad\qquad$ working from left
$\qquad\qquad$ to right.

f) $24 \div \dfrac{3^2}{9-6} + 5 = 24 \div \dfrac{9}{9-6} + 5$

$\qquad\qquad$ Apply the exponent.

$\qquad = 24 \div \dfrac{9}{3} + 5$

$\qquad\qquad$ Simplify the
$\qquad\qquad$ denominator.

$\qquad = 24 \div 3 + 5$ $\;$ Divide.

$\qquad = 8 + 5$ $\qquad$ Divide.

$\qquad = 13$ $\qquad\quad$ Add.

g) $2(3+1) + \left[3(4-2)+4\right]$

$\quad = 2(4) + \left[3(2)+4\right]$ $\quad$ Add inside
$\qquad\qquad\qquad\qquad\qquad$ parentheses.

$\quad = 8 + \left[6+4\right]$ $\qquad\quad$ Multiply from left
$\qquad\qquad\qquad\qquad\qquad$ to right.

$\quad = 8 + 10$ $\qquad\qquad$ Add inside brackets.

$\quad = 18$ $\qquad\qquad\quad$ Add.

3. Using inequality symbols:

a) Determine whether the statement is true or false: $16 \leq 5$.

b) Write the statement in symbols. *Seven is not equal to 5.*

c) Write the statement with the inequality symbol reversed: $13 > 8$

d) Write the statement in words: $5 > 1$

Solution

a) The statement $16 \leq 5$ is false because $16 > 5$.

b) $7 \neq 5$

c) $8 < 13$

d) Five is greater than one.

7.2 Variables, Expressions, and Equations

Learning Objectives:

1. Evaluate algebraic expressions.
2. Rewrite phrases as algebraic expressions.
3. Decide whether a given number is a solution to an equation.
4. Write sentences as equations.
5. Differentiate between expressions or equations.
6. Solve modeling problems.

Solved Examples:

1. Evaluate the expressions when $a = 2$, $b = 3$, and $c = 6$.

   a) $2a + 3b$

   b) $\dfrac{5(a+b)-1}{c}$

   c) $\dfrac{5ab}{6} + 3cb$

   d) $\dfrac{a+b^3-1}{7} - c$

   Solution

   a) $2a + 3b$
   $= 2 \cdot 2 + 3 \cdot 3$ Replace a with 2 and b with 3.
   $= 4 + 9$ Multiply.
   $= 13$ Add.

   b) $\dfrac{5(a+b)-1}{c}$

   $= \dfrac{5(2+3)-1}{6}$ Replace a with 2, b with 3 and c with 6.

   $= \dfrac{5(5)-1}{6}$ Add.

   $= \dfrac{25-1}{6}$ Multiply.

   $= \dfrac{24}{6}$ Subtract.

   $= 4$ Divide.

   c) $\dfrac{5ab}{6} + 3cb$

   $= \dfrac{5 \cdot 2 \cdot 3}{6} + 3 \cdot 6 \cdot 3$ Replace a with 2, b with 3 and c with 6.

   $= \dfrac{30}{6} + 54$ Multiply.

   $= 5 + 54$ Divide.

   $= 59$ Add.

   d) $c - \dfrac{a+b^3-1}{7}$

   $= 6 - \dfrac{2+3^3-1}{7}$ Replace a with 2, b with 3 and c with 6.

   $= 6 - \dfrac{2+9-1}{7}$ Apply the exponent.

   $= 6 - \dfrac{10}{7}$ Add in the numerator.

   $= \dfrac{42}{7} - \dfrac{10}{7}$ Change 6 to an improper fraction.

   $= \dfrac{32}{7}$ Subtract.

2. Write each word phrase as an algebraic expression, using x as the variable.

a) The quotient of 6 less than a number and 3.

b) Fifty more than a number.

c) The product of 8 and the total of a number and 5.

d) The square of the difference between a number and 13.

Solution

a) *Quotient* is the answer to a division problem, and *less than* indicates subtraction. The phrase translates to $\dfrac{x}{3} - 6$.

b) *More than* indicates addition. The phrase translates to $50 + x$.

c) *Product* indicates multiplication, and *total* indicates addition. The phrase translates to $8(x+5)$.

d) *Difference* indicates subtraction. The phrase translates to $(x-13)^2$.

3. Decide whether the given number is a solution of the equation.

a) Is 0 a solution of $6 - 3x = 6 - 5x$?

b) Is 5 a solution of $x + 4 = 10$?

c) Is 9 a solution of $12 = 3 + x$?

d) Is 10 a solution of $4 - \dfrac{3}{4}x + 2 = 0$?

Solution

a) $6 - 3x = 6 - 5x$

$6 - 3(0) \overset{?}{=} 6 - 5(0)$ Replace x with 0.

$6 - 0 \overset{?}{=} 6 - 0$ Multiply.

$6 = 6$ True.

0 is a solution of the equation.

c) $12 = 3 + x$

$12 \overset{?}{=} 3 + 9$ Replace x with 9.

$12 = 12$ True.

9 is a solution of the equation.

b) $x + 4 = 10$

$5 + 4 \overset{?}{=} 10$ Replace x with 5.

$9 = 10$ False.

5 is not a solution of the equation.

d) $4 - \dfrac{3}{4}x + 2 = 0$

$4 - \dfrac{3}{4}(10) + 2 \overset{?}{=} 0$ Replace x with 10.

$4 - \dfrac{15}{2} + 2 \overset{?}{=} 0$ Multiply.

$\dfrac{8}{2} - \dfrac{15}{2} + \dfrac{4}{2} \overset{?}{=} 0$ Convert 4 and 2 to improper fractions.

$-\dfrac{3}{2} = 0$ False

10 is not a solution of the equation.

4. Write each word sentence as an equation. Use x as the variable.

   a) The difference between a number and 12 is 3.
   b) The quotient of a number and 3 is 2.

   c) Twice the difference between a number and 16 is 2 times the number.
   d) Eight less than 5 times a number is 4 more than 2 times the number.

Solution

a) $x - 12 = 3$

b) $\dfrac{x}{3} = 2$

c) $2(x - 16) = 2x$

d) $5x - 8 = 4 + 2x$

7.3 Real Numbers and the Number Line

Learning Objectives:

1. Classify numbers.
2. Use an integer to represent a change.
3. Graph numbers on a number line.
4. Select the lesser number in a pair of numbers.
5. Find the opposite and absolute value of a number.
6. Simplify absolute value expressions.
7. Solve applications.

Solved Examples:

1. List all numbers from the set $\left\{ -10, \sqrt{5}, -2\frac{3}{4}, 0, -5, 10 \right\}$ that are:

   a) natural numbers   b) whole numbers   c) integers

   d) rational numbers   e) irrational numbers   f) real numbers

   Solution

   a) 10   b) 0, 10   c) $-10, 0, 10$

   d) $-10, -2\frac{3}{4}, 0, -5, 10$ e) $\sqrt{5}$ f) $-10, \sqrt{5}, -2\frac{3}{4}, 0, -5, 10$

2. Graph each of the numbers on a number line: $-2, 5, -4\frac{1}{3}, 1\frac{1}{2}$.

   Solution

3. Select the lesser number in each pair.

   a) $-3, 8$ b) $-\frac{4}{5}, -\frac{1}{3}$ c) $10, -12$

   Solution

   a) -3 b) $-\frac{4}{5}$ c) -12

4. For each number, find its opposite

   a) 6   b) -5 c) $-\frac{5}{6}$

   Solution

   a) 6   b) 5   c) $\frac{5}{6}$

5. Simplify.

a) $|-3|$ b) $-|-6|$ c) $|13-16|$

Solution

a) 3   b) -6 c) 3

6. Decide whether the statement is true or false.

     a) $|-10| > 2$ b) $|-3| \leq |-5|$ c) $-|-7| > 1$

Solution

a) True
$|-10| = 10$ and $10 > 2$.

b) True
$|-3| = 3,\ |-5| = 5,$ and $3 < 5$.

c) False
$-|-7| = 7$ and $-7 < 1$.

7.4 Adding Real Numbers

Learning Objectives:

1. Add signed numbers.
2. Evaluate whether statements are true or false.
3. Write a numerical expression for a phrase and simplify it.
4. Solve applications.

Solved Examples:

1. Find each sum.

   a) $9 + 12$

   b) $-6 + (-10)$

   c) $-14 + (-26)$

   d) $-6.3 + 5.2$

   e) $-12 + (-6) + 17$

   f) $-2 + [6 + (-1) + 7]$

   g) $\dfrac{7}{10} + \left(-\dfrac{2}{5}\right)$

   Solution

   a) Add as usual since the signs of the addends are both positive.

   $9 + 12 = 21$

   b) Add the absolute values of the numbers. Give the sum the same sign as the numbers being added.

   $-6 + (-10) = -\left(|-6| + |-10|\right) = -(6 + 10) = -14$

   c) Add the absolute values of the numbers. Give the sum the same sign as the numbers being added.

   $\begin{aligned} -14 + (-26) &= -\left(|-14| + |-26|\right) \\ &= -(14 + 26) = -40 \end{aligned}$

   d) Find the absolute values of the numbers, and subtract the lesser absolute value from the greater. Give the answer the sign of the number having the greater absolute value.

   $-6.3 + 5.2 = -(6.3 - 5.2) = -1.1$

   e) $-12 + (-6) + 17 = -18 + 17 = -1$

   f) $\begin{aligned} -2 + [6 + (-1) + 7] &= -2 + [5 + 7] \\ &= -2 + 12 = 10 \end{aligned}$

   g) $\dfrac{7}{10} + \left(-\dfrac{2}{5}\right) = \dfrac{7}{10} + \left(-\dfrac{4}{10}\right) = \dfrac{7}{10} - \dfrac{4}{10} = \dfrac{3}{10}$

2. Write a numerical expression for each phrase, and simplify the expression.

   a) The sum of −6 and 3 and −1

   b) The sum of −10 and −15, increased by 12

   c) 0.29 more than the sum of 3.56 and −2.41

   Solution

   a) $-6 + 3 + (-1) = -3 + (-1) = -4$

   b) $\begin{aligned} [-10 + (-15)] + 12 &= (-25) + 12 \\ &= -(25 - 12) = -13 \end{aligned}$

   c) $\begin{aligned} 0.29 + [3.56 + (-2.41)] &= 0.29 + (3.56 - 2.41) \\ &= 0.29 + 1.14 \\ &= 1.43 \end{aligned}$

3. Solve: A scuba diver is at a depth of 16 feet below the surface. He descends another 8 feet. What is his new depth?

   Solution

   Depth below the surface and number of feet descended are both represented by negative numbers.

   $-16 + (-8) = -24$

   The diver is 24 feet below the surface.

4. Solve: On January 14, in New Market, Indiana, the temperature rose 17° F in three hours. If the starting temperature was –5° F, what was the temperature three hours later?

   Solution

   Temperature rising is represented by a positive number.

   $-5 + 17 = 12$

   The temperature was 12°F three hours later.

7.5 Subtracting Real Numbers

Learning Objectives:

1. Subtract real numbers.
2. Write numerical expressions and simplify.
3. Solve applications.

Solved Examples:

1. Rewrite the following as addition problems using the definition of subtraction.

   a) $7-2$

   b) $-3-6$

   c) $13-(-1)$

   Solution

   a) $7-2 = 7+(-2)$

   b) $-3-6 = -3+(-6)$

   c) $13-(-1) = 13+1$

2. Find each difference.

   a) $3-7$

   b) $-2-8$

   c) $-6.3-(-4.1)$

   d) $-16-(-30)-12$

   e) $14-(8-19)$

   f) $-2+\left[(-5-11)-(4+6)\right]$

   Solution

   a) $3-7 = 3+(-7) = -4$

   b) $-2-8 = -2+(-8) = -10$

   c) $-6.3-(-4.1) = -6.3+4.1 = -2.2$

   d) $-16-(-30)-12 = -16+30+(-12) = 2$

   e) $14-(8-19) = 14-[8+(-19)] = 14-(-11)$
   $= 14+11 = 25$

   f) $-2+\left[(-5-11)-(4+6)\right]$
   $= -2+\left[(-5+(-11))-10\right]$
   $= -2+\left[-16+(-10)\right]$
   $= -2+(-26) = -28$

3. Write a numerical expression for each phrase and simplify.

   a) The difference between 3 and -6

   b) 13 less than -5

   c) Two less than the difference between -5 and -7

   Solution

   a) $3-(-6) = 3+6 = 9$

   b) $-5-13 = -5+(-13) = -18$

   c) $[-5-(-7)]-2$
   $= [-5+7]-2$
   $= -2-2$
   $= -2+(-2) = -4$

4. Solve: A scuba diver was at a depth of 17 feet below the surface. A wrecked ship was 12 feet lower than the diver. What was the depth of the wrecked ship?

    Solution

    Depth below the surface is represented by a negative number.

    $-17 - 12 = -17 + (-12) = -29$

    The wrecked ship was 29 feet below the surface.

5. Solve: The Terre Haute YWCA showed a profit of $72,000 in the year 2007, while it had a loss of $19,000 in the year 2008. Find the difference between the amounts.

    Solution

    A profit is represented by a positive number. A loss is represented by a negative number.

    $72,000 - (-19,000) = 72,000 + 19,000 = 91,000$

    There was a difference of $91,000.

7.6 Multiplying and Dividing Real Numbers

Learning Objectives:

1. Multiply real numbers.
2. Divide real numbers.
3. Simplify expressions using order of operations.
4. Evaluate algebraic expressions.
5. Write numerical expressions and simplify.
6. Translate sentences to equations.

Solved Examples:

1. Multiply. Write answers in simplest form.

   a) $-2(0)$

   b) $4(-15)$

   c) $(-30)(-5)$

   d) $(2.2)(-3.3)$

   e) $\left(-\dfrac{3}{4}\right)\left(-\dfrac{8}{9}\right)$

   Solution

   a) The product of any number and 0 is 0, so $-2(0) = 0$.

   b) The product of two numbers having different signs is negative, so $4(-15) = -60$.

   c) The product of two numbers having the same sign is positive, so $(-30)(-5) = 150$.

   d) $(2.2)(-3.3) = -7.26$

   e) $\left(-\dfrac{3}{4}\right)\left(-\dfrac{8}{9}\right)$

2. Divide.

   a) $-16 \div 8$

   b) $\dfrac{-9}{-3}$

   c) $\dfrac{-4.6}{0}$

   d) $-\dfrac{3}{5} \div \dfrac{15}{20}$

   e) $\dfrac{0}{-4}$

   Solution

   a) The quotient of two numbers having different signs is negative., so $-16 \div 8 = -2$.

   b) The quotient of two numbers having the same sign is positive, so $\dfrac{-9}{-3} = 3$.

   c) Division by zero is undefined, so $\dfrac{-4.6}{0}$ is undefined

   d) $-\dfrac{3}{5} \div \dfrac{15}{20} = -\dfrac{3}{5} \cdot \dfrac{20}{15} = -\dfrac{4}{5}$

   e) Zero divided by any nonzero number is 0 so, $\dfrac{0}{-4} = 0$.

3. Perform each indicated operation.

a) $\dfrac{4(-7)}{-28}$

b) $\dfrac{9(-8)+5(-1)}{12-1}$

c) $\dfrac{6^2-4^2}{-2[5-(-1)]}$

Solution

a) $\dfrac{4(-7)}{-28}=\dfrac{-28}{-28}=1$

b) $\dfrac{9(-8)+5(-1)}{12-1}=\dfrac{-72+(-5)}{11}$

$=\dfrac{-77}{11}=-7$

c) $\dfrac{6^2-4^2}{-2[5-(-1)]}=\dfrac{36-16}{-2(6)}$

$=\dfrac{20}{-12}=-\dfrac{5}{3}$

4. Evaluate each expression if $x=-2$, $y=4$, $a=-1$, and $b=3$.

a) $(y+b)^2-4x$

b) $\dfrac{y-2x}{b-ya^2}$

Solution

a) $(y+b)^2-4x=(4+3)^2-4(-2)$

$=7^2-4(-2)$

$=49+8=57$

b) $\dfrac{y-2x}{b-ya^2}=\dfrac{4-2(-2)}{3-4(-1)^2}$

$=\dfrac{4+4}{3-4(1)}=\dfrac{8}{3-4}$

$=\dfrac{8}{-1}=-8$

5. Write a numerical expression for each phrase and simplify.

a) The product of –5 and the difference between 4 and –9.

b) The quotient of –18 and the sum of –7 and –2.

Solution

a) $-5[4-(-9)]=-5[4+9]$

$=-5(13)=-65$

b) $\dfrac{-18}{-7+(-2)}=\dfrac{-18}{-9}=2$

6. Write each sentence with symbols, using x to represent the number.

a) Seven times a number is –35.

b) $\frac{1}{3}$ less than a number is 6.

Solution

a) $7x=-35$

b) $x-\frac{1}{3}=6$

7.7 Properties of Real Numbers

Learning Objectives:

1. Identify properties of real numbers.
2. Write an equivalent expression using the given property.
3. Simplify expressions using properties of real numbers.
4. Rewrite expressions using the distributive property.

Solved Examples:

1. Name the property (commutative, associative, identity property, inverse property, or distributive property) illustrated by each statement.

   a) $3 + (-7) = -7 + 3$

   b) $\left(\dfrac{2}{3}\right)\left(\dfrac{3}{2}\right) = 1$

   c) $-3 + \left[6 + (-2)\right] = (-3 + 6) + (-2)$

   d) $-\dfrac{1}{5} + \dfrac{1}{5} = 0$

   e) $-5(c + d) = -5c - 5d$

   f) $(-5 + 7) + 10 = 10 + (-5 + 7)$

   g) $7\left(-\dfrac{3}{7}\right) = \left(-\dfrac{3}{7}\right)7$

   h) $-56 \cdot 1 = -56$

   Solution

   a) commutative property

   b) inverse property

   c) associative property

   d) identity property

   e) distributive property

   f) associative property

   g) commutative property

   h) identity property

2. Use the distributive property to rewrite each expression. Simplify if possible.

   a) $3(k + 6)$

   b) $-5(h - 2)$

   c) $3 \cdot y + 3 \cdot z$

   d) $-2(5a - 2b - c)$

   e) $-2 \cdot 7 + (-2) \cdot 3$

   Solution

   a) $3(k + 6) = 3k + 3(6)$
   $= 3k + 18$

   b) $-5(h - 2) = -5h - 5(-2)$
   $= -5h + 10$

   c) $3 \cdot y + 3 \cdot z = 3(y + z)$

   d) $-2(5a - 2b - c) = -2(5a) - 2(-2b) - 2(-c)$
   $= -10a + 4b + 2c$

   e) $-2 \cdot 7 + (-2) \cdot 3 = -2(7 + 3)$
   $= -2(10)$
   $= -20$

7.8 Simplifying Expressions

Learning Objectives:

1. Simplify expressions by combining terms.
2. Identify numerical coefficients.
3. Identify terms as like or unlike.
4. Simplify expressions.
5. Write phrases as mathematical expressions.

Solved Examples:

1. Simplify each expression.

   a) $8x - 3 + 5x$
   b) $3(2x - 7)$
   c) $-4 - (3m + 5)$

   Solution

   a) $8x - 3 + 5x = 8x + 5x - 3$
   $= 13x - 3$

   b) $3(2x - 7) = 3(2x) + 3(-7)$
   $= 6x - 21$

   c) $-4 - (3m + 5) = -4 - 3m - 5$
   $= -4 - 5 - 3m$
   $= -9 - 3m$

2. Give the numerical coefficient of each term.

   a) $4m$
   b) $-16x^2$
   c) $-p^2 q^{-1}$
   d) $\dfrac{-4r}{5}$

   Solution

   a) 4
   b) -16
   c) -1
   d) $-\dfrac{4}{5}$

3. Simplify each expression.

   a) $-2(3m + 6) + 4m$
   b) $3b + 6b$
   c) $\frac{2}{3}b - b$

   d) $7c - (5c + 8)$
   e) $3(e^2 - 4e) - 2(5e^2 - 1)$

   Solution

   a) $-2(3m + 6) + 4m = -2(3m) - 2(6) + 4m = -6m - 12 + 4m = -6m + 4m - 12$
   $= (-6 + 4)m - 12 = -2m - 12$

   b) $3b + 6b = (3 + 6)b = 9b$
   c) $\frac{2}{3}b - b = \left(\frac{2}{3} - 1\right)b = -\frac{1}{3}b$

   d) $7c - (5c + 8) = 7c - 5c - 8 = (7 - 5)c - 8 = 2c - 8$

   e) $3(e^2 - 4e) - 2(5e^2 - 1) = 3e^2 + 3(-4e) - 2(5e^2) - 2(-1) = 3e^2 - 12e - 10e^2 + 2$
   $= 3e^2 - 10e^2 - 12e + 2 = (3 - 10)e^2 - 12e + 2 = -7e^2 - 12e + 2$

4. Write each phrase as a mathematical expression. Use x to represent the number. Combine like terms when possible.

a) A number increased by the difference between 3 and the number.

b) Six plus the product of 4 more than a number and 2.

c) A number plus 7 added to the difference between 6 and twice the number.

Solution

a) $x + (3 - x) = x + 3 - x$
$$= x - x + 3 = 3$$

b) $6 + (4 + x)2$
$$= 6 + 4(2) + x(2)$$
$$= 6 + 8 + 2x$$
$$= 14 + 2x$$

c) $(x + 7) + (6 - 2x)$
$$= x + 7 + 6 - 2x$$
$$= x - 2x + 7 + 6$$
$$= (1 - 2)x + 13$$
$$= -x + 13$$

8.1 The Addition Property of Equality

Learning Objectives:

1. Solve equations using the addition property.
2. Simplify first, and then solve.

Solved Examples:

1. Decide whether each is an expression or an equation. If it is an expression, simplify it. If it is an equation, solve it.

   a) $3x + 10 - x - 3$

   b) $-7y + 13 + 8y = -6$

   Solution

   a) There is no equals sign, so this is an expression.
   $$3x + 10 - x - 3 = 3x - x + 10 - 3$$
   $$= 2x + 7$$

   b) There is an equals sign, so this is an equation.

$-7y + 13 + 8y = -6$	
$-7y + 8y + 13 = -6$	Commutative property
$y + 13 = -6$	Combine like terms.
$y + 13 - 13 = -6 - 13$	Subtract 13 from both sides.
$y = -19$	

2. Which of the pairs of equations are equivalent equations?

   a) $x + 5 = 8$ and $x = 3$

   b) $-12 = x - 10$ and $x = -2$

   c) $x - 15 = 5$ and $x = 6$

   d) $x + 23 = 30$ and $x = -9$

   Solution

   a) $x + 5 = 8$
   $x + 5 - 5 = 8 - 5$ Subtract 5 from both sides.
   $x = 3$
   The equations are equivalent.

   b) $-12 = x - 10$
   $-12 + 10 = x - 10 + 10$
   Add 10 to both sides.
   $-2 = x$
   The equations are equivalent.

   c) $x - 15 = 5$
   $x - 15 = 5 + 15$ Add 15 to both sides.
   $x = 20$
   The equations are not equivalent.

   d) $x + 23 = 30$
   $x + 23 - 23 = 30 - 23$ Subtract 23 from both sides.
   $x = 7$
   The equations are not equivalent.

3. Which of the following are linear equations in one variable?

a) $4x - 7 = 3x + 2$

b) $x^2 = 2x + 5$

c) $2x - 8 = 0$

d) $x^3 = 8x$

Solution

a) $4x - 7 = 3x + 2$
    This is a linear equation in one variable.

b) $x^2 = 2x + 5$
    This is not a linear equation because there is an x^2 term in the equation.

c) $2x - 8 = 0$
    This is a linear equation in one variable.

d) $x^3 = 8x$
    This is not a linear equation because there is an x^3 term in the equation.

4. Solve for x. Check your answers.

a) $x - 4 = 16$

b) $14 = x - 12$

c) $-19 = x + 16$

d) $21 = -16 + x$

e) $x - (-6) = 18$

f) $19 - 6 + x = 15 - 2$

g) $\dfrac{1}{4} + x = \dfrac{3}{4}$

h) $x - \dfrac{9}{10} = -\dfrac{2}{3} + \dfrac{1}{15}$

i) $-2.2 + x = 16$

j) $6 = 19.2 + x - 3.2$

Solution

a)
$x - 4 = 16$
$x - 4 + 4 = 16 + 4$ Add 4 to both sides.
$x = 20$
Check:
$x - 4 = 16$
$20 - 4 \overset{?}{=} 16$
$16 = 16$ [a]
Solution set: {20}

b)
$14 = x - 12$
$14 + 12 = x - 12 + 12$ Add 12 to both sides.
$26 = x$
Check:
$14 = x - 12$
$14 \overset{?}{=} 26 - 12$
$14 = 14$ [a]
Solution set: {26}

c)
$-19 = x + 16$
$-19 - 16 = x + 16 - 16$ Subtract 16 from both sides.
$-35 = x$
Check:
$-19 = x + 16$
$-19 \overset{?}{=} -35 + 16$
$-19 = -19$ [a]
Solution set: {-19}

d)
$21 = -16 + x$
$21 + 16 = -16 + x + 16$ Add 16 to both sides.
$37 = x$
Check:
$21 = -16 + x$
$21 \overset{?}{=} -16 + 37$
$21 = 21$ [a]
Solution set: {-19}

e)
$$x - (-6) = 18$$
$$x - (-6) + (-6) = 18 + (-6) \quad \text{Add } (-6) \text{ to both sides.}$$
$$x = 12$$

Check:
$$x - (-6) = 18$$
$$12 - (-6) \overset{?}{=} 18$$
$$12 + 6 \overset{?}{=} 18$$
$$18 = 18^{\text{a}}$$

Solution set: $\{12\}$

f)
$$19 - 6 + x = 15 - 2$$
$$13 + x = 13 \quad \text{Combine like terms.}$$
$$13 + x - 13 = 13 - 13 \quad \text{Subtract 13 from both sides.}$$
$$x = 0$$

Check:
$$19 - 6 + x = 15 - 2$$
$$19 - 6 + 0 \overset{?}{=} 15 - 2$$
$$13 = 13^{\text{a}}$$

Solution set: $\{0\}$

g)
$$\frac{1}{4} + x = \frac{3}{4}$$
$$\frac{1}{4} + x - \frac{1}{4} = \frac{3}{4} - \frac{1}{4} \quad \text{Subtract } \frac{1}{4} \text{ from both sides.}$$
$$x = \frac{2}{4} = \frac{1}{2}$$

Check:
$$\frac{1}{4} + x = \frac{3}{4}$$
$$\frac{1}{4} + \frac{1}{2} \overset{?}{=} \frac{3}{4}$$
$$\frac{1}{4} + \frac{2}{4} \overset{?}{=} \frac{3}{4}$$
$$\frac{3}{4} = \frac{3}{4}^{\text{a}}$$

Solution set: $\left\{\frac{3}{4}\right\}$

h)
$$x - \frac{9}{10} = -\frac{2}{3} + \frac{1}{15}$$
$$x - \frac{9}{10} + \frac{9}{10} = -\frac{2}{3} + \frac{1}{15} + \frac{9}{10} \quad \text{Add } \frac{9}{10} \text{ to both sides.}$$
$$x = -\frac{20}{30} + \frac{2}{30} + \frac{27}{30}$$
$$= \frac{9}{30} = \frac{3}{10}$$

Check:
$$x - \frac{9}{10} = -\frac{2}{3} + \frac{1}{15}$$
$$\frac{3}{10} - \frac{9}{10} \overset{?}{=} -\frac{2}{3} + \frac{1}{15}$$
$$-\frac{6}{10} \overset{?}{=} -\frac{10}{15} + \frac{1}{15}$$
$$-\frac{6}{10} \overset{?}{=} -\frac{9}{15}$$
$$-\frac{3}{5} = -\frac{3}{5}^{\text{a}}$$

Solution set: $\left\{\frac{3}{10}\right\}$

i)
$$-2.2 + x = 16$$
$$-2.2 + x + 2.2 = 16 + 2.2 \quad \text{Add 2.2 to both sides.}$$
$$x = 18.2$$

Check:
$$-2.2 + x = 16$$
$$-2.2 + 18.2 \overset{?}{=} 16$$
$$16 = 6^{\text{a}}$$

Solution set: $\{12\}$

j)
$$6 = 19.2 + x - 3.2$$
$$6 = 16 + x \quad \text{Combine like terms.}$$
$$6 - 16 = 16 + x - 16 \quad \text{Subtract 16 from both sides.}$$
$$-10 = x$$

Check:
$$6 = 19.2 + x - 3.2$$
$$6 \overset{?}{=} 19.2 + (-10) - 3.2$$
$$6 = 6^{\text{a}}$$

Solution set: $\{6\}$

8.2 The Multiplication Property of Equality

Learning Objectives:

1. Use the multiplication property of equality.
2. Simplify first, and then use the multiplication property.
3. Translate a sentence to an equation, and solve.

Solved Examples:

1. By what number is it necessary to multiply each side of each equation in order to isolate x on the left side? Do not solve.

   a) $\dfrac{1}{5}x = 3$ b) $\dfrac{x}{6} = -2$ c) $-\dfrac{3}{4}x = 21$ d) $-x = 41$

   Solution

   a) Multiply each side by 5.   b) Multiply each side by 6.   c) Multiply each side by $-\dfrac{4}{3}$.   d) Multiply each side by -1.

2. Solve for x. Be sure to reduce your answers. Check your answers.

   a) $\dfrac{1}{5}x = 6$ b) $\dfrac{1}{4}x = -25$

   c) $\dfrac{x}{12} = 5$ d) $-9 = \dfrac{x}{9}$

   e) $3x = 9$ f) $-11 = 2x$

   g) $1.2x = 90$ h) $-16 = -x$

   i) $-9.8x = -211$ j) $\dfrac{5}{6}x = -9$

   k) $8x - 4x - x = 6 - 4$ l) $-7 = 3x - x$

   Solution

   a) $\dfrac{1}{5}x = 6$

   $5 \cdot \dfrac{1}{5}x = 5 \cdot 6$ Multiply both sides by 5.

   $\quad x = 30$

   Check:

   $\dfrac{1}{5}x = 6$

   $\dfrac{1}{5} \cdot 30 \overset{?}{=} 6$

   $\quad 6 = 6^{\text{a}}$

   Solution set: $\{30\}$

   b) $\dfrac{1}{4}x = -25$

   $4 \cdot \dfrac{1}{4}x = 4 \cdot (-25)$ Multiply both sides by 4.

   $\quad x = -100$

   Check:

   $\dfrac{1}{4}x = -25$

   $\dfrac{1}{4}(-100) \overset{?}{=} -25$

   $\quad -25 = -25^{\text{a}}$

   Solution set: $\{-100\}$

c) $\dfrac{x}{12} = 5$

$12 \cdot \dfrac{x}{12} = 12 \cdot 5$ Multiply both sides by 12.

$x = 60$

Check:

$\dfrac{x}{12} = 5$

$\dfrac{60}{12} \overset{?}{=} 5$

$5 = 5^{a}$

Solution set: $\{60\}$

d) $-9 = \dfrac{x}{9}$

$-9 \cdot 9 = \dfrac{x}{9} \cdot 9$ Multiply both sides by 9.

$-81 = x$

Check:

$-9 = \dfrac{x}{9}$

$-9 \overset{?}{=} \dfrac{-81}{9}$

$-9 = -9^{a}$

Solution set: $\{-81\}$

e) $3x = 9$

$\dfrac{3x}{3} = \dfrac{9}{3}$ Divide each side by 3.

$x = 3$

Check:

$3x = 9$

$3 \cdot 3 \overset{?}{=} 9$

$9 = 9^{a}$

Solution set: $\{3\}$

f) $-11 = 2x$

$\dfrac{-11}{2} = \dfrac{2x}{2}$ Divide each side by 2.

$-\dfrac{11}{2} = x$

Check:

$-11 = 2x$

$-11 \overset{?}{=} 2 \cdot \left(-\dfrac{11}{2}\right)$

$-11 = -11^{a}$

Solution set: $\left\{-\dfrac{11}{2}\right\}$

g) $1.2x = 90$

$\dfrac{1.2x}{1.2} = \dfrac{90}{1.2}$ Divide each side by 1.2.

$x = 75$

Check:

$1.2x = 90$

$1.2(70) \overset{?}{=} 90$

$90 = 90^{a}$

Solution set: $\{75\}$

h) $-16 = -x$

$-16(-1) = -x(-1)$ Multiply each side by -1.

$16 = x$

Check:

$-16 = -x$

$-16 = -16^{a}$

Solution set: $\{16\}$

i) $-9.8x = -200.9$

$\dfrac{-9.8x}{-9.8} = \dfrac{-200.9}{-9.8}$ Divide each side by -9.8.

$x = 20.5$

Check:

$-9.8x = -200.9$

$-9.8(20.5) \overset{?}{=} -200.9$

$-200.9 = -200.9^{a}$

Solution set: $\{20.5\}$

j) $\dfrac{5}{6}x = -9$

$\dfrac{6}{5} \cdot \dfrac{5}{6}x = \dfrac{6}{5} \cdot (-9)$ Multiply each side by $\dfrac{6}{5}$.

$x = -\dfrac{54}{5}$

Check:

$\dfrac{5}{6}x = -9$

$\dfrac{5}{6}\left(-\dfrac{54}{5}\right) \overset{?}{=} -9$

$-9 = -9^{a}$

Solution set: $\left\{-\dfrac{54}{5}\right\}$

k) $8x - 4x - x = 6 - 4$

$\qquad 3x = 2$ Combine like terms.

$\qquad \dfrac{3x}{3} = \dfrac{2}{3}$ Divide each side by 3.

$\qquad x = \dfrac{2}{3}$

Check:

$$8x - 4x - x = 6 - 4$$

$$8\left(\dfrac{2}{3}\right) - 4\left(\dfrac{2}{3}\right) - \left(\dfrac{2}{3}\right) \overset{?}{=} 6 - 4$$

$$\dfrac{16}{3} - \dfrac{8}{3} - \dfrac{2}{3} \overset{?}{=} 2$$

$$\dfrac{6}{3} \overset{?}{=} 2$$

$$2 = 2^{\text{a}}$$

Solution set: $\left\{ \dfrac{2}{3} \right\}$

$-7 = 3x - x$

$-7 = 2x$ Combine like terms.

l) $\quad \dfrac{-7}{2} = \dfrac{2x}{2}$ Divide each term by 2.

$\qquad -\dfrac{7}{2} = x$

Check:

$$-7 = 3x - x$$

$$-7 \overset{?}{=} 3\left(-\dfrac{7}{2}\right) - \left(-\dfrac{7}{2}\right)$$

$$-7 \overset{?}{=} -\dfrac{21}{2} + \dfrac{7}{2}$$

$$-7 \overset{?}{=} -\dfrac{14}{2}$$

$$-7 = -7^{\text{a}}$$

Solution set: $\left\{ -\dfrac{7}{2} \right\}$

8.3 More on Solving Linear Equations

Learning Objectives:

1. Solve linear equations.
2. Write the answer as an algebraic expression.

Solved Examples:

1. Solve each equation. Check your solution.

   a) $10x + 7 = 107$

   b) $33 = 6x - 3$

   c) $164 = 15x + 14$

   d) $\dfrac{1}{2}x - 8 = -2$

   e) $-\dfrac{2}{3}x - 8 = -32$

   f) $8x - 6 = 3 + 9x$

   g) $-9x + 4 + 7x = -3x + 9$

   h) $6(2x - 1) = 30$

   i) $6(x - 8) = 6x - 48$

   j) $3x + 6(x + 9) = 9x - 15$

   k) $0.4x - 0.2(3 - x) = 6.6$

   l) $7x - 3(x - 8) = 3x + 24$

Solution

a)
$$10x + 7 = 107$$
$$10x + 7 - 7 = 107 - 7 \quad \text{Subtract 7.}$$
$$10x = 100$$
$$\frac{10x}{10} = \frac{100}{10} \quad \text{Divide by 10.}$$
$$x = 10$$
Check:
$$10x + 7 = 107$$
$$\overset{?}{10(10) + 7 = 107}$$
$$10 = 10^{a}$$
Solution set: $\{10\}$

b)
$$33 = 6x - 3$$
$$33 + 3 = 6x - 3 + 3 \quad \text{Add 3.}$$
$$36 = 6x$$
$$\frac{36}{6} = \frac{6x}{6} \quad \text{Divide by 6.}$$
$$6 = x$$
Check:
$$33 = 6x - 3$$
$$\overset{?}{33 = 6(6) + 3}$$
$$33 = 33^{a}$$
Solution set: $\{6\}$

c)
$$164 = 15x + 14$$
$$164 - 14 = 15x + 14 - 14 \quad \text{Add 14.}$$
$$150 = 15x$$
$$\frac{150}{15} = \frac{15x}{15} \quad \text{Divide by 15.}$$
$$10 = x$$
Check:
$$164 = 15x + 14$$
$$\overset{?}{164 = 15(10) + 14}$$
$$164 = 164^{a}$$
Solution set: $\{10\}$

d)
$$\frac{1}{2}x - 8 = -2$$
$$\frac{1}{2}x - 8 + 8 = -2 + 8 \quad \text{Add 8.}$$
$$\frac{1}{2}x = 6$$
$$2 \cdot \frac{1}{2}x = 2 \cdot 6 \quad \text{Multiply by 2.}$$
$$x = 12$$
Check:
$$\frac{1}{2}x - 8 = -2$$
$$\overset{?}{\frac{1}{2}(12) - 8 = -2} \quad \text{*Solution set*: } \{12\}$$
$$-2 = -2^{a}$$

e)
$$-\frac{2}{3}x - 8 = -32$$
$$-\frac{2}{3}x - 8 + 8 = -32 + 8 \qquad \text{Add 8.}$$
$$-\frac{2}{3}x = -24$$
$$\left(-\frac{3}{2}\right)\left(-\frac{2}{3}x\right) = \left(-\frac{3}{2}\right)(-24) \quad \text{Multiply by } -\frac{3}{2}.$$
$$x = 36$$

Check:
$$-\frac{2}{3}x - 8 = -32$$
$$-\frac{2}{3}(36) - 8 \overset{?}{=} -32$$
$$-32 = -32^{a}$$

Solution set: $\{36\}$

g)
$$-9x + 4 + 7x = -3x + 9$$
$$-2x + 4 = -3x + 9 \qquad \text{Combine like terms.}$$
$$-2x + 4 + 3x = -3x + 9 + 3x \quad \text{Add } 3x.$$
$$x + 4 = 9$$
$$x + 4 - 4 = 9 - 4 \qquad \text{Subtract 4.}$$
$$x = 5$$

Check:
$$-9x + 4 + 7x = -3x + 9$$
$$-9(5) + 4 + 7(5) \overset{?}{=} -3(5) + 9$$
$$-6 = -6^{a}$$

Solution set: $\{5\}$

i)
$$6(x - 8) = 6x - 48$$
$$6x - 48 = 6x - 48 \quad \text{Distributive property}$$

Both sides of the equations are the same, so any real number will satisfy the equation.

Solution set: $\{$all real numbers$\}$

f)
$$8x - 6 = 3 + 9x$$
$$8x - 6 + 6 = 3 + 9x + 6 \qquad \text{Add 6.}$$
$$8x = 9 + 9x$$
$$8x - 9x = 9 + 9x - 9x \qquad \text{Subtract } 9x.$$
$$-x = 9$$
$$x = -9 \qquad \text{Multiply by } -1.$$

Check:
$$8x - 6 = 3 + 9x$$
$$8(-9) - 6 \overset{?}{=} 3 + 9(-9)$$
$$-78 = -78^{a}$$

Solution set: $\{-9\}$

h)
$$6(2x - 1) = 30$$
$$12x - 6 = 30 \qquad \text{Distributive property}$$
$$12x - 6 + 6 = 30 + 6 \quad \text{Add 6.}$$
$$12x = 36$$
$$\frac{12x}{12} = \frac{36}{12} \qquad \text{Divide by 12.}$$
$$x = 3$$

Check:
$$6(2x - 1) = 30$$
$$6[2(3) - 1] \overset{?}{=} 30$$
$$30 = 30^{a}$$

Solution set: $\{3\}$

j)
$$3x + 6(x + 9) = 9x - 15$$
$$3x + 6x + 54 = 9x - 15 \qquad \text{Distributive property}$$
$$9x + 54 = 9x - 15 \qquad \text{Combine like terms}$$
$$9x + 54 - 9x = 9x - 15 - 9x \quad \text{Subtract } 9x.$$
$$54 = -15 \qquad \text{False}$$

The equation is a contradiction and has no solution.

Solution set: $\varnothing$

k) $0.4x - 0.2(3 - x) = 6.6$

$0.4x - 0.6 + 0.2x = 6.6$ Distributive property

$0.6x - 0.6 = 6.6$ Combine like terms.

$0.6x - 0.6 + 0.6 = 6.6 + 0.6$ Add 0.6.

$0.6x = 7.2$

$\dfrac{0.6x}{0.6} = \dfrac{7.2}{0.6}$ Divide by 0.6.

$x = 12$

Check:

$0.4x - 0.2(3 - x) = 6.6$

$0.4(12) - 0.2(3 - 12) \overset{?}{=} 6.6$

$6.6 = 6.6\,^{a}$

Solution set: $\{12\}$

l) $7x - 3(x - 8) = 3x + 24$

$7x - 3x + 24 = 3x + 24$ Distributive property

$4x + 24 = 3x + 24$ Combine like terms.

$4x + 24 - 3x = 3x + 24 - 3x$ Subtract $3x$.

$x + 24 = 24$

$x + 24 - 24 = 24 - 24$ Subtract 24.

$x = 0$

Check:

$7x - 3(x - 8) = 3x + 24$

$7(0) - 3(0 - 8) \overset{?}{=} 3(0) + 24$

$24 = 24\,^{a}$

Solution set: $\{0\}$

2. Write the answers to each problem in terms of the variable.

a) Two numbers have a sum of 11. One number is y. What expression represents the other number?

b) Monica is 18 years old. What expression represents her age d years from now?

c) Enrico has k nickels. Express the value of the nickels in cents.

Solution

a) $11 - y$

b) $18 + d$

c) $5k$

8.4 An Introduction to Applications of Linear Equations

Learning Objectives:

1. Solve problems involving unknown numbers.

Solved Examples:

1. Solve each problem using the six steps for solving applied problems.

    a) The product of 6, and a number increased by 7, is 72. What is the number?

    b) If 5 is added to a number, and the sum is tripled, the result is 11 more than the number. Find the number.

    c) What number minus 310 gives 822?

    d) When six is subtracted from half of a number, the result is −18. What is the original number?

    e) One number is 24 greater than another number. The sum of the two numbers is −72. Fine each number.

    f) In 2008, Pierre's Pizza and Burger Palace together had revenue totaling $470,000. If Burger Palace took in $90,000 less than Pierre's Pizza, how much did each take in as revenue?

    g) Find the measure of an angle whose complement is two times its measure.

    h) Find the measure of an angle such that the sum of the measures of its complement and its supplement is 142°.

    i) The sum of two consecutive integers is 59. Find the integers.

    j) When the lesser of two consecutive odd integers is added to twice the greater, the result is 188. Find the integers.

Solution

The six steps for solving applied problems are (1) read the problem, (2) assign a variable to represent the unknown value, (3) write an equation , (4) solve the equation, (5) state the answer, and (6) check the answer in the words of the original problem.

a) Let x = the number. Then the equation is $6(x+7) = 72$.

$$6(x+7) = 72$$
$$6x + 42 = 72 \qquad \text{Distributive property}$$
$$6x + 42 = 72 - 42 \quad \text{Subtract 42.}$$
$$6x = 30$$
$$\frac{6x}{6} = \frac{30}{6} \qquad \text{Divide by 6.}$$
$$x = 5$$

The number is 5.

Check:
Five increased by seven is 12. The product of 6 and 12 is 72, so the answer is correct.

b) Let x = the number. Then the equation is $(5+x)3 = 11 + x$.

$$(5+x)3 = 11 + x$$
$$15 + 3x = 11 + x \qquad \text{Distributive property}$$
$$15 + 3x - x = 11 + x - x \quad \text{Subtract } x.$$
$$15 + 2x = 11$$
$$15 + 2x - 15 = 11 - 15 \qquad \text{Subtract 15.}$$
$$2x = -4$$
$$\frac{2x}{2} = \frac{-4}{2} \qquad \text{Divide by 2.}$$
$$x = -2$$

The number is −2.

Check:
Five added to −2 is 3. When 3 is tripled, the product is 9. Eleven more than −2 is 9. The answer is correct.

c) Let n = the number. Then the equation is $n - 310 = 822$.

$n - 310 = 822$ *Check*:

$n - 310 + 310 = 822 + 310$ Add 310. $1132 - 310 = 822$, so the answer is correct.

$n = 1132$

The number is 1132.

d) Let n = the number. Then the equation is $\frac{1}{2}n - 6 = -18$.

$\frac{1}{2}n - 6 = -18$ *Check*:

One-half of the number is −12. When 6 is subtracted

$\frac{1}{2}n - 6 + 6 = -18 + 6$ Add 6.   from − 12, the result is −18. The answer is correct.

$\frac{1}{2}n = -12$

$2 \cdot \frac{1}{2}n = -12 \cdot 2$ Multiply by 2.

$n = -24$

The number is −24.

e) Let x = the smaller number. Then $x + 24$ = the larger number. The equation is $x + (x + 24) = 72$.

$x + (x + 24) = 72$ *Check*:

$2x + 24 = 72$ Combine like terms. $24 + 48 = 72$, so the answer is correct.

$2x + 24 - 24 = 72 - 24$ Subtract 24.

$2x = 48$

$\frac{2x}{2} = \frac{48}{2}$ Divide by 2.

$x = 24$

$x + 24 = 24 + 24 = 48$

The numbers are 24 and 48.

f) Let x = Pierre's Pizza revenue. Then $x - 90,000$ = Burger Palace revenue.

The equation is $x + (x - 90,000) = 470,000$.

$x + (x - 90,000) = 470,000$ *Check*:

$2x - 90,000 = 470,000$ $\$280,000 + \$190,000 = \$470,000$, so

Combine like terms.   the answer is correct.

$2x - 90,000 + 90,000 = 470,000 + 90,000$

Subtract 90,000

$2x = 560,000$

$\frac{2x}{2} = \frac{560,000}{2}$

Divide by 2.

$x = 280,000$

$x - 90,000 = 280,000 - 90,000 = 190,000$

Pierre's Pizza took in $280,000 and Burger Palace took in $190,000.

g) Let x = the measure of the angle. Then $90 - x$ = the measure of its complement.
The equation is $90 - x = 2x$.

$$90 - x = 2x$$
$$90 - x + x = 2x + x \quad \text{Add } x.$$
$$90 = 3x$$
$$\frac{90}{3} = \frac{3x}{3} \quad \text{Divide by 3.}$$
$$30 = x$$
$$90 - x = 90 - 30 = 60$$

Check:
The complement of the angle is 60°, and 60° is twice 30°. The answer is correct.

The measure of the angle is 30°.

h) Let x = the measure of the angle. Then $90 - x$ = the measure of its complement and
$180 - x$ = the measure of its supplement. The equation is $(90 - x) + (180 - x) = 142$.

$$(90 - x) + (180 - x) = 142$$
$$90 - x + 180 - x = 142$$
$$270 - 2x = 142$$
$$270 - 2x - 270 = 142 - 270$$
$$-2x = -128$$
$$\frac{-2x}{-2} = \frac{-128}{-2}$$
$$x = 64$$
$$90 - x = 90 - 64 = 26$$
$$180 - x = 180 - 64 = 116$$

Check:
The complement of the angle is 26°, and the supplement of the angle is 116°. 26° + 116° = 142°, so the answer is correct.

The measure of the angle is 64°.

i) Let n = the first integer. Then $n + 1$ = the second integer. The equation is $n + (n + 1) = 59$.

$$n + (n + 1) = 59$$
$$2n + 1 = 59 \quad \text{Combine like terms.}$$
$$2n + 1 - 1 = 59 - 1 \quad \text{Subtract 1.}$$
$$2n = 58$$
$$\frac{2n}{2} = \frac{58}{2} \quad \text{Divide by 2.}$$
$$n = 29$$
$$n + 1 = 29 + 1 = 30$$

Check:
29 + 30 = 59, so the answer is correct.

The integers are 29 and 30.

j) Let n = the smaller odd integer. Then $n + 2$ = the larger odd integer.
The equation is $n + 2(n + 2) = 187$.

$$n + 2(n + 2) = 187$$
$$n + 2n + 4 = 187 \quad \text{Combine like terms.}$$
$$3n + 4 = 187$$
$$3n + 4 - 4 = 187 - 4 \quad \text{Subtract 4.}$$
$$3n = 183$$
$$\frac{3n}{3} = \frac{183}{3} \quad \text{Divide by 3.}$$
$$n = 61$$
$$n + 2 = 61 + 2 = 63$$

Check:
$61 + 2(63) = 61 + 126 = 187$, so the answer is correct.

The integers are 61 and 63.

8.5 Formulas and Additional Applications from Geometry

Learning Objectives:

1. Determine whether to use perimeter or area.
2. Solve formulas, given the values of some variables.
3. Use a formula to solve problems.
4. Solve problems involving vertical and straight angles.
5. Solve a formula for a specified variable.
6.

Solved Examples:

1. Substitute values into the given formula and solve.

   a) The formula for the perimeter of a rectangle is $P = 2L + 2W$. If the length, L, is 9 meters and the width, W, is 5 meters, find the perimeter, P, of the rectangle.

   b) The area of a triangle is given by $A = \frac{1}{2}bh$. If the base, b, is 19 in. and the height, h, is 17 in., find the area.

   Solution

   a) $P = 2 \cdot 9 + 2 \cdot 5 = 18 + 10 = 28$
   The perimeter of the rectangle is 28 meters.

   b) $A = \frac{1}{2} \cdot 19 \cdot 17 = \frac{323}{2} = 161\frac{1}{2}$

   The area of the triangle is $161\frac{1}{2}$ sq. in.

2. The perimeter of a triangle is 37 feet. One side of the triangle is 2 feet longer than the second side. The third side is 5 feet longer than the second side. Find the length of each side.

   Solution

   Let x = the length of the second side of the triangle. Then $x + 2$ = the length of the first side, and $x + 5$ = the length of the third side. The equation is $x + (x + 2) + (x + 5) = 37$.

   $x + (x + 2) + (x + 5) = 37$

   $3x + 7 = 37$ Combine like terms.

   $3x + 7 - 7 = 37 - 7$ Subtract 7.

   $3x = 30$

   $\dfrac{3x}{3} = \dfrac{30}{3}$ Divide by 10.

   $x = 10$

   $x + 2 = 10 + 2 = 12$

   $x + 5 = 10 + 5 = 15$

   The three sides have lengths 10 feet, 12 feet, and 15 feet.

   Check:
   If the sides of the triangle are 10 ft, 12 ft, and 15 ft, then the perimeter of the triangle = 10 + 12 + 15 = 37, so the answer is correct.

3. Find the measure of each marked angle.

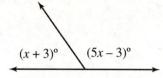

$(x+3)°$ $(5x-3)°$

Solution

The sum of the two marked angles is 180°.

$(x+3)+(5x-3) = 180$

$6x = 180$ Combine like terms.

$\dfrac{6x}{6} = \dfrac{180}{6}$ Divide by 6.

$x = 30$

$x+3 = 30+3 = 33$

$5x-3 = 5\cdot30-3 = 150-3 = 147$

The measures of the angles are 33° and 147°.

4. Solve for the indicated variable.

a) $d = rt$, for r

b) $V = \dfrac{1}{3}Bh$, for h

c) $P = 2L+2W$, for W

d) $F = \dfrac{9}{5}C+32$, for C

Solution

a) $d = rt$

$\dfrac{d}{t} = r$ Divide by t.

b) $V = \dfrac{1}{3}Bh$

$3V = 3\cdot\dfrac{1}{3}Bh$ Multiply by 3.

$3V = Bh$

$\dfrac{3V}{B} = \dfrac{Bh}{B}$ Divide by B.

$\dfrac{3V}{B} = h$

c) $P = 2L+2W$

$P-2L = 2L+2W-2L$ Subtract 2L.

$P-2L = 2W$

$\dfrac{P-2L}{2} = \dfrac{2W}{2}$ Divide by 2.

$\dfrac{P-2L}{2} = W$

d) $F = \dfrac{9}{5}C+32$

$F-32 = \dfrac{9}{5}C+32-32$ Subtract 32.

$F-32 = \dfrac{9}{5}C$

$\dfrac{5}{9}(F-32) = \dfrac{5}{9}\cdot\dfrac{9}{5}C$ Multiply by $\dfrac{5}{9}$.

$\dfrac{5}{9}(F-32) = C$

8.6 Solving Linear Inequalities

Learning Objectives:

1. Graph solutions on a number line.
2. Use the addition property of inequality.
3. Use the multiplication property of inequality.
4. Use both properties of inequality.
5. Solve applications.

Solved Examples:

1. Graph each inequality on a number line.

 a) $x > 1$ b) $x \geq -4$ c) $x < \dfrac{2}{3}$ d) $x \leq -2.5$

 Solution

 a)

 b)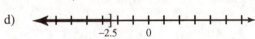

 c)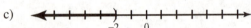

 d)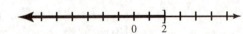

2. Translate each graph to an inequality using the variable x.

 a) 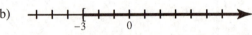 b)

 c)   d)

 Solution

 a) $x > 4$ b) $x \geq -3$ c) $x < -2$ d) $x \leq 2$

3. Solve each inequality and write the answer in interval notation.

 a) $4x - 10 > 5x + 4$ b) $5x - 2 \leq 6x + 4$

 c) $-4x - 7 < 5(2x + 7)$ d) $7(x - 3) > 2(5x - 8)$

 e) $1 > 3x - 2 > -5$ f) $1 \leq 2x - 5 < 5$

 Solution

 a)
$$4x - 10 > 5x + 4$$
$$4x - 10 - 4x > 5x + 4 - 4x \quad \text{Subtract } 4x.$$
$$-10 > x + 4 \quad \text{Combine like terms.}$$
$$-10 - 4 > x + 4 - 4 \quad \text{Subtract 4.}$$
$$-14 > x$$
$$x < -14 \quad \text{Reverse the inequality.}$$

 Solution set: $(-\infty, -14)$

 b)
$$5x - 2 \leq 6x + 4$$
$$5x - 2 - 5x \leq 6x + 4 - 5x \quad \text{Subtract } 5x.$$
$$-2 \leq x + 4 \quad \text{Combine like terms.}$$
$$-2 - 4 \leq x + 4 - 4 \quad \text{Subtract 4.}$$
$$-6 \leq x$$
$$x \geq -6 \quad \text{Reverse the inequality.}$$

 Solution set: $[-6, \infty)$

c)
$$-4x - 7 < 5(2x + 7)$$
$$-4x - 7 < 10x + 35 \quad \text{Distributive property}$$
$$-4x - 7 + 4x < 10x + 35 + 4x \quad \text{Add } 4x.$$
$$-7 < 14x + 35 \quad \text{Combine like terms.}$$
$$-7 - 35 < 14x + 35 - 35 \quad \text{Subtract 35.}$$
$$-42 < 14x \quad \text{Combine terms.}$$
$$\frac{-42}{14} < \frac{14x}{14} \quad \text{Divide by 14.}$$
$$-3 < x$$
$$x > -3 \quad \text{Reverse the inequality.}$$

Solution set: $(-3, \infty)$

d)
$$7(x - 3) > 2(5x - 8)$$
$$7x - 21 > 10x - 16 \quad \text{Distributive property}$$
$$7x - 21 - 7x > 10x - 16 - 7x \quad \text{Subtract } 7x.$$
$$-21 > 3x - 16 \quad \text{Combine terms.}$$
$$-21 + 16 > 3x - 16 + 16 \quad \text{Subtract 16.}$$
$$-5 > 3x$$
$$\frac{-5}{3} > \frac{3x}{3} \quad \text{Divide by 3.}$$
$$-\frac{5}{3} > x$$
$$x < -\frac{5}{3} \quad \text{Reverse the inequality.}$$

Solution set: $\left(-\infty, -\frac{5}{3}\right)$

e)
$$1 > 3x - 2 > -5$$
$$1 + 2 > 3x - 2 + 2 > -5 + 2 \quad \text{Add 2.}$$
$$3 > 3x > -3$$
$$\frac{3}{3} > \frac{3x}{3} > \frac{-3}{3} \quad \text{Divide by 3.}$$
$$1 > x > -1$$

Solution set: $(-1, 1)$

f)
$$1 \le 2x - 5 < 5$$
$$1 + 5 \le 2x - 5 + 5 < 5 + 5 \quad \text{Add 5.}$$
$$6 \le 2x < 10$$
$$\frac{6}{2} \le \frac{2x}{2} < \frac{10}{2} \quad \text{Divide by 2.}$$
$$3 \le x < 5$$

Solution set: $[3, 5)$

4. Solve using an inequality.

a) A certain car has a weight limit for all passengers and cargo of 1129 pounds. The four passengers in the car weight an average of 165 pounds. Find the maximum weight of the cargo that the car can handle.

b) A certain store has a fax machine available for use by its customers. The store charges $1.85 to send the first page and $0.45 for each subsequent page. Find the maximum number of pages that can be faxed for $7.25.

i) An archery set containing a bow and three arrows costs $68. Additional arrows can be purchased for $9 each. Jerry has $230 to spend on the set and additional arrows. Including the arrows in the set, what is the maximum total number of arrows Jerry can purchase?

Solution

a) Let x = the amount of cargo. There are four passengers, so the total weight of the passengers is $4(165)$ pounds. The inequality is $4(165) + x \le 1129$.

$$4(165) + x \le 1129$$
$$660 + x \le 1129$$
$$660 + x - 660 \le 1129 - 660 \quad \text{Subtract 660.}$$
$$x \le 469$$

The maximum weight of the cargo is 660 pounds.

b) Let x = the number of pages. Then the inequality is $1.85 + 0.45x \le 7.25$.

$$1.85 + 0.45x \le 7.25$$
$$1.85 + 0.45x - 1.85 \le 7.25 - 1.85 \quad \text{Subtract } 1.85.$$
$$0.45x \le 5.40$$
$$\frac{0.45x}{0.45} \le \frac{5.40}{0.45} \qquad \text{Divide by } 0.45.$$
$$x \le 12$$

The maximum number of pages that can be faxed for \$7.25 is 12 pages..

c) Let x = the maximum number of arrows Jerry can purchase. The inequality is $68 + 9x \le 230$.

$$68 + 9x \le 230$$
$$68 + 9x - 68 \le 230 - 68 \quad \text{Subtract } 68.$$
$$9x \le 162$$
$$\frac{9x}{9} \le \frac{162}{9} \qquad \text{Divide by } 9.$$
$$x \le 18$$

Jerry can buy a maximum of 18 additional arrows, or 21 arrows in total.

9.1 Reading Graphs: Linear Equations in Two Variables

Learning Objectives:

1. Read and interpret graphs.
2. Decide whether an ordered pair is a solution to an equation.
3. Complete ordered pairs.
4. Find coordinates of a point on a plane, and plot points.
5. Complete tables of values and plot ordered pairs.
6. Solve applications.

Solved Examples:

1. The line graph shows the price per share for Compu-Tech stock for the first four months of year 2008. Use the line graph to answer the following questions.

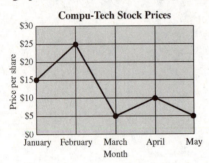

a) For what months did the price of the stock rise?

b) How much did the value of the stock change from the beginning of February until the beginning of May?

Solution

a) The stock rose in January and in March.

b) At the beginning of February, the stock price was $25 per share. At the beginning of May, the stock price was $5 per share. The stock dropped $25 – $5 = $20 per share.

2. Identify and label the following features on the rectangular coordinate system shown.

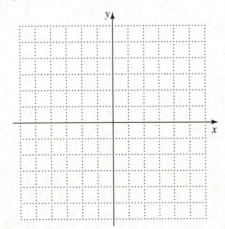

a) *x*-axis

b) *y*-axis

c) *x*-values –6 to 6

d) *y*-values –6 to 6

e) origin (0, 0)

f) ordered pair (3, 4)

g) ordered pair (–2, –5)

(continued on next page)

Solution

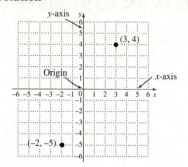

3. Decide whether each ordered pair is a solution of the equation $3x - y = 5$.

    a) $(0, -5)$ b) $(2, -1)$ c) $(-1, -8)$

    Solution

    a) Replace x with 0 and y with -5.

    $$3(0) - (-5) \overset{?}{=} 5$$
    $$5 = 5 \checkmark$$

    Yes, $(0, -5)$ is a solution of the equation.

    b) Replace x with 2 and y with -1.

    $$3(2) - (-1) \overset{?}{=} 5$$
    $$7 \neq 5$$

    No, $(2, -1)$ is not a solution of the equation.

    c) Replace x with -1 and y with -8.

    $$3(-1) - (-8) \overset{?}{=} 5$$
    $$5 = 5 \checkmark$$

    Yes, $(-1, -8)$ is a solution of the equation.

4. Find the missing coordinate to complete the ordered-pair solution to the given linear equation.

    a) $y = -x \quad (-4, \ )$ b) $x + y = 6 \quad (\ , 5)$ c) $3x - 2y = 9 \quad (0, \ )$

    Solution

    a) Replace x with -4, then solve for y.

    $$y = -(-4) = 4$$

    The missing coordinate is 4.

    b) Replace y with 5, then solve for x.

    $$x + 5 = 6$$
    $$x + 5 - 5 = 6 - 5$$
    $$x = 1$$

    The missing coordinate is 1.

    c) Replace x with 0, then solve for y.

    $$3(0) - 2y = 9$$
    $$-2y = 9$$
    $$\frac{-2y}{-2} = \frac{9}{-2}$$
    $$y = -\frac{9}{2}$$

    The missing coordinate is $-\frac{9}{2}$.

5. A hardware store rents lawn aerators for a flat fee of $25, plus $3 per hour. Therefore, the cost to rent an aerator for x hours is given by $y = 25 + 3x$, where y is in dollars. Express the following as an ordered pair:

    When an aerator is rented for 4 hours, the cost is $37.

    Solution

    $x = 4$ and $y = 37$, so the ordered pair is $(4, 27)$.

9.2 Graphing Linear Equations in Two Variables

Learning Objectives:

1. Complete ordered pairs for equations.
2. Find intercepts.
3. Graph linear equations.
4. Solve applications.

Solved Examples:

1. Find five ordered-pair solutions to the linear equation $x + y = 6$.

   Solution

   Choose values for x, then solve for y:
   If $x = 0$, then $0 + y = 6$ leads to $y = 6$.
   If $x = 1$, then $1 + y = 6$ leads to $y = 5$.
   If $x = 2$, then $2 + y = 6$ leads to $y = 4$.
   If $x = 3$, then $3 + y = 6$ leads to $y = 3$.
   If $x = 4$, then $4 + y = 6$ leads to $y = 2$.

   Thus, the ordered pairs are (0, 6), (1, 5), (2, 3), (3, 3), and (4, 2). (Note that answers may vary depending on what values are chosen for x.)

2. Complete the ordered-pair solutions. Then plot each solution and graph the equation by connecting the points by a straight line.

   a) $y = x - 3$
   (5,) (–2,) (0,)

   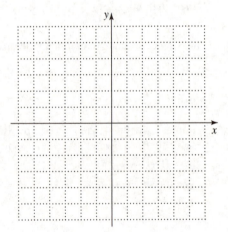

   b) $y = -3x + 2$
   (0,) (1,) (–1,)

   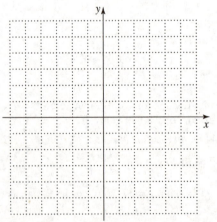

SE-168

c) $4x + 3y = 0$

$(-3, \ ) \ (0, \ ) \ (3, \ )$

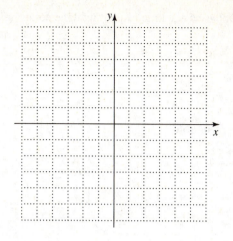

Solution

a) $y = x - 3$
$y = 5 - 3 = 2$
$y = -2 - 3 = -5$
$y = 0 - 3 = -3$

The points are $(5, 2) \ (-2, -5), \ (0, -3)$.

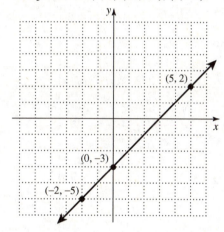

b) $y = -3x + 2$
$y = -3(0) + 2 = 2$
$y = -3(1) + 2 = -1$
$y = -3(-1) + 2 = 5$

The points are $(0, 2), \ (1, -1), \ (-1, 5)$

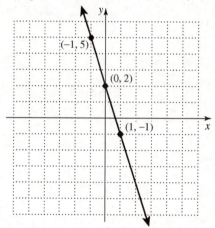

c) $4x + 3y = 0$

$4(-3) + 3y = 0 \qquad 4(0) + 3y = 0$
$-12 + 3y = 0 \qquad\quad 3y = 0$
$\qquad 3y = 12 \qquad\qquad y = 0$
$\qquad\quad y = 4$

$4(3) + 3y = 0$
$12 + 3y = 0$
$\quad 3y = -12$
$\qquad y = -4$

The points are $(-3, 4), \ (0, 0), \ (3, -4)$

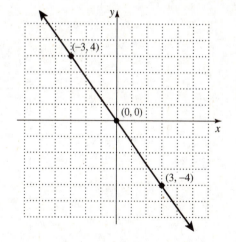

3. Graph each equation by plotting three points or by plotting the intercepts and one additional point.

a) $y = 3x + 4$

b) $2x + 3y = 6$

c) $y = -\dfrac{1}{2}x + 4$

d) $\dfrac{2}{5}x + y = 3$

e) $x = 4$

f) $y = -3$

Solution

a)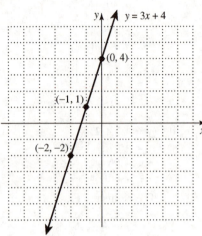

b)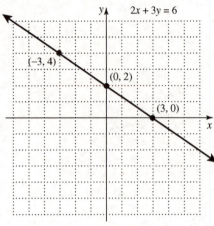

c)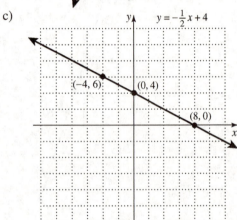

d)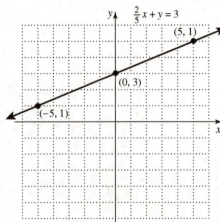

e)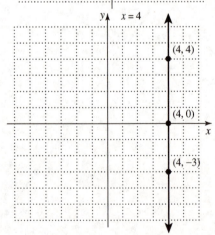

f)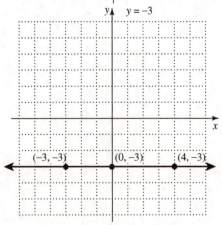

4. The population of an endangered species of fish living in a controlled habitat is given by the equation $P = 7t + 46$, where P is the population and t is the time in months since the population was moved to the habitat. Graph the equation for $t = 0$, 6, 12, and 18.

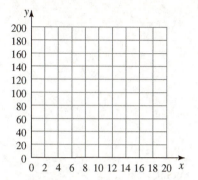

Solution

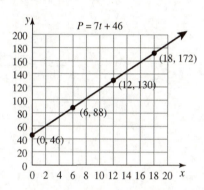

9.3 Slope of a Line

Learning Objectives:

1. Find the slope from a graph.
2. Find the slope given two points on a line.
3. Find the slope given the equation of a line.
4. Decide whether a slope is positive, negative, or zero.
5. Tell whether lines are parallel, perpendicular, or neither.
6. Solve applications.

Solved Examples:

1. Find the slope of a straight line that passes through the given pair of points.

   a) (5, 3) and (9, 6)   b) (−5, 8) and (−3, −9)   c) (−8, −5) and (−3, −5).

   Solution

   a) $m = \dfrac{y_2 - y_1}{x_2 - x_1} = \dfrac{6-3}{9-5} = \dfrac{3}{4}$

   b) $m = \dfrac{y_2 - y_1}{x_2 - x_1} = \dfrac{-9-8}{-3-(-5)}$
   $= \dfrac{-17}{2} = -\dfrac{17}{2}$

   c) $m = \dfrac{y_2 - y_1}{x_2 - x_1} = \dfrac{-5-(-5)}{-3-(-8)}$
   $= \dfrac{0}{5} = 0$

2. Find the slope and the y-intercept.

   a) $y = 3x + 4$ b) $y = -2x + 9$ c) $y = \dfrac{4}{3}x - \dfrac{3}{5}$ d) $2x + 5y = 3$

   Solution

   An equation written in the form $y = mx + b$ is in slope-intercept form, where the slope is m and the y-intercept is $(0, b)$.

   a) $y = 3x + 4$
   slope = 3; y-intercept = (0, 4)

   b) $y = -2x + 9$
   slope = −2; y-intercept = (0, 9)

   c) $y = \dfrac{4}{3}x - \dfrac{3}{5}$
   slope = $\dfrac{4}{3}$; y-intercept = $\left(0, -\dfrac{3}{5}\right)$

   d) $2x + 5y = 3$
   Rewrite the equation in slope-intercept form.
   $2x + 5y = 3$
   $5y = -2x + 3$ Subtract $2x$.
   $y = -\dfrac{2}{5}x + \dfrac{3}{5}$ Divide by 5.

   slope = $-\dfrac{2}{5}$; y-intercept = $\left(0, \dfrac{3}{5}\right)$

3. Write the equation of the line in slope–intercept form.

a) $m = 2$, y–intercept $= (0, -4)$

b) $m = -\dfrac{3}{2}$, y–intercept $= \left(0, \dfrac{5}{6}\right)$

Solution

a) $y = 2x - 4$

b) $y = -\dfrac{3}{2}x + \dfrac{5}{6}$

4. Find the slope of a line parallel to and a line perpendicular to the given line.

a) A line of slope $\dfrac{2}{3}$

b) A line of slope -6

c) The line $y = \dfrac{1}{2}x - 3$

Solution

The slopes of parallel lines are equal. The product of the slopes of two perpendicular lines, neither of which is vertical, is always -1. In other words, the slopes are negative reciprocals of each other.

a) A line parallel to a line of slope $\dfrac{2}{3}$ has slope $\dfrac{2}{3}$.

A line perpendicular to a line of slope $\dfrac{2}{3}$ has slope $-\dfrac{3}{2}$.

b) A line parallel to a line of slope -6 has slope -6.

A line perpendicular to a line of slope -6 has slope $\dfrac{1}{6}$.

c) The line $y = \dfrac{1}{2}x - 3$ has slope $\dfrac{1}{2}$, so a line parallel to it also has slope $\dfrac{1}{2}$.

A line perpendicular to the line $y = \dfrac{1}{2}x - 3$ has slope -2.

9.4 Equations of Lines

Learning Objectives:

1. Write equations of lines, given slope and y-intercept.
2. Graph equations, given slope and y-intercept.
3. Graph a line given point and slope.
4. Write an equation given a point and slope.
5. Write an equation given two points.
6. Write equations of parallel and perpendicular lines.
7. Solve applications.

Solved Examples:

1. Write the equation of the line that has the given slope and passes through the given point. Write each answer in slope-intercept form.

   a) $m = 3$, $(0, 4)$　　　b) $m = 2$, $(4, -4)$　　　c) $m = -3$, $\left(\dfrac{3}{2}, 2\right)$　　　d) $m = \dfrac{2}{3}$, $(4, -2)$

   Solution

   The slope-intercept form of a linear equation is $y = mx + b$. The point-slope form of a linear equation is $y - y_1 = m(x - x_1)$.

   a) We are given the slope and the y-intercept, so use the slope-intercept form.
   $y = 3x + 4$

   b) We are given the slope and a point, so start with the point-slope form.
   $$y - (-4) = 2(x - 4)$$
   $$y + 4 = 2x - 8 \quad \text{Distributive property}$$
   $$y = 2x - 12 \quad \text{Subtract 4.}$$

   c) We are given the slope and a point, so start with the point-slope form.
   $$y - 2 = -3\left(x - \frac{3}{2}\right)$$
   $$y - 2 = -3x + \frac{9}{2} \quad \text{Distributive property}$$
   $$y = -3x + \frac{13}{2} \quad \text{Add 2.}$$

   d) We are given the slope and a point, so start with the point-slope form.
   $$y - (-2) = \frac{2}{3}(x - 4)$$
   $$y + 2 = \frac{2}{3}x - \frac{8}{3} \quad \text{Distributive property}$$
   $$y = \frac{2}{3}x - \frac{14}{3} \quad \text{Subtract 2.}$$

2. Write the equation of the line passing through the given points. Write each answer in slope-intercept form.

   a) $(2, 14)$ and $(-4, -4)$　　　b) $(1, -16)$ and $(-2, -1)$　　　c) $\left(2, \dfrac{4}{3}\right)$ and $\left(\dfrac{2}{3}, \dfrac{2}{3}\right)$

   Solution

   a) First find the slope:
   $$m = \frac{y_2 - y_1}{x_2 - x_1} = \frac{-4 - 14}{-4 - 2} = \frac{-18}{-6} = 3$$
   Now use either point to write the equation in point-slope form:

   $$y - 14 = 2(x - 2)$$
   $$y - 14 = 2x - 4 \quad \text{Distributive property}$$
   $$y = 2x - 10 \quad \text{Add 14.}$$

SE-174

b) First find the slope:

$$m = \frac{y_2 - y_1}{x_2 - x_1} = \frac{-1 - (-16)}{-2 - 1} = \frac{15}{-3} = -5$$

Now use either point to write the equation in point-slope form:

c) First find the slope:

$$m = \frac{y_2 - y_1}{x_2 - x_1} = \frac{\dfrac{2}{3} - \dfrac{4}{3}}{\dfrac{2}{3} - 2} = \frac{-\dfrac{2}{3}}{-\dfrac{4}{3}} = \frac{1}{2}$$

Now use either point to write the equation in point-slope form:

$$y - (-16) = -5(x - 1)$$
$$y + 16 = -5x + 5 \quad \text{Distributive property}$$
$$y = -5x - 11 \quad \text{Subtract 16.}$$

$$y - \frac{2}{3} = \frac{1}{2}\left(x - \frac{2}{3}\right)$$
$$y - \frac{2}{3} = \frac{1}{2}x - \frac{1}{3} \quad \text{Distributive property}$$
$$y = \frac{1}{2}x + \frac{1}{3} \quad \text{Add } \frac{2}{3}.$$

3. Write the equation of each line in slope-intercept form.

a)

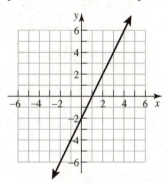

b)

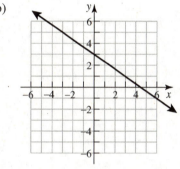

c)

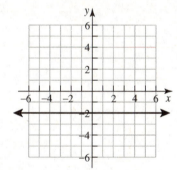

d)

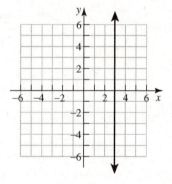

Solution

a) Find the graph, we see that the line goes through the points $(0, -2)$ and $(1, 0)$.

$$m = \frac{y_2 - y_1}{x_2 - x_1} = \frac{0 - (-2)}{1 - 0} = 2$$

The y-intercept is $(0, -2)$, so the equation is $y = 2x - 2$.

b) Find the graph, we see that the line goes through the points $(0, 3)$ and $(3, 1)$.

$$m = \frac{y_2 - y_1}{x_2 - x_1} = \frac{1 - 3}{3 - 0} = \frac{-2}{3} = -\frac{2}{3}$$

The y-intercept is $(0, 3)$, so the equation is

$$y = -\frac{2}{3}x + 3..$$

c) The line is a horizontal line passing through $(0, -2)$, so the equation is $y = -2$.

d) The line is a vertical line passing through $(3, 0)$, so the equation is $x = 3..$

4. Let the ordered pairs $(3, 125)$ and $(8, 220)$ represent the number of fish caught in a lake where 3 represents March and 8 represents August, and 125 and 220 represent the numbers of fish caught. Use these ordered pairs to write an equation of a line that approximates this data.

   Solution

   First find the slope of the line: $m = \dfrac{y_2 - y_1}{x_2 - x_1} = \dfrac{220 - 125}{8 - 3} = \dfrac{95}{5} = 19$

   Now, using the point-slope form with either point, write the equation of the line: $y - 125 = 19(x - 3)$

   Solve for y in order to write the equation in slope-intercept form:

   $y - 125 = 19(x - 3)$
   $y - 125 = 19x - 57$ Distributive property
   $\quad\quad y = 19x + 68$ Add 68.

   Note that we obtain the same answer if we used the point (8, 220) to start.

9.5 Graphing Linear Inequalities in Two Variables

Learning Objectives:

1. Analyze linear inequalities.
2. Graph linear inequalities.
3. Solve applications.

Solved Examples:

1. Determine whether the ordered pair satisfies $y \geq x + 2$.

   a) $(0, 2)$ b) $(1, 4)$ c) $(-1, -2)$

   Solution

   a) Substitute 0 for x and 2 for y in the inequality.
   $$2 \overset{?}{\geq} 0 + 2$$
   $$2 \geq 2 \checkmark$$
   $(0, 2)$ is a solution of the inequality.

   b) Substitute 1 for x and 4 for y in the inequality.
   $$4 \overset{?}{\geq} 1 + 2$$
   $$4 \geq 3 \checkmark$$
   $(1, 4)$ is a solution of the inequality.

   c) Substitute -1 for x and -2 for y in the inequality.
   $$-2 \overset{?}{\geq} -1 + 2$$
   $$-2 \geq 1 \quad \text{False}$$
   $(-1, -2)$ is not a solution of the inequality.

2. Graph the region described by the inequality.

   a) $y < x$

   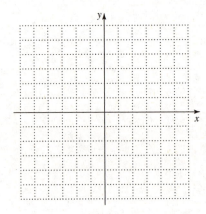

   b) $y \geq x + 2$

   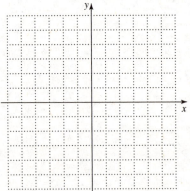

c) $y \le -x - 3$

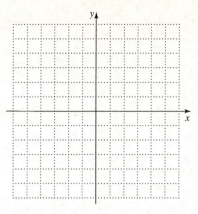

d) $y > -5x + 3$

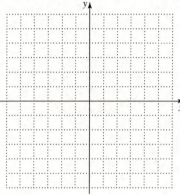

e) $x + 2y > -2$

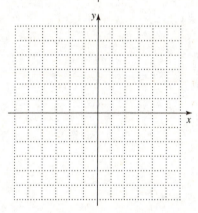

f) $-2x - 5y \le 10$

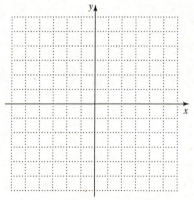

g) $y > \dfrac{1}{2}x$

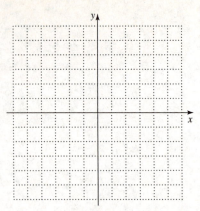

h) $y \le 2$

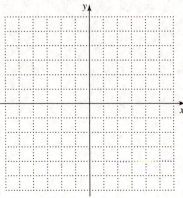

i) $x \ge -2$

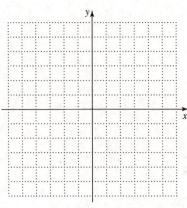

j) $2x < -3y$

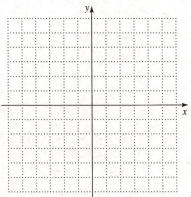

k) $x > -3y$

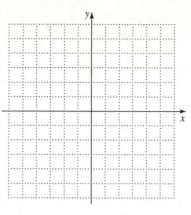

l) $2x \le 3 + y$

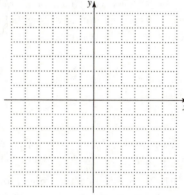

Solution

a) First graph the line $y = x$. Since the inequality is $y < x$, the line should be dashed. Then choose a test point, for example (1, 2). This point makes the inequality false, so it is not in the region to be shaded.

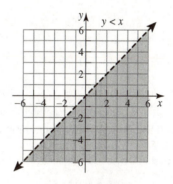

b) First graph the line $y = x + 2$. Since the inequality is $y \ge x + 2$, the line should be solid. Then choose a test point, for example (1, 2). This point makes the inequality false, so it is not in the region to be shaded.

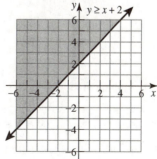

c) First graph the line $y = -x - 3$. Since the inequality is $y \le -x - 3$, the line should be solid. Then choose a test point, for example $(0, 0)$. This point makes the inequality false, so it is not in the region to be shaded.

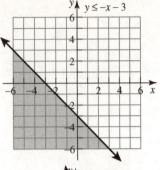

d) First graph the line $y = -5x + 3$. Since the inequality is $y > -5x + 3$, the line should be dashed. Then choose a test point, for example $(0, 0)$. This point makes the inequality false, so it is not in the region to be shaded.

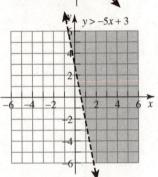

e) First graph the line $x + 2y = -2$. Since the inequality is $x + 2y > -2$, the line should be dashed. Then choose a test point, for example $(0, 0)$. This point makes the inequality true, so it is in the region to be shaded.

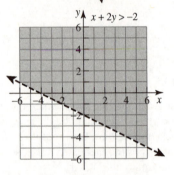

f) First graph the line $-2x - 5y = 10$. Since the inequality is $-2x - 5y \le 10$, the line should be solid. Then choose a test point, for example $(0, 0)$. This point makes the inequality true, so it is in the region to be shaded.

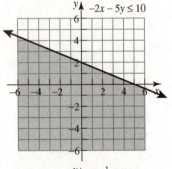

g) First graph the line $y > \dfrac{1}{2}x$. Since the inequality is $y > \dfrac{1}{2}x$, the line should be dashed. Then choose a test point, for example $(0, 1)$. This point makes the inequality true, so it is in the region to be shaded.

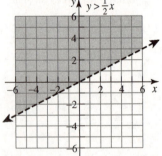

h) First graph the line $y = 2$. Since the inequality is $y \leq 2$, the line should be solid. Then choose a test point, for example $(0, 0)$. This point makes the inequality true, so it is in the region to be shaded.

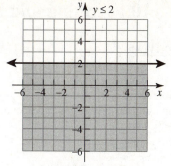

i) First graph the line $x = -2$. Since the inequality is $x \geq -2$, the line should be solid. Then choose a test point, for example $(0, 0)$. This point makes the inequality true, so it is in the region to be shaded.

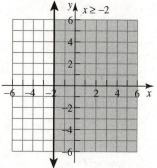

j) First graph the line $2x = -3y$. Since the inequality is $2x < -3y$, the line should be dashed. Then choose a test point, for example $(2, 0)$. This point makes the inequality false, so it is not in the region to be shaded.

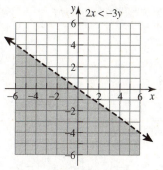

k) First graph the line $2x = 3 + y$. Since the inequality is $2x \leq 3 + y$, the line should be solid. Then choose a test point, for example $(0, 0)$. This point makes the inequality true, so it is in the region to be shaded.

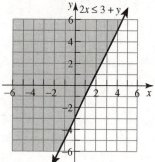

10.1 Adding and Subtracting Polynomials

Learning Objectives:

1. Know the vocabulary for polynomials.
2. Combine like terms.
3. Simplify and write polynomials in descending order.
4. Evaluate polynomials.
5. Add or subtract polynomials.
6. Solve problems.
7. Perform more calculations

Solved Examples:

1. State the degree of the polynomial and whether it is a monomial, binomial, or trinomial.

   a) $2xy^7$

   b) $12x^5 - 3x^2 + 2$

   c) $9x^3y + 10x^4y^4$

   Solution

   a) The degree is 8 because the sum of the exponents is 8. The polynomial is a monomial because it has just one term.

   b) The degree is 5 because the degree of a polynomial is the greatest degree of any nonzero term of the polynomial. There are three terms, so the polynomial is a trinomial.

   c) The degree is 8 because the degree of a polynomial is the greatest degree of any nonzero term of the polynomial. There are two terms, so the polynomial is a biinomial.

2. Add.

   a) $(2x - 12) + (-3x + 22)$

   b) $(5x^2 - 3x - 6) + (2x^2 - 8x - 3)$

   c) $\left(\dfrac{3}{4}x^2 - \dfrac{1}{3}x - 12\right) + \left(-\dfrac{1}{3}x^2 + \dfrac{2}{9}x + 3\right)$

   d) $(-2.2x^3 + 5.4x - 0.1) + (6.4x^2 - 3.4)$

   Solution

   a) $(2x - 12) + (-3x + 22)$

   $\quad = 2x + (-3x) + (-12) + 22$ Commutative property

   $\quad = -x + 10$ Combine like terms.

   b) $\left(5x^2 - 3x - 6\right) + \left(2x^2 - 8x - 3\right)$

   $\quad = 5x^2 + 2x^2 + (-3x) + (-8x) + (-6) + (-3)$ Commutative property

   $\quad = 7x^2 - 11x - 9$ Combine like terms.

   c) $\left(\dfrac{3}{4}x^2 - \dfrac{1}{3}x - 12\right) + \left(-\dfrac{1}{3}x^2 + \dfrac{2}{9}x + 3\right)$

   $\quad = \dfrac{3}{4}x^2 + \left(-\dfrac{1}{3}x^2\right) + \left(-\dfrac{1}{3}x\right) + \dfrac{2}{9}x + (-12) + 3$

   $\quad = \dfrac{5}{12}x^2 - \dfrac{1}{9}x - 9$

   d) $\left(-2.2x^3 + 5.4x - 0.1\right) + \left(6.4x^2 - 3.4\right)$

   $\quad = -2.2x^3 + 6.4x^2 + 5.4x - 0.1 + (-3.4)$

   $\quad = -2.2x^3 + 6.4x^2 + 5.4x - 3.5$

3. Subtract.

   a) $(4x - 8) - (3x - 4)$

   b) $(6x^3 + 5x^2 - 3) - (-2x^3 + 3x^2 - x + 1)$

   c) $\left(\dfrac{5}{8}x^2 - \dfrac{1}{3}x - 7\right) - \left(\dfrac{2}{3}x^2 - \dfrac{3}{4}x + 5\right)$

   d) $(2.3x^4 - 4x^3 + 5x) - (x^4 - 6.2x^2 + 2.2x)$

   Solution

   a) $(4x - 8) - (3x - 4) = 4x - 8 + (-3x) + 4$
   $$= 4x + (-3x) - 8 + 4$$
   $$= x - 4$$

   b)
   $$\left(6x^3 + 5x^2 - 3\right) - \left(-2x^3 + 3x^2 - x + 1\right)$$
   $$= 6x^3 + 5x^2 - 3 + 2x^3 - 3x^2 + x - 1$$
   $$= 6x^3 + 2x^3 + 5x^2 - 3x^2 + x - 3 - 1$$
   $$= 8x^3 + 2x^2 + x - 4$$

   c) $\left(\dfrac{5}{8}x^2 - \dfrac{1}{3}x - 7\right) - \left(\dfrac{2}{3}x^2 - \dfrac{3}{4}x + 5\right)$
   $$= \dfrac{5}{8}x^2 - \dfrac{1}{3}x - 7 - \dfrac{2}{3}x^2 + \dfrac{3}{4}x - 5$$
   $$= \dfrac{5}{8}x^2 - \dfrac{2}{3}x^2 - \dfrac{1}{3}x + \dfrac{3}{4}x - 7 - 5$$
   $$= -\dfrac{1}{24}x^2 + \dfrac{5}{12}x - 12$$

   d) $\left(2.3x^4 - 4x^3 + 5x\right) - \left(x^4 - 6.2x^2 + 2.2x\right)$
   $$= 2.3x^4 - x^4 - 4x^3 + 6.2x^2 + 5x - 2.2x$$
   $$= 1.3x^4 - 4x^3 + 6.2x^2 + 2.8x$$

4. Perform the indicate operations.

   a) $(x^3 + 8x^2 + 5) + (6x - 3) - (3x^2 - x - 9)$

   b) $(7x^3 + 4x^2y + 7xy^2 - 3y^3) + (2x^3 + 5x^2y + 8xy^2) - (5x^2y + 2xy^2 + y^3)$

   Solution

   a) $\left(x^3 + 8x^2 + 5\right) + \left(6x - 3\right) - \left(3x^2 - x - 9\right) = x^3 + 8x^2 + 5 + 6x - 3 - 3x^2 + x + 9$
   $$= x^3 + 8x^2 - 3x^2 + 6x + x + 5 - 3 + 9$$
   $$= x^3 + 5x^2 + 7x + 11$$

   b) $\left(7x^3 + 4x^2y + 7xy^2 - 3y^3\right) + \left(2x^3 + 5x^2y + 8xy^2\right) - \left(5x^2y + 2xy^2 + y^3\right)$
   $$= 7x^3 + 4x^2y + 7xy^2 - 3y^3 + 2x^3 + 5x^2y + 8xy^2 - 5x^2y - 2xy^2 - y^3$$
   $$= 7x^3 + 2x^3 + 4x^2y + 5x^2y - 5x^2y + 7xy^2 + 8xy^2 - 2xy^2 - 3y^3 - y^3$$
   $$= 9x^3 + 4x^2y + 13xy^2 - 4y^3$$

10.2 The Product Rule and Power Rule for Exponents

Learning Objectives:

1. Use exponents.
2. Use the product rule for exponents.
3. Use the power rules for exponents.
4. Use a combination of rules for exponents.
5. Solve problems.

Solved Examples:

1. Write in simplest exponent form.

    a) $3 \cdot 3 \cdot 3 \cdot x \cdot x$ b) $(-4)(m)(m)(n)(m)(n)$ c) $(5)(5)(x)(y)(x)(z)(y)(z)$

Solution

    a) $3 \cdot 3 \cdot 3 \cdot x \cdot x = 3^2 x^2 = 27x^2$ b) $(-4)(m)(m)(n)(m)(n)$ c) $(5)(5)(x)(y)(x)(z)(y)(z)$

                                                    $= -4m^3 n^2$ $= 5^2 x^2 y^2 z^2$

                                                                                 $= 25x^2 y^2 z^2$

2. Multiply. Leave your answer in exponent form.

    a) $4^2 \cdot 4^3$ b) $x^8 \cdot x^6$ c) $\left(2p^4\right)\left(6p^5\right)$ d) $(-3y)\left(5y^7\right)\left(2y^3\right)$

    e) $\left(7x^4 y^3\right)\left(3xy^6\right)$ f) $\left(\dfrac{2}{3}m^4 n^3\right)\left(-\dfrac{3}{4}mn^6\right)$ g) $(2.3abc)\left(8.8a^2 b^5\right)$

Solution

For any positive integers m and n, $a^m \cdot a^n = a^{m+n}$.

    a) $4^2 \cdot 4^3 = 4^5$ b) $x^8 \cdot x^6 = x^{14}$

    c) $\left(2p^4\right)\left(6p^5\right) = 2 \cdot 6 \cdot p^4 \cdot p^5 = 12p^9$ d) $(-3y)\left(5y^7\right)\left(2y^3\right) = (-3)(5)(2)(y)\left(y^7\right)\left(y^3\right)$

                                                                                      $= -30y^{11}$

    e) $\left(7x^4 y^3\right)\left(3xy^6\right) = 7 \cdot 3 \cdot x^4 \cdot x \cdot y^3 \cdot y^6$ f) $\left(\dfrac{2}{3}m^4 n^3\right)\left(-\dfrac{3}{4}mn^6\right) = \left(\dfrac{2}{3}\right)\left(-\dfrac{3}{4}\right)m^4 n^3 mn^6$

                                  $= 21x^5 y^9$ $= -\dfrac{1}{2}m^5 n^9$

    g) $(2.3abc)\left(8.8a^2 b^5\right) = (2.3)(8.8)abca^2 b^5$

                                  $= 20.24a^3 b^6 c$

3. Raise exponential expressions to a power.

   a) $\left(x^2\right)^3$ b) $\left(a^2b^3\right)^4$ c) $\left(2xy^3\right)^5$ d) $\left(\dfrac{-6x}{y^3}\right)^3$

   Solution

   For any positive integers m and n, $\left(a^m\right)^n = a^{mn}$, $(ab)^m = a^m b^m$, and $\left(\dfrac{a}{b}\right)^m = \dfrac{a^m}{b^m}$.

   a) $\left(x^2\right)^3 = x^6$

   b) $\left(a^2b^3\right)^4 = \left(a^2\right)^4\left(b^3\right)^4 = a^8 b^{12}$

   c) $\left(2xy^3\right)^5 = 2^5 x^5 \left(y^3\right)^5 = 32 x^5 y^{15}$

   d) $\left(\dfrac{-6x}{y^3}\right)^3 = \dfrac{(-6)^3 x^3}{\left(y^3\right)^3} = -\dfrac{216 x^3}{y^9}$

4. Simplify each expression.

   a) $\left(4xy^3\right)^3 xy$ b) $\left(\dfrac{a^2b^3}{c^5 d}\right)^4$ c) $\left(-3a^2b^3\right)^2\left(ab^3\right)$

   Solution

   a) $\left(4xy^3\right)^3 xy = 4^3 x^3 \left(y^3\right)^3 xy$
   $= 64 x^3 y^9 xy = 64 x^4 y^{10}$

   b) $\left(\dfrac{a^2b^3}{c^5 d}\right)^4 = \dfrac{\left(a^2\right)^4\left(b^3\right)^4}{\left(c^5\right)^4 d^4} = \dfrac{a^8 b^{12}}{c^{20} d^4}$

   c) $\left(-3a^2b^3\right)^2\left(ab^3\right) = (-3)^2\left(a^2\right)^2\left(b^3\right)^2 ab^3$
   $= 9a^4 b^6 ab^3 = 9a^5 b^9$

10.3 Multiplying Polynomials

Learning Objectives:

1. Find products using the rectangular method.
2. Multiply a monomial and a polynomial.
3. Multiply two polynomials.

Solved Examples:

1. Multiply.

a) $2x^2(6x-3)$

b) $4x(-2x^4+6x)$

c) $5x^3(-2x^3+7x-1)$

d) $\dfrac{2}{3}(5x+6x^2-7x^3)$

e) $(3x^3+x^2-5x)(3xy)$

f) $(2x^2y^2-4xy+8)(-3xy)$

Solution

a) $2x^2(6x-3) = 2x^2(6x)+2x^2(-3)$
$\qquad\qquad = 12x^3-6x^2$

b) $4x(-2x^4+6x) = 4x(-2x^4)+4x(6x)$
$\qquad\qquad\qquad = -8x^5+24x^2$

c) $5x^3(-2x^3+7x-1)$
$\qquad = 5x^3(-2x^3)+5x^3(7x)+5x^3(-1)$
$\qquad = -10x^6+35x^4-5x^3$

d) $\dfrac{2}{3}(5x+6x^2-7x^3)$
$\qquad = \dfrac{2}{3}(5x)+\dfrac{2}{3}(6x^2)+\dfrac{2}{3}(-7x^3)$
$\qquad = \dfrac{10}{3}x+4x^2-\dfrac{14}{3}x^3$

e) $(3x^3+x^2-5x)(3xy)$
$\qquad = 3x^3(3xy)+x^2(3xy)-5x(3xy)$
$\qquad = 9x^4y+3x^3y-15x^2y$

f) $(2x^2y^2-4xy+8)(-3xy)$
$\qquad = 2x^2y^2(-3xy)-4xy(-3xy)+8(-3xy)$
$\qquad = -6x^3y^3+12x^2y^2-24xy$

2. Multiply.

a) $(x+2)(x+3)$

b) $(x+2)(x-6)$

c) $(x-4)(x-5)$

d) $(3x-2)(-5x-6)$

e) $(4x-3y)(3x-4y)$

f) $(6x-3)^2$

g) $(4y-3z)(5y-7z)$

h) $(0.5x+2)(6x-0.2)$

i) $\left(\dfrac{1}{2}x-\dfrac{1}{3}\right)\left(\dfrac{1}{3}x+\dfrac{1}{4}\right)$

Solution

Use the FOIL method to multiply the binomials: First terms, Outer product, Inner product, Last terms

a) $(x+2)(x+3) = x^2 + 3x + 2x + 2 \cdot 3$
$= x^2 + 5x + 6$

b) $(x+2)(x-6) = x^2 - 6x + 2x + 2(-6)$
$= x^2 - 4x - 12$

c) $(x-4)(x-5) = x^2 - 5x - 4x - 4(-5)$
$= x^2 - 9x + 20$

d) $(3x-2)(-5x-6)$
$= (3x)(-5x) + 3x(-6) - 2(-5x) - 2(-6)$
$= -15x^2 - 18x + 10x + 12$
$= -15x^2 - 8x + 12$

e) $(4x-3y)(3x-4y)$
$= 4x(3x) + 4x(-4y) - 3y(3x) - 3y(-4y)$
$= 12x^2 - 16xy - 9xy + 12y^2$
$= 12x^2 - 25xy + 12y^2$

f) $(6x-3)^2 = (6x-3)(6x-3)$
$= (6x)^2 + 6x(-3) - 3(6x) - 3(-3)$
$= 36x^2 - 18x - 18x + 9$
$= 36x^2 - 36x + 9$

g) $(4y-3z)(5y-7z)$
$= 4y(5y) + 4y(-7z) - 3z(5y) - 3z(-7z)$
$= 20y^2 - 28yz - 15yz + 21z^2$
$= 20y^2 - 43yz + 21z^2$

h) $(0.5x+2)(6x-0.2)$
$= 0.5x(6x) + 0.5x(-0.2) + 2(6x) + 2(-0.2)$
$= 3x^2 - 0.1x + 12x - 0.4$
$= 3x^2 + 11.9x - 0.4$

i) $\left(\frac{1}{2}x - \frac{1}{3}\right)\left(\frac{1}{3}x + \frac{1}{4}\right)$
$= \frac{1}{2}x\left(\frac{1}{3}x\right) + \frac{1}{2}x\left(\frac{1}{4}\right) - \frac{1}{3}\left(\frac{1}{3}x\right) - \frac{1}{3}\left(\frac{1}{4}\right)$
$= \frac{1}{6}x^2 + \frac{1}{8}x - \frac{1}{9}x - \frac{1}{12}$
$= \frac{1}{6}x^2 + \frac{1}{72}x - \frac{1}{12}$

3. Multiply.

a) $\left(x^3 + 2x^2 - 3x + 1\right)(x+2)$

b) $(2x-3)\left(2x^3 - 3x^2 + 2x - 1\right)$

Solution

Use the FOIL method to multiply the binomials: First terms, Outer product, Inner product, Last terms

a) $\left(x^3 + 2x^2 - 3x + 1\right)(x+2) = x^3(x+2) + 2x^2(x+2) - 3x(x+2) + 1(x+2)$
$= x^4 + 2x^3 + 2x^3 + 4x^2 - 3x^2 - 6x + x + 2$
$= x^4 + 4x^3 + x^2 - 5x + 2$

b) $(2x-3)\left(2x^3 - 3x^2 + 2x - 1\right) = 2x\left(2x^3 - 3x^2 + 2x - 1\right) - 3\left(2x^3 - 3x^2 + 2x - 1\right)$
$= 4x^4 - 6x^3 + 4x^2 - 2x - 6x^3 + 9x^2 - 6x + 3$
$= 4x^4 - 6x^3 - 6x^3 + 4x^2 + 9x^2 - 2x - 6x + 3$
$= 4x^4 - 12x^3 + 13x^2 - 8x + 3$

10.4 Special Products

Learning Objectives:

1. Square binomials.
2. Find products of sum and differences of two terms.
3. Find greater powers of binomials.
4. Solve problems.

Solved Examples:

1. Use the formula $(a+b)(a-b) = a^2 - b^2$ to multiply.

   a) $(x+2)(x-2)$

   b) $(x+7)(x-7)$

   c) $(3x-5)(3x+5)$

   d) $(4x-3)(4x+3)$

   e) $(6a-b)(6a+b)$

   f) $(3x-0.2)(3x+0.2)$

   Solution

   a) $(x+2)(x-2) = x^2 - 2^2 = x^2 - 4$

   b) $(x+7)(x-7) = x^2 - 7^2 = x^2 - 49$

   c) $(3x-5)(3x+5) = (3x)^2 - 5^2 = 9x^2 - 25$

   d) $(4x-3)(4x+3) = (4x)^2 - 3^2 = 16x^2 - 9$

   e) $(6a-b)(6a+b) = (6a)^2 - b^2 = 36a^2 - b^2$

   f) $(3x-0.2)(3x+0.2) = (3x)^2 - 0.2^2$
   $= 9x^2 - .04$

2. Multiply

   a) $(3x-1)^2$

   b) $(4x+5)^2$

   c) $(7x-3)^2$

   d) $(3x+2y)^2$

   e) $\left(\dfrac{2}{3}x + \dfrac{1}{4}\right)^2$

   f) $(3y-4xz)^2$

   Solution

   To square a binomial, use the formulas $(a+b)^2 = a^2 + 2ab + b^2$ and $(a-b)^2 = a^2 - 2ab + b^2$.

   a) $(3x-1)^2 = (3x)^2 - 2(3x)(1) + 1^2$
   $= 9x^2 - 6x + 1$

   b) $(4x+5)^2 = (4x)^2 + 2(4x)(5) + 5^2$
   $= 16x^2 + 40x + 25$

   c) $(7x-3)^2 = (7x)^2 - 2(7x)(3) + 3^2$
   $= 49x^2 - 42x + 9$

   d) $(3x+2y)^2 = (3x)^2 + 2(3x)(2y) + (2y)^2$
   $= 9x^2 + 12xy + 4y^2$

   e) $\left(\dfrac{2}{3}x + \dfrac{1}{4}\right)^2 = \left(\dfrac{2}{3}x\right)^2 + 2\left(\dfrac{2}{3}x\right)\left(\dfrac{1}{4}\right) + \left(\dfrac{1}{4}\right)^2$
   $= \dfrac{4}{9}x^2 + \dfrac{1}{3}x + \dfrac{1}{16}$

   f) $(3y-4xz)^2 = (3y)^2 - 2(3y)(4xz) + (4xz)^2$
   $= 9y^2 - 24xyz + 16x^2z^2$

3. Multiply.

    a) $x(2x+5)(2x-5)$ b) $\left(7p^2+1\right)\left(7p^2-1\right)$

    c) $(k+6)^3$ d) $(4m-3)^4$

Solution

a) $\begin{aligned} x(2x+5)(2x-5) &= x\left[(2x+5)(2x-5)\right] \\ &= x\left[(2x)^2-5^2\right] \\ &= x\left(4x^2-25\right) \\ &= 4x^3-25 \end{aligned}$

b) $\begin{aligned} \left(7p^2+1\right)\left(7p^2-1\right) &= \left(7p^2\right)^2-1^2 \\ &= 49p^4-1 \end{aligned}$

c) $\begin{aligned} (k+6)^3 &= (k+6)^2(k+6) = \left[k^2+2(k)(6)+6^2\right](k+6) = \left(k^2+12k+36\right)(k+6) \\ &= k^2(k+6)+12k(k+6)+36(k+6) \\ &= k^3+6k^2+12k^2+72k+36k+108 = k^3+18k^2+108k+108 \end{aligned}$

d) $\begin{aligned} (4m-3)^4 &= \left[(4m-3)^2\right]^2 = \left[(4m)^2-2(4m)(3)+3^2\right]^2 \\ &= \left(16m^2-24m+9\right)\left(16m^2-24m+9\right) \\ &= 16m^2\left(16m^2-24m+9\right)-24m\left(16m^2-24m+9\right)+9\left(16m^2-24m+9\right) \\ &= 256m^4-384m^3+144m^2-384m^2+576m^2-216m+144m^2-216m+81 \\ &= 256m^4-768m^3+864m^2-432m+81 \end{aligned}$

10.5 Integers, Exponents, and The Quotient Rule

Learning Objectives:

1. Work with zero and negative exponents.
2. Use the quotient rule to simplify.
3. Use combinations of rules.

Solved Examples:

1. Evaluate each expression.

   a) $(-1)^{-3}$ b) 4^{-2} c) $(-11)^0$ d) $\dfrac{0^5}{5^0}$

   Solution

   Remember that for all numbers except 0, $a^0 = 1$ and $a^{-m} = \dfrac{1}{a^m}$.

   a) $(-1)^{-3} = \dfrac{1}{(-1)^3}$ b) $4^{-2} = \dfrac{1}{4^2} = \dfrac{1}{16}$ c) $(-11)^0 = 1$ d) $\dfrac{0^5}{5^0} = \dfrac{0}{1} = 1$

          $= \dfrac{1}{-1} = -1$

2. Simplify. Express your answer with positive exponents.

   a) x^{-3} b) $2x^{-3}$ c) $\dfrac{1}{a^{-7}}$ d) $\dfrac{2}{3y^{-2}}$

   e) $\left(x^{-3}\right)\left(x^{-5}\right)$ f) $\dfrac{a^{-3}}{a^{-2}}$ g) $\left(x^{-3}y^{-4}\right)^5$ h) $\left(-2a^{-4}b\right)^{-3}$

   Solution

   Remember that for all numbers except 0, $a^0 = 1$, $a^{-m} = \dfrac{1}{a^m}$, $\dfrac{1}{a^{-m}} = a^m$, and $\dfrac{a^m}{a^n} = a^{m-n}$. Also recall

   that for any positive integers m and n, $a^m \cdot a^n = a^{m+n}$, $\left(a^m\right)^n = a^{mn}$, and $(ab)^m = a^m b^m$.

   a) $x^{-3} = \dfrac{1}{x^3}$ b) $2x^{-3} = \dfrac{2}{x^3}$ c) $\dfrac{1}{a^{-7}} = a^7$ d) $\dfrac{2}{3y^{-2}} = \dfrac{2y^2}{3}$

   e) $\left(x^{-3}\right)\left(x^{-5}\right) = x^{-8}$ f) $\dfrac{a^{-3}}{a^{-2}} = a^{-3-(-2)}$ g) $\left(x^{-3}y^{-4}\right)^5 = \left(x^{-3}\right)^5\left(y^{-4}\right)^5$

          $= \dfrac{1}{x^8}$ $= a^{-1} = \dfrac{1}{a}$ $= x^{-15}y^{-20} = \dfrac{1}{x^{15}y^{20}}$

   h) $\left(-2a^{-4}b\right)^{-3} = (-2)^{-3}\left(a^{-4}\right)^{-3}b^{-3} = \dfrac{1}{(-2)^3}\left(a^{12}\right)\left(\dfrac{1}{b^3}\right) = -\dfrac{a^{12}}{8b^3}$

4. Simplify. Express your answer with positive exponents.

a) $a^6 b^{-2} c^{-4}$

b) $\dfrac{3x^{-4}y^{-2}}{y^3}$

c) $\dfrac{\left(3xy^2\right)^{-3}}{\left(3xy^2\right)^{-4}}$

d) $\left(4^2\right)\left(2^{-3}\right)$

e) $\left(\dfrac{3x^3 y^0}{z^2}\right)^{-3}$

f) $\dfrac{a^{-3}b^{-4}}{a^5 b^{-3}}$

Solution

a) $a^6 b^{-2} c^{-4} = \dfrac{a^6}{b^2 c^4}$

b) $\dfrac{3x^{-4}y^{-2}}{y^3} = \dfrac{3}{x^4 y^2 y^3} = \dfrac{3}{x^4 y^5}$

c) $\dfrac{\left(3xy^2\right)^{-3}}{\left(3xy^2\right)^{-4}} = \left(3xy^2\right)^{-3-(-4)} = \left(3xy^2\right)^{1} = 3xy^2$

d) $\left(4^2\right)\left(2^{-3}\right) = \dfrac{4^2}{2^3} = \dfrac{16}{8} = 2$

e) $\left(\dfrac{3x^3 y^0}{z^2}\right)^{-3} = \left(\dfrac{z^2}{3x^3 y^0}\right)^3 = \dfrac{\left(z^2\right)^3}{\left(3x^3\left(1\right)\right)^3}$

$= \dfrac{z^6}{3^3 x^9} = \dfrac{z^6}{27x^9}$

f) $\dfrac{a^{-3}b^{-4}}{a^5 b^{-3}} = a^{-3-4}b^{-4-(-3)} = a^{-7}b^{-1}$

$= \dfrac{1}{a^7 b}$

10.6 Dividing a Polynomial by a Monomial

Learning Objectives:

1. Divide a polynomial by a monomial.

Solved Examples:

1. In the statement $\dfrac{5x^2 - 10}{5}$, the dividend is _____ and the divisor is _____ .

   Solution

   In the statement $\dfrac{5x^2 - 10}{5}$, the dividend is $\underline{5x^2 - 10}$ and the divisor is $\underline{5}$.

2. The expression $\dfrac{-7x + 21}{7x}$ is undefined if _____ .

   Solution

   The expression $\dfrac{-7x + 21}{7x}$ is undefined if $\underline{7x = 0 \text{ or } x = 0}$.

3. The problem $\dfrac{6x^3 - 3x^2 + 12x}{3x} = 2x^2 - x + 4$ can be checked by multiplying _____ and _____ to show that the product is _____.

   Solution

   The problem $\dfrac{6x^3 - 3x^2 + 12x}{3x} = 2x^2 - x + 4$ can be checked by multiplying $\underline{2x^2 - x + 4}$ and $\underline{3x}$ to show that the product is $\underline{6x^3 - 3x^2 + 12x}$. .

4. Divide.

   a) $\dfrac{20x^4 - 25x^2 + 20x}{5x}$

   b) $\dfrac{27b^5 - 15b^3 - 9b^2}{3b^2}$

   c) $\left(64x^7 - 32x^5 + 16x^3\right) \div 8x^3$

   Solution

   a) $\dfrac{20x^4 - 25x^2 + 20x}{5x} = \dfrac{20x^4}{5x} + \dfrac{-25x^2}{5x} + \dfrac{20x}{5x} = 4x^3 - 5x + 4$

   b) $\dfrac{27b^5 - 15b^3 - 9b^2}{3b^2} = \dfrac{27b^5}{3b^2} + \dfrac{-15b^3}{3b^2} + \dfrac{-9b^2}{3b^2} = 9b^3 - 5b - 3$

c) $\left(64x^7 - 32x^5 + 16x^3\right) \div 8x^3 = \dfrac{64x^7 - 32x^5 + 16x^3}{8x^3} = \dfrac{64x^7}{8x^3} + \dfrac{-32x^5}{8x^3} + \dfrac{16x^3}{8x^3} = 8x^4 - 4x^2 + 2$

10.7 Dividing a Polynomial by a Polynomial

Learning Objectives:

1. Divide a polynomial by a polynomial.
2. Solve problems.

Solved Examples:

1. Divide. Check your answers by multiplication.

   a) $\dfrac{x^2 + 18x + 81}{x + 9}$ b) $\dfrac{2m^2 + 11m - 40}{m + 8}$ c) $\dfrac{6m^3 + 49m^2 - 36m + 81}{m + 9}$

   Solution

   a) $\dfrac{x^2 + 18x + 81}{x + 9}$ $\quad x+9\overline{\smash{)}x^2 + 18x + 81}$ Check: $(x+9)(x+9) = x^2 + 2x(9) + 9^2 = x^2 + 18x + 81$

   $$\begin{array}{r} x +9 \\ x+9\overline{\smash{)}x^2 + 18x + 81} \\ \underline{x^2 + 9x} \\ 9x + 81 \\ \underline{9x + 81} \\ 0 \end{array}$$

   b) $\dfrac{2m^2 + 11m - 40}{m + 8}$ $\quad m+8\overline{\smash{)}2m^2 + 11m - 40}$ Check: $(2m-5)(m+8) = 2m^2 + 2m(8) - 5m - 5(8)$

   $$\begin{array}{r} 2m -5 \\ m+8\overline{\smash{)}2m^2 + 11m - 40} \\ \underline{2m^2 + 16m} \\ -5m - 40 \\ \underline{-5m - 40} \\ 0 \end{array}$$
   $ = 2m^2 + 11m - 40$

   c) $\dfrac{6m^3 + 49m^2 - 36m + 81}{m + 9}$ $\quad m+9\overline{\smash{)}6m^3 + 49m^2 - 36m + 81}$

   $$\begin{array}{r} 6m^2 - 5m + 9 \\ m+9\overline{\smash{)}6m^3 + 49m^2 - 36m + 81} \\ \underline{6m^3 + 54m^2} \\ -5m^2 - 36m \\ \underline{-5m^2 - 45m} \\ 9m + 81 \\ \underline{9m + 81} \\ 0 \end{array}$$

   Check: $(6m^2 - 5m + 9)(m + 9) = 6m^2(m+9) - 5m(m+9) + 9(m+9)$
   $$= 6m^3 + 54m^2 - 5m^2 - 45m + 9m + 81$$
   $$= 6m^3 + 49m^2 - 36m + 81$$

2. Divide. Check your answers by multiplication.

   a) $\dfrac{x^2 + 2x - 5}{x + 4}$ b) $\dfrac{x^2 + 17x + 66}{x + 8}$ c) $\dfrac{4x^2 - 22x + 32}{2x + 3}$

d) $\dfrac{9x^2 + 6x + 10}{3x - 2}$ e) $\dfrac{3y^4 + y^2 + 2}{y + 1}$ f) $\dfrac{y^3 + 1}{y + 1}$

Solution

a) $\dfrac{x^2 + 2x - 5}{x + 4}$

$$x + 4 \overline{)x^2 + 2x - 5}$$

with quotient $x - 2$

$$\begin{array}{r} x - 2 \\ x+4\overline{)x^2 + 2x - 5} \\ \underline{x^2 + 4x} \\ -2x - 5 \\ \underline{-2x - 8} \\ 3 \end{array}$$

$\dfrac{x^2 + 2x - 5}{x + 4} = x - 2 + \dfrac{3}{x + 4}$

Check: $(x + 4)\left(x - 2 + \dfrac{3}{x + 4}\right) = (x + 4)x + (x + 4)(-2) + (x + 4)\left(\dfrac{3}{x + 4}\right)$

$\qquad = x^2 + 4x - 2x - 8 + 3 = x^2 + 2x - 5$

b) $\dfrac{x^2 + 17x + 66}{x + 8}$

$$\begin{array}{r} x + 9 \\ x+8\overline{)x^2 + 17x + 66} \\ \underline{x^2 + 8x} \\ 9x + 66 \\ \underline{9x + 72} \\ -6 \end{array}$$

$\dfrac{x^2 + 17x + 66}{x + 8} = x + 9 - \dfrac{6}{x + 8}$

Check: $(x + 8)\left(x + 9 - \dfrac{6}{x + 8}\right) = (x + 8)x + (x + 8)(9) + (x + 8)\left(-\dfrac{6}{x + 8}\right)$

$\qquad = x^2 + 8x + 9x + 72 - 6 = x^2 + 17x + 66$

c) $\dfrac{4x^2 - 22x + 32}{2x + 3}$

$$\begin{array}{r} 2x - 14 \\ 2x+3\overline{)4x^2 - 22x + 32} \\ \underline{4x^2 + 6x} \\ -28x + 32 \\ \underline{-28x - 42} \\ 74 \end{array}$$

$\dfrac{4x^2 - 22x + 32}{2x + 3} = 2x - 14 + \dfrac{74}{2x + 3}$

Check: $(2x + 3)\left(2x - 14 + \dfrac{74}{2x + 3}\right) = (2x + 3)(2x) + (2x + 3)(-14) + (2x + 3)\left(\dfrac{74}{2x + 3}\right)$

$\qquad = 4x^2 + 6x - 28x - 42 + 74 = 4x^2 - 22x + 32$

d) $\dfrac{9x^2 + 6x + 10}{3x - 2}$

$$\begin{array}{r} 3x + 4 \\ 3x-2\overline{)9x^2 + 6x + 10} \\ \underline{9x^2 - 6x} \\ 12x + 10 \\ \underline{12x - 8} \\ 18 \end{array}$$

$\dfrac{9x^2 + 6x + 10}{3x - 2} = 3x + 4 + \dfrac{18}{3x - 2}$

Check: $(3x - 2)\left(3x + 4 + \dfrac{18}{3x - 2}\right) = (3x - 2)(3x) + (3x - 2)(4) + (3x - 2)\left(\dfrac{18}{3x - 2}\right)$

$\qquad = 9x^2 - 6x + 12x - 8 + 18 = 9x^2 + 6x + 10$

e) $\dfrac{3y^4 + y^2 + 2}{y+1}$

$$y+1\overline{\smash{\big)}\,3y^4 + 0y^3 + y^2 + 0y + 2}$$

$$
\begin{array}{r}
3y^3 - 3y^2 + 4y - 4 \\
\underline{3y^4 + 3y^3} \\
-3y^3 + y^2 \\
\underline{-3y^3 - 3y^2} \\
4y^2 + 0y \\
\underline{4y^2 + 4y} \\
-4y + 2 \\
\underline{-4y - 4} \\
6
\end{array}
$$

$\dfrac{3y^4 + y^2 + 2}{y+1} = 3y^3 - 3y^2 + 4y - 4 + \dfrac{6}{y+1}$

Check: $(y+1)\left(3y^3 - 3y^2 + 4y - 4 + \dfrac{6}{y+1}\right)$

$= (y+1)(3y^3) + (y+1)(-3y^2) + (y+1)(4y) + (y+1)(-4) + (y+1)\left(\dfrac{6}{y+1}\right)$

$= 3y^4 + 3y^3 - 3y^3 - 3y^2 + 4y^2 + 4y - 4y - 4 + 6 = 3y^4 + y^2 + 2$

f) $\dfrac{y^3 + 1}{y+1}$

$$y+1\overline{\smash{\big)}\,y^3 + 0y^2 + 0y + 1}$$

$$
\begin{array}{r}
y^2 - y + 1 \\
\underline{y^3 + y^2} \\
-y^2 + 0y \\
\underline{-y^2 - y} \\
y + 1 \\
\underline{y + 1} \\
0
\end{array}
$$

$\dfrac{y^3 + 1}{y+1} = y^2 - y + 1$

Check: $(y+1)(y^2 - y + 1) = (y+1)(y^2) + (y+1)(-y) + (y+1)(1) = y^3 + y^2 - y^2 + 1 = y^3 + 1$

3. Give the width of the rectangle.

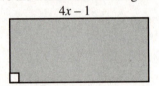

4x − 1

Area = $12x^2 - 23x + 5$
Length = $4x - 1$

Solution

Area = length × width, so width = $\dfrac{\text{area}}{\text{length}} = \dfrac{12x^2 - 23x + 5}{4x - 1}$.

$$4x+1\overline{\smash{\big)}\,12x^2 - 23x + 5}$$

$$
\begin{array}{r}
3x + 5 \\
\underline{12x^2 + 3x} \\
20x + 5 \\
\underline{20x + 5} \\
0
\end{array}
$$

The width is $3x + 5$.

10.8 An Application of Exponents: Scientific Notation

Learning Objectives:

1. Express numbers in scientific notation.
2. Convert from scientific to decimal notation.
3. Perform calculations using scientific notation.

Solved Examples:

1. Write in scientific notation.

    a) 125   b) 3442   c) 0.022   d) 0.00000453

Solution

    a) Move the decimal two places to the left.

$$125 = 1.25 \times 10^2$$

    b) Move the decimal three places to the left.

$$3442 = 3.442 \times 10^3$$

    c) Move the decimal two places to the right.

$$0.022 = 2.2 \times 10^{-2}$$

    d) Move the decimal six places to the right.

$$0.00000453 = 4.53 \times 10^{-6}$$

2. Write in standard notation.

    a) 2.04×10^3 b) 1.9902×10^7 c) 9.311×10^{-4}

Solution

    a) The decimal moves three places to the right.

$$2.04 \times 10^3 = 2040$$

    b) The decimal moves seven places to the right.

$$1.9902 \times 10^7 = 19,902,000$$

    c) The decimal moves four places to the left.

$$9.311 \times 10^{-4} = 0.0009311$$

3. Evaluate by using scientific notation and the laws of exponents. Write answers in scientific notation.

    a) $(43,000,000)(560,000)$ b) $(1,500,000)(0.00045) \div 1200$

    c) $(7 \times 10^{-3})(5 \times 10^{-9})$ d) $(9 \times 10^6)(4 \times 10^7)$

    e) $\dfrac{8 \times 10^{-3}}{2 \times 10^{-4}}$ f) $\dfrac{(2.3 \times 10^5)(4.1 \times 10^{-1})}{6.8 \times 10^3}$

Solution

    a) $(43,000,000)(560,000)$

$$= (4.3 \times 10^7)(5.6 \times 10^5)$$
$$= 4.3 \times 5.6 \times 10^7 \times 10^5 = 24.08 \times 10^{12}$$
$$= 2.408 \times 10^{13}$$

    b) $(1,500,000)(0.00045) \div 1200$

$$= (1.5 \times 10^6)(4.5 \times 10^{-4}) \div (1.2 \times 10^3)$$
$$= (1.5 \times 4.5 \div 1.2) \times (10^6 \times 10^{-4} \div 10^3)$$
$$= 5.625 \times 10^{-1}$$

c) $\left(7\times10^{-3}\right)\left(5\times10^{-9}\right)=\left(7\times5\right)\times\left(10^{-3}\times10^{-9}\right)$
$$=35\times10^{-12}$$
$$=3.5\times10^{-11}$$

d) $\left(9\times10^{6}\right)\left(4\times10^{7}\right)=\left(9\times4\right)\times\left(10^{6}\times10^{7}\right)$
$$=36\times10^{13}=3.6\times10^{12}$$

e) $\dfrac{8\times10^{-3}}{2\times10^{-4}}=4\times10^{-3-(-4)}=4\times10^{1}$

f) $\dfrac{\left(2.3\times10^{5}\right)\left(4.1\times10^{-1}\right)}{6.8\times10^{3}}$
$$=\dfrac{2.3\times4.1}{6.8}\times\dfrac{10^{5}\times10^{-1}}{10^{3}}\approx1.39\times10^{1}$$

11.1 The Greatest Common Factor

Learning Objectives:

1. Find the greatest common factor of a list of numbers.
2. Find the greatest common factor of a list of variable terms.
3. Factor out the greatest common factor.
4. Factor by grouping.

Solved Examples:

1. Find the greatest common factor for each list of numbers.

   a) 15, 27

   c) 30, 18, 24

   Solution

   First write each number in prime factored form. Then use each prime the least number of time it appears in all the factored form as a factor in the GCF.

   a) $15 = 3 \cdot 5$
   $27 = 3 \cdot 3 \cdot 3$
   $\text{GCF} = 3$

   b) $30 = 2 \cdot 3 \cdot 5$
   $18 = 2 \cdot 3 \cdot 3$
   $24 = 2 \cdot 2 \cdot 2 \cdot 3$
   $\text{GCF} = 2 \cdot 3 = 6$

2. Factor out the greatest common factor.

   a) $3x^2 + 6x$

   b) $4y^2 + 4y$

   c) $8a^2b^2 - 32ab$

   d) $15xy - 18yz - 27xz$

   e) $36a^2 - 24ab - 16a$

   f) $15x^2y - 25xy^2 + 20xy$

   Solution

   a) $3x^2 + 6x = 3x(x + 2)$

   b) $4y^2 + 4y = 4y(y + 1)$

   c) $8a^2b^2 - 32ab = 8ab(ab - 4)$

   d) $15xy - 18yz - 27xz = 3(5xy - 6yz - 9xz)$

   e) $36a^2 - 24ab - 16a = 4a(9a - 6b - 4)$

   f) $15x^2y - 25xy^2 + 20xy = 5xy(3x - 5y + 4)$

3. Factor completely.

   a) $3(2x + y) - z(2x + y)$

   b) $7x(x - 5) + 3(x - 5)$

   c) $4x(2y + 3z) - 7t(2y + 3z)$

   d) $12x^3 + 8x^2 - 15x - 10$

   e) $(x + 2d) + 5c^2(x + 2d)$

   f) $x(5y - 2) - (5y - 2)$

   g) $pq + 3q + 5p + 15$

   h) $2b(y^2 - x) - 3a(y^2 - x) + 6c(y^2 - x)$

Solution

a) $3(2x+y)-z(2x+y)=(3-z)(2x+y)$

b) $7x(x-5)+3(x-5)=(7x+3)(x-5)$

c) $4x(2y+3z)-7t(2y+3z)$
$=(4x-7t)(2y+3z)$

d) $12x^3+8x^2-15x-10$
$=4x^2(3x+2)-5(3x+2)$
$=(4x^2-5)(3x+2)$

e) $(x+2d)+5c^2(x+2d)$
$=1(x+2d)+5c^2(x+2d)$
$=(1+5c^2)(x+2d)$

f) $x(5y-2)-(5y-2)=x(5y-2)-1(5y-2)$
$=(x-1)(5y-2)$

g) $pq+3q+5p+15=q(p+3)+5(p+3)$
$=(q+5)(p+3)$

h) $2b(y^2-x)-3a(y^2-x)+6c(y^2-x)$
$=(2b-3a+6c)(y^2-x)$

11.2 Factoring Trinomials

Learning Objectives:

1. Factor trinomials with a coefficient of 1 for the squared term.
2. Factor trinomials after factoring out the greatest common factor.

Solved Examples:

1. Factor.

a) $x^2 + 3x + 2$

b) $x^2 + 6x + 8$

c) $x^2 - 6x + 8$

d) $x^2 - 10x + 9$

e) $x^2 + x - 2$

f) $x^2 + 7x - 8$

g) $x^2 - 2x - 8$

h) $x^2 - 3x - 10$

i) $x^2 + 12x + 35$

j) $x^2 + 2x - 48$

k) $x^2 - 11x + 24$

l) $x^2 - 4x - 21$

Solution

a) $x^2 + 3x + 2$

Factors of 2	Sums of Factors
1, 2	3

$x^2 + 3x + 2 = (x+1)(x+2)$

b) $x^2 + 6x + 8$

Factors of 8	Sums of Factors
1, 8	9
2, 4	6

$x^2 + 6x + 8 = (x+2)(x+4)$

c) $x^2 - 6x + 8$

Factors of 2	Sums of Factors
−1, −8	−9
−2, −4	−6

$x^2 - 6x + 8 = (x-2)(x-4)$

d) $x^2 - 10x + 9$

Factors of 9	Sums of Factors
−1, −9	−10
−3, −3	−6

$x^2 - 10x + 9 = (x-1)(x-9)$

e) $x^2 + x - 2$

Factors of −2	Sums of Factors
−1, 2	1
1, −2	−1

$x^2 + x - 2 = (x-1)(x+2)$

f) $x^2 + 7x - 8$

Factors of −8	Sums of Factors
1, −8	−7
−1, 8	7
2, −4	−2
−2, 4	2

$x^2 + 7x - 8 = (x-1)(x+7)$

g) $x^2 - 2x - 8$

Factors of −8	Sums of Factors
1, −8	−7
−1, 8	7
2, −4	−2
−2, 4	2

$x^2 - 2x - 8 = (x+2)(x-4)$

h) $x^2 - 3x - 10$

Factors of −10	Sums of Factors
1, −10	−9
−1, 10	9
−2, 5	3
2, −5	−3

$x^2 - 3x - 10 = (x-5)(x+2)$

i) $x^2 + 12x + 35$

Factors of 35	Sums of Factors
1, 35	36
2, 18	20
5, 7	12

$$x^2 + 3x + 2 = (x+1)(x+2)$$

j) $x^2 + 2x - 48$

Factors of –48	Sums of Factors
–1, 48	47
–2, 24	22
–3, 16	13
–4, 12	8
–6, 8	2

$$x^2 + 2x - 48 = (x-6)(x+8)$$

k) $x^2 - 11x + 24$

Factors of 24	Sums of Factors
–1, –24	–25
–2, –12	–14
–3, –8	–11

$$x^2 - 11x + 24 = (x-3)(x-8)$$

l) $x^2 - 4x - 21$

Factors of –21	Sums of Factors
1, –21	–20
3, –7	–4

$$x^2 - 4x - 21 = (x+3)(x-7)$$

2. Factor out the greatest common factor. Then factor the remaining polynomial.

a) $3x^2 + 21x + 30$ b) $5x^2 + 20x + 15$ c) $4x^2 - 8x - 96$

d) $6x^2 + 6x - 72$ e) $3x^2 - 30x + 48$ f) $5x^2 - 20x - 105$

Solution

a) $3x^2 + 21x + 30 = 3\left(x^2 + 7x + 10\right)$
$$= 3(x+2)(x+5)$$

b) $5x^2 + 20x + 15 = 5\left(x^2 + 4x + 3\right)$
$$= 5(x+1)(x+3)$$

c) $4x^2 - 8x - 96 = 4\left(x^2 - 2x - 24\right)$
$$= 4(x-6)(x+4)$$

d) $6x^2 + 6x - 72 = 6\left(x^2 + x - 12\right)$
$$= 6(x+4)(x-3)$$

e) $3x^2 - 30x + 48 = 3\left(x^2 - 10x + 16\right)$
$$= 3(x-2)(x-8)$$

f) $5x^2 - 20x - 105 = 5\left(x^2 - 4x - 21\right)$
$$= 5(x+3)(x-7)$$

11.3 Factoring Trinomials by Grouping

Learning Objectives:

1. Factor trinomials by grouping.

Solved Examples:

1. The middle term of each trinomial has been rewritten. Now factor by grouping.

    a) $x^2 + 7x + 10 = x^2 + 5x + 2x + 10$

    b) $b^2 + 5b - 24 = b^2 - 3b + 8b - 24$

    c) $6c^2 - 13c - 5 = 6c^2 + 2c - 15c - 5$

Solution

a) $x^2 + 7x + 10 = x^2 + 5x + 2x + 10$
$= (x^2 + 5x) + (2x + 10)$
$= x(x+5) + 2(x+5)$
$= (x+2)(x+5)$

b) $b^2 + 5b - 24 = b^2 - 3b + 8b - 24$
$= (b^2 - 3b) + (8b - 24)$
$= b(b-3) + 8(b-3)$
$= (b+8)(b-3)$

c) $6c^2 - 13c - 5 = 6c^2 + 2c - 15c - 5$
$= (6c^2 + 2c) - (15c - 5)$
$= 2c(3c+1) - 5(3c+1)$
$= (2c-5)(3c+1)$

2. Factor each trinomial by grouping.

    a) $2y^2 - y - 10$

    b) $6m^2 + 35m - 6$

    c) $12 - s - s^2$

Solution

Look for two integers whose product is the coefficient of the first term times the third term, and whose sum is the coefficient of the middle term.

a) $2y^2 - y - 10$
Look for two integers whose product is $2(-10) = -20$ and whose sum is -1. $4(-5) = -20$ and $4 + (-5) = -1$.
$2y^2 - y - 10 = 2y^2 + 4y - 5y - 10$
$= (2y^2 + 4y) - (5y - 10)$
$= 2y(y+2) - 5(y+2)$
$= (2y-5)(y+2)$

b) $6m^2 + 35m - 6$
Look for two integers whose product is $6(-6) = -36$ and whose sum is 35. $-1(36) = -36$ and $-1 + 36 = 35$.
$6m^2 + 35m - 6 = 6m^2 - m + 36m - 6$
$= (6m^2 - m) + (36m - 6)$
$= m(6m-1) + 6(6m-1)$
$= (m+6)(6m-1)$

c) $12 - s - s^2$

Look for two integers whose product
is $12(-1) = -12$ and whose sum is -1.
$-4(3) = -12$ and $-4 + 3 = -1$.

$$12 - s - s^2 = 12 - 4s + 3s - s^2$$
$$= (12 - 4s) + (3s - s^2)$$
$$= 4(3 - s) + s(3 - s)$$
$$= (4 + s)(3 - s)$$

3. Factor the following expression as described below. $3x^2 - x + 9x - 3$

   a) Factor by grouping the first two terms together   b) Factor by grouping the first and third terms
       and the last two terms together.   together and the second and fourth terms
                                                             together.

   c) Do both methods give the same final answer?

   Solution

   a) $3x^2 - x + 9x - 3 = (3x^2 - x) + (9x - 3)$ b) $3x^2 - x + 9x - 3 = 3x^2 + 9x - x - 3$
       $\qquad\qquad\qquad = x(3x - 1) + 3(3x - 1)$ $\qquad\qquad\qquad = (3x^2 + 9x) + (-x - 3)$
       $\qquad\qquad\qquad = (x + 3)(3x - 1)$ $\qquad\qquad\qquad = 3x(x + 3) - 1(x + 3)$
                                                                $\qquad\qquad\qquad = (3x - 1)(x + 3)$

   c) Yes, both methods give the same answer.

11.4 Factoring Trinomials Using FOIL

Learning Objectives:

1. Factor trinomials using FOIL.

Solved Examples:

1. Factor by the trial–and–error method. Check your answers using FOIL.

  a) $2x^2 + 7x + 3$ b) $3x^2 - 2x - 8$ c) $5x^2 - 17x + 6$

  d) $6x^2 + 19x - 20$ e) $9x^2 + 29x + 6$ f) $8x^2 + 18x + 9$

  g) $10y^2 + 23y + 12$ h) $6z^2 + 5z - 6$ i) $15z^2 - 4z - 4$

Solution

  a) $2x^2 + 7x + 3$ b) $3x^2 - 2x - 8$ c) $5x^2 - 17x + 6$
   $= (2x+1)(x+3)$ $= (3x+4)(x-2)$ $= (5x-2)(x-3)$

  d) $6x^2 + 19x - 20$ e) $9x^2 + 29x + 6$ f) $8x^2 + 18x + 9$
   $= (6x-5)(x+4)$ $= (x+3)(9x+2)$ $= (4x+3)(2x+3)$

  g) $10y^2 + 23y + 12$ h) $6z^2 + 5z - 6$ i) $15z^2 - 4z - 4$
   $= (5x+4)(2x+3)$ $= (3x-2)(2x+3)$ $= (5x+2)(3x-2)$

2. Factor out the greatest common factor from each term. Then factor the remaining trinomial.

  a) $18x^2 - 78x - 60$ b) $10x^2 - 35x - 20$ c) $8y^2 + 44y + 20$

  d) $-4x^2 + 10x + 6$ e) $9x^3 - 6x^2 - 24x$ f) $-45x^3 + 96x^2 - 48x$

Solution

  a) $18x^2 - 78x - 60 = 6(3x^2 - 13x - 10)$ b) $10x^2 - 35x - 20 = 5(2x^2 - 7x - 4)$
        $= 6(3x+2)(x-5)$ $= 5(2x+1)(x-4)$

  c) $8y^2 + 44y + 20 = 4(2y^2 + 11y + 5)$ d) $-4x^2 + 10x + 6 = -2(2x^2 - 5x - 3)$
       $= 4(2y+1)(y+5)$ $= -2(2x+1)(2x-3)$

  e) $9x^3 - 6x^2 - 24x = 3x(3x^2 - 2x - 8)$ f) $-45x^3 + 96x^2 - 48x = -3x(15x^2 - 32x + 16)$
        $= 3x(3x+4)(x-2)$ $= -3x(5x-4)(3x-4)$

11.5 Special Factoring Techniques

Learning Objectives:

1. Find the square of a number.
2. Factor a difference of squares.
3. Factor a perfect square trinomial.
4. Evaluate cubes and recognize sums and differences of cubes.
5. Factor sums and differences of cubes.

Solved Examples:

1. Factor by using the difference of squares formula, $a^2 - b^2 = (a+b)(a-b)$.

   a) $x^2 - 4$ b) $x^2 - 49$ c) $9x^2 - 25$ d) $25x^2 - 49y^2$

   e) $100 - x^2$ f) $81x^4 - 1$ g) $81x^2 - y^2$ h) $x^4 - 4$

   Solution

   a) $x^2 - 4 = (x+2)(x-2)$ b) $x^2 - 49 = (x+7)(x-7)$

   c) $9x^2 - 25 = (3x+5)(3x-5)$ d) $25x^2 - 49y^2 = (5x+7)(5x-7)$

   e) $100 - x^2 = (10+x)(10-x)$ f) $81x^4 - 1 = (9x^2+1)(9x^2-1)$
   $$= (9x^2+1)(3x+1)(3x-1)$$

   g) $81x^2 - y^2 = (9x+y)(9x-y)$ h) $x^4 - 4 = (x^2+2)(x^2-2)$

2. Factor by using the perfect–square trinomial formula, $a^2 + 2ab + b^2 = (a+b)^2$ or $a^2 - 2ab + b^2 = (a-b)^2$.

   a) $x^2 + 18x + 81$ b) $x^2 + 2x + 1$ c) $x^2 + 12x + 36$

   d) $4x^2 - 28x + 49$ e) $49x^2 - 42xy + 9y^2$ f) $16x^2 + 72xy + 81y^2$

   Solution

   a) $x^2 + 18x + 81 = (x+9)^2$ b) $x^2 + 2x + 1 = (x+1)^2$

   c) $x^2 + 12x + 36 = (x+6)^2$ d) $4x^2 - 28x + 49 = (2x-7)^2$

   e) $49x^2 - 42xy + 9y^2 = (7x-3y)^2$ f) $16x^2 + 72xy + 81y^2 = (4x+9y)^2$

3. Factor by first looking for a greatest common factor.

   a) $4x^2 - 16$ b) $72x^2 - 98y^2$ c) $ab^2 - 16a$

   d) $18x^2 + 12x + 2$ e) $75x^2 + 90x + 27$ f) $125x^2 - 150x + 45$

Solution

a) $4x^2 - 16 = 4(x^2 - 4) = 4(x+2)(x-2)$

b) $72x^2 - 98y^2 = 2(36x^2 - 49y^2)$
$$= 2(6x+7y)(6x-7y)$$

c) $ab^2 - 16a = a(b^2 - 16) = a(b+4)(b-4)$

d) $18x^2 + 12x + 2 = 2(9x^2 + 6x + 1) = 2(3x+1)^2$

e) $75x^2 + 90x + 27 = 3(25x^2 + 30x + 9)$
$$= 3(5x+3)^2$$

f) $125x^2 - 150x + 45 = 5(25x^2 - 30x + 9)$
$$= 5(5x-3)^2$$

4. Factor completely.

   a) $2x^2 + 6x + 4$

   b) $3x^2 - 2x - 8$

   c) $25x^2 - 100$

   d) $10x^2 - 35x - 20$

   e) $9x^2 + 24xy + 16y^2$

   f) $16x^2 + 72xy + 81y^2$

Solution

a) $2x^2 + 6x + 4 = 2(x^2 + 3x + 2)$
$$= 2(x+2)(x+1)$$

b) $3x^2 - 2x - 8 = (3x+4)(x-2)$

c) $25x^2 - 100 = (5x+10)(5x-10)$

d) $10x^2 - 35x - 20 = 5(2x^2 - 7x - 4)$
$$= 5(2x+1)(x-4)$$

e) $9x^2 + 24xy + 16y^2 = (3x+4y)^2$

f) $16x^2 + 72xy + 81y^2 = (4x+9y)^2$

11.6 Solving Quadratic Equations by Factoring

Learning Objectives:

1. Solve quadratic equations by factoring.
2. Solve other equations by factoring.
3. Solve application problems.

Solved Examples:

1. Solve for x using the zero factor property.

    a) $3 \cdot x = 0$

    b) $2(x-5) = 0$

    c) $x(x+6) = 0$

    d) $(2x-3)(4x+5) = 0$

    Solution

    a) $3 \cdot x = 0$
       $x = 0$

    b) $2(x-5) = 0$
       $x-5 = 0$
       $x = 5$

    c) $x(x+6) = 0$
       $x = 0$ or $x+6 = 0$
       $x = -6$

    d) $(2x-3)(4x+5) = 0$
       $2x-3 = 0$ or $4x+5 = 0$
       $2x = 3$ $4x = -5$
       $x = \dfrac{3}{2}$ $x = -\dfrac{5}{4}$

2. Using the factoring method, solve for the roots of each quadratic equation. Be sure to put the equation in standard form before factoring. Check your answers.

    a) $x^2 + 9x - 36 = 0$

    b) $x^2 - x = 6$

    c) $3x^2 - 21x + 30 = 0$

    d) $5x^2 - 3x - 8 = 0$

    e) $9x^2 - 2x = 0$

    f) $x^2 - 64 = 63x$

    g) $x(3x+13) = 10$

    h) $(x-3)(x+4) = -4(x+4)$

    i) $2 + \dfrac{x^2}{3} = 5x + 2$

    j) $\dfrac{x^2 - 7x}{2} = 9$

    Solution

    Remember to check your work by substituting each solution in the original equation.

    a) $x^2 + 9x - 36 = 0$
       $(x+12)(x-3) = 0$ Factor
       Using the zero-factor property, we have
       $x+12 = 0$ or $x-3 = 0$
       $x = -12$ $x = 3$
       Solution set: $\{-12, 3\}$

    b) $x^2 - x = 6$
       $x^2 - x - 6 = 0$ Standard form
       $(x-3)(x+2) = 0$ Factor.
       Using the zero-factor property, we have
       $x-3 = 0$ or $x+2 = 0$
       $x = 3$ $x = -2$
       Solution set: $\{3, -2\}$

c) $3x^2 - 21x + 30 = 0$

$3\left(x^2 - 7x + 10\right) = 0$ Factor out 3.

$x^2 - 7x + 10 = 0$ Divide by 3.

$(x - 5)(x - 2) = 0$ Factor.

Using the zero-factor property, we have

$x - 5 = 0$ or $x - 2 = 0$

$x = 5$ $x = 2$

Solution set: $\{5, 2\}$

d) $5x^2 - 3x - 8 = 0$

$(5x - 8)(x + 1) = 0$ Factor.

Using the zero-factor property, we have

$5x - 8 = 0$ or $x + 1 = 0$

$5x = 8$ $x = -1$

$x = \dfrac{8}{5}$

Solution set: $\left\{\dfrac{8}{5}, -1\right\}$

e) $9x^2 - 2x = 0$

$x(9x - 2) = 0$

Using the zero-factor property, we have

$x = 0$ or $9x - 2 = 0$

$9x = 2$

$x = \dfrac{2}{9}$

Solution set: $\left\{\dfrac{2}{9}, 0\right\}$

f) $x^2 - 64 = 63x$

$x^2 - 63x - 64 = 0$ Standard form

$(x - 64)(x + 1) = 0$ Factor.

Using the zero-factor property, we have

$x - 64 = 0$ or $x + 1 = 0$

$x = 64$ $x = -1$

Solution set: $\{64, -1\}$

g) $x(3x + 13) = 10$

$3x^2 + 13x = 10$ Distributive property

$3x^2 + 13x - 10 = 0$ Standard form

$(3x - 2)(x + 5) = 0$ Factor.

Using the zero-factor property, we have

$3x - 2 = 0$ or $x + 5 = 0$

$3x = 2$ $x = -5$

$x = \dfrac{2}{3}$

Solution set: $\left\{\dfrac{2}{3}, -5\right\}$

h) $(x - 3)(x + 4) = -4(x + 4)$

$x^2 + x - 12 = -4x - 16$ Multiply.

$x^2 + 5x + 4 = 0$ Standard form

$(x + 4)(x + 1) = 0$ Factor.

Using the zero-factor property, we have

$x + 4 = 0$ or $x + 1 = 0$

$x = -4$ $x = -1$

Solution set: $\{-4, -1\}$

i) $2 + \dfrac{x^2}{3} = 5x + 2$

$6 + x^2 = 15x + 6$ Multiply by 3.

$x^2 - 15x = 0$ Standard form

$x(x - 15) = 0$ Factor.

Using the zero-factor property, we have

$x = 0$ or $x - 15 = 0$

$x = 15$

Solution set: $\{0, 15\}$

j) $\dfrac{x^2 - 7x}{2} = 9$

$x^2 - 7x = 18$ Multiply by 2.

$x^2 - 7x - 18 = 0$ Standard form

$(x - 9)(x + 2) = 0$ Factor.

Using the zero-factor property, we have

$x - 9 = 0$ or $x + 2 = 0$

$x = 9$ $x = -2$

Solution set: $\{9, -2\}$

11.7 Applications of Quadratic Equations

Learning Objectives:

1. Solve problems about geometric figures.
2. Solve problems about consecutive integers.
3. Solve problems using the Pythagorean formula.
4. Solve problems using given quadratic models.

Solved Examples:

1. The area of a circle is 144π square meters. Find its radius.

   Solution

   The area of a circle is given by $A = \pi r^2$, so the equation is $144\pi = \pi r^2$.

   $$144\pi = \pi r^2$$
   $$144 = r^2 \quad \text{Divide by } \pi.$$
   $$144 - r^2 = 0 \quad \text{Standard form}$$
   $$(12 - r)(12 + r) = 0 \quad \text{Factor.}$$
   $$12 - r = 0 \quad \text{or} \quad 12 + r = 0 \quad \text{Zero factor property}$$
   $$12 = r \qquad\qquad r = -12$$

   Since radius cannot be negative, discard -12. The radius of the circle is 12 meters.

2. The width of a rectangle is 6 kilometers less than twice its length. If its area is 56 square kilometers, find the dimensions of the rectangle.

   Solution

   Let l = the length of the rectangle. Then $2l + 6$ = the width of the rectangle.
   Since the area of a rectangle is given by $A = lw$, the equation is $l(2l + 6) = 56$.

   $$l(2l + 6) = 56$$
   $$2l^2 + 6l = 56 \quad \text{Multiply.}$$
   $$2l^2 + 6l - 56 = 0 \quad \text{Standard form}$$
   $$2(l^2 + 3l - 28) = 0 \quad \text{Factor out 2.}$$
   $$l^2 + 3l - 28 = 0 \quad \text{Divide by 2.}$$
   $$(l + 7)(l - 4) = 0 \quad \text{Factor.}$$
   $$l + 7 = 0 \quad \text{or} \quad l - 4 = 0 \quad \text{Zero factor property}$$
   $$l = -7 \qquad\qquad l = 4$$

   Since length cannot be negative, discard -7. If the length = 4 km, then the width = $2(4) + 6 = 14$ km.
   The rectangle is 4 km $\times$ 14 km.

3. The product of two consecutive even integers is 14 more than 7 times their sum. Find the integers.

Solution

Let n = the first even integer. Then $n + 2$ = the next even integer. The equation is $n(n+2) = 14 + 7[n + (n+2)]$.

$$n(n+2) = 14 + 7[n + (n+2)]$$

$n^2 + 2n = 14 + 7(2n + 2)$ Multiply and combine like terms.

$n^2 + 2n = 14 + 14n + 14$ Multiply.

$n^2 + 2n = 14n + 28$ Combine like terms.

$n^2 - 12n - 28 = 0$ Standard form

$(n - 14)(n + 2) = 0$ Factor.

$n - 14 = 0$ or $n + 2 = 0$ Zero factor property

$n = 14$ $n = -2$

If $n = 14$, then the next even integer is 16. If $n = -2$, then the next even integer is 0. There are two sets of integers: 14 and 16, and -2 and 0.

4. Find the lengths of the sides of the triangle.

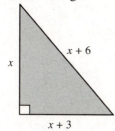

$x + 6$

x

$x + 3$

Solution

Using the Pythagorean theorem, we have $x^2 + (x+3)^2 = (x+6)^2$.

$$x^2 + (x+3)^2 = (x+6)^2$$

$x^2 + x^2 + 6x + 9 = x^2 + 12x + 36$ Expand.

$2x^2 + 6x + 9 = x^2 + 12x + 36$ Combine like terms.

$x^2 - 6x - 27 = 0$ Standard form

$(x - 9)(x + 3) = 0$ Factor.

$x - 9 = 0$ or $x + 3 = 0$ Zero factor property

$x = 9$ $x = -3$

The length of a side of a triangle cannot be negative, so discard -3. If one side is 9, then the other sides are $9 + 3 = 12$ and $9 + 6 = 15$.

5. A window washer accidentally drops a bucket from the top of a 144-foot building. The height, h, of the bucket after t seconds is given by $h = -16t^2 + 144$. When will the bucket hit the ground?

Solution

When the bucket hits the ground, $h = 0$. So we have $-16t^2 + 144 = 0$.

$$-16t^2 + 144 = 0$$
$$-16\left(t^2 - 9\right) = 0 \qquad \text{Factor out } -16.$$
$$t^2 - 9 = 0 \qquad \text{Divide by } -16.$$
$$(t+3)(t-3) = 0 \qquad \text{Factor.}$$
$$t + 3 = 0 \quad \text{or} \quad t - 3 = 0 \quad \text{Zero factor property}$$
$$t = -3 \qquad\qquad t = 3$$

Time cannot be negative, so discard -3. The bucket hits the ground 3 seconds after it is dropped.

12.1 The Fundamental Property of Rational Expressions

Learning Objectives:

1. Find the values of the variable for which a rational expression is undefined.
2. Find the numerical value of a rational expression.
3. Write rational expressions in lowest terms.
4. Recognize equivalent forms of rational expressions.
5. Solve applications using rational expressions.

Solved Examples:

1. Find any value(s) for which each rational expression is undefined.

   a) $\dfrac{3}{x}$ b) $\dfrac{5x}{x-2}$ c) $\dfrac{x-3}{x^2+x-12}$

   Solution

   Remember that the denominator cannot equal zero, so find the values for which the expression is undefined by setting the denominator equal to zero and solving for the variable.

   a) $\dfrac{3}{x}$

   The expression is not defined for $x = 0$.

   b) $\dfrac{5x}{x-2}$

   $x - 2 = 0$

   $x = 2$

   The expression is not defined for $x = 2$.

   c) $\dfrac{x-3}{x^2+x-12}$

   $x^2 + x - 12 = 0$

   $(x+4)(x-3) = 0$

   $x+4 = 0 \quad$ or $\quad x-3 = 0$

   $x = -4 \qquad\qquad x = 3$

   The expression is not defined for $x = -4$ or $x = 3$.

2. Find the numerical value of each rational expression for the given value.

   a) $\dfrac{x^2-9}{3x+1}; \; x = 2$

   b) $\dfrac{3-8x}{2x^2-5x+3}; \; x = -3$

   Solution

   a) $\dfrac{2^2-9}{3(2)+1} = \dfrac{4-9}{7} = -\dfrac{5}{7}$

   b) $\dfrac{3-8(-3)}{2(-3)^2-5(-3)+3} = \dfrac{3+24}{18+15+3} = \dfrac{27}{36} = \dfrac{3}{4}$

3. Simplify.

   a) $\dfrac{2x-12}{3x-18}$

   b) $\dfrac{9x^2+27x^3}{7x+21x^2}$

   c) $\dfrac{12-6x}{4x-8}$

   d) $\dfrac{12xy^2}{9x^2y^2(y+3x)}$

   e) $\dfrac{x^2-3x-10}{x^2-4x-12}$

   f) $\dfrac{-4x-4}{20x^2+28x+8}$

   g) $\dfrac{25-x^2}{2x^2-7x-15}$

   h) $\dfrac{x^2-xy+4x-4y}{x+4}$

   i) $\dfrac{x^2-2ax-3x+6a}{9x^3-18ax^2-4x+8a}$

Solution

To simplify, factor the numerator and denominator completely, then divide out any common factors.

a) $\dfrac{2x-12}{3x-18} = \dfrac{2(x-6)}{3(x-6)} = \dfrac{2}{3}$

b) $\dfrac{9x^2+27x^3}{7x+21x^2} = \dfrac{9x^2(1+3x)}{7x(1+3x)} = \dfrac{9x}{7}$

c) $\dfrac{12-6x}{4x-8} = \dfrac{-6(x-2)}{4(x-2)} = -\dfrac{3}{2}$

d) $\dfrac{12xy^2}{9x^2y^2(y+3x)} = \dfrac{4}{3x(y+3x)}$

e) $\dfrac{x^2-3x-10}{x^2-4x-12} = \dfrac{(x-5)(x+2)}{(x-6)(x+2)}$

$= \dfrac{x-5}{x-6}$

f) $\dfrac{-4x-4}{20x^2+28x+8} = \dfrac{-4(x+1)}{4(5x^2+7x+2)}$

$= \dfrac{-4(x+1)}{4(5x+2)(x+1)}$

$= -\dfrac{1}{5x+2}$

g) $\dfrac{25-x^2}{2x^2-7x-15} = \dfrac{(5-x)(5+x)}{(2x-3)(x+5)}$

$= \dfrac{5-x}{2x-3} = -\dfrac{x-5}{2x-3}$

h) $\dfrac{x^2-xy+4x-4y}{x+4} = \dfrac{(x^2-xy)+(4x-4y)}{x+4}$

$= \dfrac{x(x-y)+4(x-y)}{x+4}$

$= \dfrac{(x+4)(x-y)}{x+4}$

$= x-y$

i) $\dfrac{x^2-2ax-3x+6a}{9x^3-18ax^2-4x+8a}$

$= \dfrac{(x^2-2ax)-(3x-6a)}{(9x^3-18ax^2)-(4x-8a)}$

$= \dfrac{x(x-2a)-3(x-2a)}{9x^2(x-2a)-4(x-2a)}$

$= \dfrac{(x-3)(x-2a)}{(9x^2-4)(x-2a)} = \dfrac{x-3}{9x^2-4}$

4. Write four equivalent forms for the rational expression: $-\dfrac{x+6}{x-1}$.

Solution

$-\dfrac{x+6}{x-1} = \dfrac{-(x+6)}{x-1} = \dfrac{-x-6}{x-1}$

$-\dfrac{x+6}{x-1} = \dfrac{x+6}{-(x-1)} = \dfrac{x+6}{-x+1}$

12.2 Multiplying and Dividing Rational Expressions

Learning Objectives:

1. Multiply rational expressions.
2. Find reciprocals.
3. Divide rational expressions.
4. Multiply or divide the rational expression.

Solved Examples:

1. Multiply the numerical fractions, and write each answer in lowest terms.

   a) $\dfrac{3}{4} \cdot \dfrac{5}{7}$

   b) $\dfrac{3}{4} \div \dfrac{9}{6}$

   c) $\dfrac{5}{36} \div 125$

   Solution

   a) $\dfrac{3}{4} \cdot \dfrac{5}{7} = \dfrac{15}{28}$

   b) $\dfrac{3}{4} \div \dfrac{9}{6} = \dfrac{3}{4} \cdot \dfrac{6}{9} = \dfrac{1}{2}$

   c) $\dfrac{5}{36} \div 125 = \dfrac{5}{36} \cdot \dfrac{1}{125} = \dfrac{1}{900}$

2. Multiply the rational expressions, and write each answer in lowest terms.

   a) $\dfrac{2p-2}{p} \cdot \dfrac{3p^2}{9p-9}$

   b) $\dfrac{x^2+11x+24}{x^2+12x+27} \cdot \dfrac{x^2+9x}{x^2+4x-32}$

   c) $\dfrac{x^2-4x+3}{x^2-12x+20} \cdot \dfrac{x^2-19x+90}{x^2-10x+9}$

   d) $\dfrac{12x^2}{9x+45} \cdot \dfrac{3x^3-75x}{36x^4}$

   Solution

   a) $\dfrac{2p-2}{p} \cdot \dfrac{3p^2}{9p-9} = \dfrac{2(p-1)}{p} \cdot \dfrac{3p^2}{9(p-1)}$

   $= \dfrac{2p}{3}$

   b) $\dfrac{x^2+11x+24}{x^2+12x+27} \cdot \dfrac{x^2+9x}{x^2+4x-32}$

   $= \dfrac{(x+3)(x+8)}{(x+3)(x+9)} \cdot \dfrac{x(x+9)}{(x+8)(x-4)}$

   $= \dfrac{x}{x-4}$

   c) $\dfrac{x^2-4x+3}{x^2-12x+20} \cdot \dfrac{x^2-19x+90}{x^2-10x+9}$

   $= \dfrac{(x-1)(x-3)}{(x-2)(x-10)} \cdot \dfrac{(x-9)(x-10)}{(x-1)(x-9)}$

   $= \dfrac{x-3}{x-2}$

   d) $\dfrac{12x^2}{9x+45} \cdot \dfrac{3x^3-75x}{36x^4}$

   $= \dfrac{12x^2}{9(x+5)} \cdot \dfrac{3x(x^2-25)}{36x^4}$

   $= \dfrac{12x^2}{9(x+5)} \cdot \dfrac{3x(x-5)(x+5)}{36x^4}$

   $= \dfrac{x-5}{9x}$

3. Divide the rational expressions, and write each answer in lowest terms.

   a) $\dfrac{6p-6}{p} \div \dfrac{7p-7}{9p^2}$

   b) $\dfrac{x^2+5x+6}{x^2+10x+21} \div \dfrac{x^2+2x}{x^2+16x+63}$

c) $\dfrac{(x-11)^2}{2} \div \dfrac{2x-22}{4}$

d) $\dfrac{x^2-16x+60}{x-10} \div (x-6)$

e) $\dfrac{3x^2+6xy+3y^2}{x^2+4xy+3y^2} \div \dfrac{4x+4y}{x+3y}$

Solution

a) $\dfrac{6p-6}{p} \div \dfrac{7p-7}{9p^2} = \dfrac{6p-6}{p} \cdot \dfrac{9p^2}{7p-7}$

$\qquad = \dfrac{6(p-1)}{p} \cdot \dfrac{9p^2}{7(p-1)} = \dfrac{54p}{7}$

b) $\dfrac{x^2+5x+6}{x^2+10x+21} \div \dfrac{x^2+2x}{x^2+16x+63}$

$\qquad = \dfrac{x^2+5x+6}{x^2+10x+21} \cdot \dfrac{x^2+16x+63}{x^2+2x}$

$\qquad = \dfrac{(x+2)(x+3)}{(x+3)(x+7)} \cdot \dfrac{(x+7)(x+9)}{x(x+2)}$

$\qquad = \dfrac{x+9}{x}$

c) $\dfrac{(x-11)^2}{2} \div \dfrac{2x-22}{4} = \dfrac{(x-11)^2}{2} \cdot \dfrac{4}{2x-22}$

$\qquad = \dfrac{(x-11)^2}{2} \cdot \dfrac{4}{2(x-11)}$

$\qquad = x-11$

d) $\dfrac{x^2-16x+60}{x-10} \div (x-6)$

$\qquad = \dfrac{x^2-16x+60}{x-10} \cdot \dfrac{1}{x-6}$

$\qquad = \dfrac{(x-6)(x-10)}{x-10} \cdot \dfrac{1}{x-6} = 1$

e) $\dfrac{3x^2+6xy+3y^2}{x^2+4xy+3y^2} \div \dfrac{4x+4y}{x+3y}$

$\qquad = \dfrac{3x^2+6xy+3y^2}{x^2+4xy+3y^2} \cdot \dfrac{x+3y}{4x+4y}$

$\qquad = \dfrac{3(x^2+2xy+y^2)}{x^2+4xy+3y^2} \cdot \dfrac{x+3y}{4(x+y)}$

$\qquad = \dfrac{3(x+y)^2}{(x+y)(x+3y)} \cdot \dfrac{x+3y}{4(x+y)} = \dfrac{3}{4}$

12.3 Least Common Denominator

Learning Objectives:

1. Find the least common denominator for a group of fractions.
2. Rewrite rational expressions with given denominators.

Solved Examples:

1. Find the least common denominator for the fractions in each list.

a) $\dfrac{3}{16}, \dfrac{5}{24}, \dfrac{1}{12}$

b) $\dfrac{7}{32m}, \dfrac{9}{48m}$

c) $\dfrac{10}{27t^3}, \dfrac{5}{36t}$

d) $\dfrac{5}{24y^2}, \dfrac{7}{14y-20}$

e) $\dfrac{3}{e-f}, \dfrac{8}{f-e}$

f) $\dfrac{1}{2c^2-9c+10}, \dfrac{6}{2c^2+c-15}$

g) $\dfrac{3}{4xy}, \dfrac{9}{5x^2y^2}$

h) $\dfrac{3}{m^2-4m}, \dfrac{9}{m^2-6m+8}$

i) $\dfrac{5}{x^2-2x-15}, \dfrac{7}{x^2-10x+25}$

Solution

To find the LCD, follow these steps.
Step 1: Factor each denominator into prime factors.
Step 2: List each different denominator factor the greatest number of times it appears in any of the denominators.
Step 3: Multiply the denominator factors from Step 2 to get the LCD.

a) $16 = 2\cdot2\cdot2\cdot2$
   $24 = 2\cdot2\cdot2\cdot3$
   $12 = 2\cdot2\cdot3$
   $\text{LCD} = 2\cdot2\cdot2\cdot2\cdot3 = 48$

b) $32m = 2\cdot2\cdot2\cdot2\cdot2\cdot m$
   $48m = 2\cdot2\cdot2\cdot2\cdot3\cdot m$
   $\text{LCD} = 2\cdot2\cdot2\cdot2\cdot2\cdot3\cdot m = 96m$

c) $27t^3 = 3\cdot3\cdot3\cdot t\cdot t\cdot t$
   $36t = 2\cdot2\cdot3\cdot3\cdot t$
   $\text{LCD} = 2\cdot2\cdot3\cdot3\cdot3\cdot t\cdot t\cdot t = 108t^3$

d) $24y^2 = 2\cdot2\cdot2\cdot3\cdot y\cdot y$
   $14y-20 = 2(7y-10)$
   $\text{LCD} = 2\cdot2\cdot2\cdot3\cdot y\cdot y\cdot(7y-10)$
   $\qquad = 24y^2(7y-10)$

e) $e-f = e-f \quad$ or $\quad -1(f-e)$
   $f-e = -(e-f)$ or $\ f-e$
   $\text{LCD} = -(e-f) \ $ or $\ -1(f-e)$

f) $2c^2-9c+10 = (2c-5)(c-2)$
   $2c^2+c-15 = (2c-5)(c+3)$
   $\qquad \text{LCD} = (2c-5)(c-2)(c+3)$

g) $4xy = 4\cdot x\cdot y$
   $5x^2y^2 = 5\cdot x\cdot x\cdot y\cdot y$
   $\text{LCD} = 4\cdot5\cdot x\cdot x\cdot y\cdot y = 20x^2y^2$

h) $m^2-4m = m(m-4)$
   $m^2-6m+8 = (m-2)(m-4)$
   $\qquad \text{LCD} = m(m-4)(m-2)$

i) $x^2 - 2x - 15 = (x-5)(x+3)$
     $x^2 - 10x + 25 = (x-5)(x-5)$
     $\text{LCD} = (x-5)(x-5)(x+3)$

2. Write each rational expression as an equivalent expression with the indicated denominator.

a) $\dfrac{7}{15} = \dfrac{}{60}$

b) $\dfrac{-3}{b} = \dfrac{}{7b}$

c) $\dfrac{5p^3}{3q} = \dfrac{}{18q^3}$

d) $\dfrac{-2m}{3m-24} = \dfrac{}{48-6m}$

e) $\dfrac{6(d-1)}{d(d+3)} = \dfrac{}{d^3 + 8d^2 + 15d}$

Solution

To write a rational expression as an equivalent expression with a specified denominator, follow these steps.

Step 1: Factor both denominators.

Step 2: Decide what factor(s) the denominator must be multiplied by in order to equal the specified denominator.

Step 3: Multiply the rational expression by that factor divided by itself. (That is, multiply by 1.)

a) $60 = 4 \cdot 15$, so multiply $\dfrac{7}{15}$ by $\dfrac{4}{4}$.

$$\dfrac{7}{15} \cdot \dfrac{4}{4} = \dfrac{28}{60}$$

b) $7b = 7 \cdot b$, so multiply $\dfrac{-3}{b}$ by $\dfrac{7}{7}$.

$$\dfrac{-3}{b} \cdot \dfrac{7}{7} = \dfrac{-21}{7b}$$

c) $18q^3 = 3q(6q)$, so multiply $\dfrac{5p^3}{3q}$ by $\dfrac{6q}{6q}$.

$$\dfrac{5p^3}{3q} \cdot \dfrac{6q}{6q} = \dfrac{30p^3q}{18q^2}$$

d) $48 - 6m = -6(m-8) = -2 \cdot 3(m-8)$ and

$3m - 24 = 3(m-8)$, so multiply $\dfrac{-2m}{3m-24}$ by

$$\dfrac{-2}{-2} \cdot \quad \dfrac{-2m}{3m-24} \cdot \dfrac{-2}{-2} = \dfrac{4m}{48m-6}$$

e) $d^3 + 8d^2 + 15d = d\left(d^2 + 8d + 15\right)$
    $\qquad\qquad\qquad = d(d+3)(d+5)$

Multiply $\dfrac{6(d-1)}{d(d+3)}$ by $\dfrac{d+5}{d+5}$.

$$\dfrac{6(d-1)}{d(d+3)} \cdot \dfrac{d+5}{d+5} = \dfrac{6(d-1)(d+5)}{d(d+3)(d+5)}$$
$$= \dfrac{6\left(d^2 + 4d - 5\right)}{d^3 + 8d^2 + 15d}$$
$$= \dfrac{6d^2 + 24d - 30}{d^3 + 8d^2 + 15d}$$

12.4 Adding and Subtracting Rational Expressions

Learning Objectives:

1. Factor trinomials using FOIL.

Solved Examples:

1. Add or subtract the numerical fractions.

    a) $\dfrac{1}{4} + \dfrac{2}{4}$

    b) $\dfrac{6}{7} - \dfrac{5}{6}$

    c) $\dfrac{1}{6} + \dfrac{4}{9}$

    Solution

    To add fractions with the same denominator, add the numerators and keep the same denominator.

    To add fractions with different denominators, follow these steps.
    Step 1: Find the least common denominator (LCD).
    Step 2: Write each fraction as an equivalent fraction with the LCD as the denominator.
    Step 3: Add the numerators to get the numerator of the sum. The LCD is the denominator of the sum.
    Step 4: Write in lowest terms

    a) $\dfrac{1}{4} + \dfrac{2}{4} = \dfrac{3}{4}$

    b) $\dfrac{6}{7} - \dfrac{5}{6} = \dfrac{36}{42} - \dfrac{35}{42} = \dfrac{1}{42}$

    c) $\dfrac{1}{6} + \dfrac{4}{9} = \dfrac{3}{18} + \dfrac{8}{18} = \dfrac{11}{18}$

2. Add or subtract as indicated.

    a) $\dfrac{7x+4}{9x+6} + \dfrac{x-2}{9x+6}$

    b) $\dfrac{m^2 - 11m}{m-5} + \dfrac{30}{m-5}$

    c) $\dfrac{6x-16}{x-2} - \dfrac{4x-7}{x-2}$

    Solution

    To add rational expressions with the same denominator, add the numerators and keep the same denominator.

    a) $\begin{aligned} \dfrac{7x+4}{9x+6} + \dfrac{x-2}{9x+6} &= \dfrac{(7x+4)+(x-2)}{9x+6} \\ &= \dfrac{8x+2}{9x+6} = \dfrac{2(4x+1)}{3(3x+2)} \end{aligned}$

    b) $\begin{aligned} \dfrac{m^2-11m}{m-5} + \dfrac{30}{m-5} &= \dfrac{m^2-11m+30}{m-5} \\ &= \dfrac{(m-5)(m-6)}{m-5} \\ &= m-6 \end{aligned}$

    c) $\begin{aligned} \dfrac{6x-16}{x-2} - \dfrac{4x-7}{x-2} &= \dfrac{(6x-16)-(4x-7)}{x-2} \\ &= \dfrac{2x-9}{x-2} \end{aligned}$

3. Add or subtract as indicated.

    a) $\dfrac{6}{x^2} + \dfrac{4}{x}$

    b) $\dfrac{4}{9} - \dfrac{6}{3x}$

    c) $\dfrac{x-4}{x^2+5x+6} + \dfrac{5x+6}{x^2+4x+3}$

    d) $\dfrac{x}{x^2-16} - \dfrac{6}{x^2+5x+4}$

    e) $\dfrac{3}{x^2-3x+2} + \dfrac{7}{x^2-1}$

Solution

To add rational expressions with the same denominator, add the numerators and keep the same denominator.

To add rational expressions with the different denominators, follow these steps.

Step 1: Find the least common denominator (LCD).

Step 2: Write each rational expression as an equivalent rational expression with the LCD as the denominator.

Step 3: Add the numerators to get the numerator of the sum. The LCD is the denominator of the sum.

Step 4: Write in lowest terms

a) $\dfrac{6}{x^2} + \dfrac{4}{x}$

The LCD is x^2.

$\dfrac{6}{x^2} + \dfrac{4}{x} = \dfrac{6}{x^2} + \dfrac{4x}{x^2} = \dfrac{6+4x}{x^2}$

b) $\dfrac{4}{9} - \dfrac{6}{3x}$

$9 = 3 \cdot 3;\ 3x = 3 \cdot x;\ \text{LCD} = 3 \cdot 3 \cdot x = 9x$

$\dfrac{4}{9} - \dfrac{6}{3x} = \dfrac{4x}{9x} - \dfrac{3 \cdot 6}{3 \cdot 3x} = \dfrac{4x}{9x} - \dfrac{18}{9x} = \dfrac{4x-18}{9x}$

c) $\dfrac{x-4}{x^2+5x+6} + \dfrac{5x+6}{x^2+4x+3}$

$x^2+5x+6 = (x+2)(x+3);\ x^2+4x+3 = (x+1)(x+3);\ \text{LCD} = (x+1)(x+2)(x+3)$

$\dfrac{x-4}{x^2+5x+6} + \dfrac{5x+6}{x^2+4x+3} = \dfrac{x-4}{(x+2)(x+3)} + \dfrac{5x+6}{(x+1)(x+3)} = \dfrac{(x+1)(x-4)}{(x+1)(x+2)(x+3)} + \dfrac{(5x+6)(x+2)}{(x+1)(x+2)(x+3)}$

$= \dfrac{x^2-3x-4}{(x+1)(x+2)(x+3)} + \dfrac{5x^2+16x+12}{(x+1)(x+2)(x+3)} = \dfrac{6x^2+13x+8}{(x+1)(x+2)(x+3)}$

d) $\dfrac{x}{x^2-16} - \dfrac{6}{x^2+5x+4}$

$x^2-16 = (x-4)(x+4);\ x^2+5x+4 = (x+1)(x+4);\ \text{LCD} = (x+1)(x+4)(x-4)$

$\dfrac{x}{x^2-16} - \dfrac{6}{x^2+5x+4} = \dfrac{x}{(x-4)(x+4)} - \dfrac{6}{(x+1)(x+4)} = \dfrac{x(x+1)}{(x+1)(x-4)(x+4)} - \dfrac{6(x-4)}{(x+1)(x+4)(x-4)}$

$= \dfrac{x^2+x}{(x+1)(x-4)(x+4)} - \dfrac{6x-24}{(x+1)(x-4)(x+4)} = \dfrac{x^2-5x+24}{(x+1)(x-4)(x+4)}$

e) $\dfrac{3}{x^2-3x+2} + \dfrac{7}{x^2-1}$

$x^2-3x+2 = (x-1)(x-2);\ x^2-1 = (x+1)(x-1);\ \text{LCD} = (x+1)(x-1)(x-2)$

$\dfrac{3}{x^2-3x+2} + \dfrac{7}{x^2-1} = \dfrac{3}{(x-1)(x-2)} + \dfrac{7}{(x+1)(x-1)} = \dfrac{3(x+1)}{(x+1)(x-1)(x-2)} + \dfrac{7(x-2)}{(x+1)(x-1)(x-2)}$

$= \dfrac{3x+3}{(x+1)(x-1)(x-2)} + \dfrac{7x-14}{(x+1)(x-1)(x-2)} = \dfrac{10x-11}{(x+1)(x-1)(x-2)}$

12.5 Complex Fractions

Learning Objectives:

1. Simplify complex fractions.

Solved Examples:

1. Simplify the complex numerical fractions, first by adding or subtracting in the numerator and denominator, and then by writing it as a division problem. Then rework using the LCD. (Both methods are shown in the textbook).

a) $\dfrac{\dfrac{2}{7}}{\dfrac{4}{9}}$

b) $\dfrac{\dfrac{1}{5}}{\dfrac{2}{3}+\dfrac{7}{15}}$

Solution

a) Method 1:

$$\frac{\frac{2}{7}}{\frac{4}{9}}=\frac{2}{7}\div\frac{4}{9}=\frac{2}{7}\cdot\frac{9}{4}=\frac{9}{14}$$

Method 2:

$$\frac{\frac{2}{7}}{\frac{4}{9}}=\frac{\frac{2}{7}(63)}{\frac{4}{9}(63)}=\frac{18}{28}=\frac{9}{14}$$

b)

$$\frac{\frac{1}{5}}{\frac{2}{3}+\frac{7}{15}}=\frac{\frac{1}{5}}{\frac{10}{15}+\frac{7}{15}}=\frac{\frac{1}{5}}{\frac{17}{15}}=\frac{1}{5}\div\frac{17}{15}$$

$$=\frac{1}{5}\cdot\frac{15}{17}=\frac{3}{17}$$

Method 2:

$$\frac{\frac{1}{5}}{\frac{2}{3}+\frac{7}{15}}=\frac{\frac{1}{5}(15)}{\left(\frac{2}{3}+\frac{7}{15}\right)(15)}=\frac{3}{\left(\frac{2}{3}\right)(15)+\left(\frac{7}{15}\right)(15)}$$

$$=\frac{3}{10+7}=\frac{3}{17}$$

2. Simplify by adding or subtracting in the numerator and denominator. Then rewrite each as a division problem and solve. (The LCD method will be used in number 3 for the exact same problems.)

a) $\dfrac{\dfrac{1}{x}-\dfrac{1}{y}}{\dfrac{1}{xy}}$

b) $\dfrac{\dfrac{5}{x}}{\dfrac{9}{x+2}}$

c) $\dfrac{4+\dfrac{2}{x}}{\dfrac{x}{3}+\dfrac{1}{6}}$

d) $\dfrac{\dfrac{5}{9x-1}-5}{\dfrac{5}{9x-1}+5}$

e) $\dfrac{\dfrac{4}{x+5}}{\dfrac{1}{x-5}-\dfrac{2}{x^2-25}}$

f) $\dfrac{\dfrac{x}{x+2}+2}{\dfrac{x^2-1}{x-1}}$

Solution

a)
$$\frac{\frac{1}{x}-\frac{1}{y}}{\frac{1}{xy}}=\frac{\frac{y}{xy}-\frac{x}{xy}}{\frac{1}{xy}}=\frac{\frac{y-x}{xy}}{\frac{1}{xy}}=\frac{y-x}{xy}\div\frac{1}{xy}$$
$$=\frac{y-x}{xy}\cdot xy = y-x$$

b)
$$\frac{\frac{5}{x}}{\frac{9}{x+2}}=\frac{5}{x}\div\frac{9}{x+2}=\frac{5}{x}\cdot\frac{x+2}{9}=\frac{5(x+2)}{9x}$$

c)
$$\frac{4+\frac{2}{x}}{\frac{x}{3}+\frac{1}{6}}=\frac{\frac{4x}{x}+\frac{2}{x}}{\frac{2x}{6}+\frac{1}{6}}=\frac{\frac{4x+2}{x}}{\frac{2x+1}{6}}=\frac{4x+2}{x}\div\frac{2x+1}{6}$$
$$=\frac{4x+2}{x}\cdot\frac{6}{2x+1}=\frac{2(2x+1)}{x}\cdot\frac{6}{2x+1}$$
$$=\frac{12}{x}$$

d)
$$\frac{\frac{5}{9x-1}-5}{\frac{5}{9x-1}+5}=\frac{\frac{5}{9x-1}-\frac{5(9x-1)}{9x-1}}{\frac{5}{9x-1}+\frac{5(9x-1)}{9x-1}}$$
$$=\frac{\frac{5}{9x-1}-\frac{45x-5}{9x-1}}{\frac{5}{9x-1}+\frac{45x}{9x-1}}=\frac{\frac{-45x+10}{9x-1}}{\frac{45x}{9x-1}}$$
$$=\frac{-45x+10}{9x-1}\cdot\frac{9x-1}{45x}$$
$$=\frac{-5(9x-2)}{9x-1}\cdot\frac{9x-1}{45x}=-\frac{9x-2}{9x}$$
$$=\frac{-9x+2}{9x}$$

e)
$$\frac{\frac{4}{x+5}}{\frac{1}{x-5}-\frac{2}{x^2-25}}$$
$$=\frac{\frac{4}{x+5}}{\frac{1}{x-5}-\frac{2}{(x-5)(x+5)}}$$
$$=\frac{\frac{4}{x+5}}{\frac{1(x+5)}{(x-5)(x+5)}-\frac{2}{(x-5)(x+5)}}$$
$$=\frac{\frac{4}{x+5}}{\frac{x+3}{(x-5)(x+5)}}=\frac{4}{x+5}\cdot\frac{(x-5)(x+5)}{x+3}$$
$$=\frac{4(x-5)}{x+3}=\frac{4x-20}{x+3}$$

f)
$$\frac{\frac{x}{x+2}+2}{\frac{x^2-1}{x-1}}=\frac{\frac{x}{x+2}+\frac{2(x+2)}{x+2}}{\frac{(x+1)(x-1)}{x-1}}$$
$$=\frac{\frac{x}{x+2}+\frac{2x+4}{x+2}}{x+1}=\frac{\frac{3x+4}{x+2}}{x+1}$$
$$=\frac{3x+4}{x+2}\div(x+1)=\frac{3x+4}{x+2}\cdot\frac{1}{x+1}$$
$$=\frac{3x+4}{(x+1)(x+2)}$$

3. Simplify by using the LCD.

a) $$\frac{\frac{1}{x}-\frac{1}{y}}{\frac{1}{xy}}$$

b) $$\frac{\frac{5}{x}}{\frac{9}{x+2}}$$

c) $$\frac{4+\frac{2}{x}}{\frac{x}{3}+\frac{1}{6}}$$

d) $\dfrac{\dfrac{5}{9x-1}-5}{\dfrac{5}{9x-1}+5}$

e) $\dfrac{\dfrac{4}{x+5}}{\dfrac{1}{x-5}-\dfrac{2}{x^2-25}}$

f) $\dfrac{\dfrac{x}{x+2}+2}{\dfrac{x^2-1}{x-1}}$

Solution

a) $\dfrac{\dfrac{1}{x}-\dfrac{1}{y}}{\dfrac{1}{xy}}=\dfrac{\left(\dfrac{1}{x}-\dfrac{1}{y}\right)(xy)}{\left(\dfrac{1}{xy}\right)(xy)}$

$=\dfrac{\left(\dfrac{1}{x}\right)(xy)-\left(\dfrac{1}{y}\right)(xy)}{\left(\dfrac{1}{xy}\right)(xy)}=y-x$

b) $\dfrac{\dfrac{5}{x}}{\dfrac{9}{x+2}}=\dfrac{\left(\dfrac{5}{x}\right)(x)(x+2)}{\left(\dfrac{9}{x+2}\right)(x)(x+2)}=\dfrac{5(x+2)}{9x}$

c) $\dfrac{4+\dfrac{2}{x}}{\dfrac{x}{3}+\dfrac{1}{6}}=\dfrac{\left(4+\dfrac{2}{x}\right)(6x)}{\left(\dfrac{x}{3}+\dfrac{1}{6}\right)(6x)}=\dfrac{(4)(6x)+\left(\dfrac{2}{x}\right)(6x)}{\left(\dfrac{x}{3}\right)(6x)+\left(\dfrac{1}{6}\right)(6x)}$

$=\dfrac{24x+12}{2x^2+x}=\dfrac{12(2x+1)}{x(2x+1)}=\dfrac{12}{x}$

d) $\dfrac{\dfrac{5}{9x-1}-5}{\dfrac{5}{9x-1}+5}=\dfrac{\left(\dfrac{5}{9x-1}-5\right)(9x-1)}{\left(\dfrac{5}{9x-1}+5\right)(9x-1)}$

$=\dfrac{\left(\dfrac{5}{9x-1}\right)(9x-1)-5(9x-1)}{\left(\dfrac{5}{9x-1}\right)(9x-1)+5(9x-1)}$

$=\dfrac{5-45x+5}{5+45x-5}=\dfrac{-45x+10}{45x}$

$=\dfrac{-5(9x-2)}{45x}=\dfrac{-9x+2}{9x}$

e) $\dfrac{\dfrac{4}{x+5}}{\dfrac{1}{x-5}-\dfrac{2}{x^2-25}}=\dfrac{\dfrac{4}{x+5}}{\dfrac{1}{x-5}-\dfrac{2}{(x-5)(x+5)}}=\dfrac{\left(\dfrac{4}{x+5}\right)(x-5)(x+5)}{\left(\dfrac{1}{x-5}-\dfrac{2}{(x-5)(x+5)}\right)(x-5)(x+5)}$

$=\dfrac{4(x-5)}{\left(\dfrac{1}{x-5}\right)(x-5)(x+5)-\left(\dfrac{2}{(x-5)(x+5)}\right)(x-5)(x+5)}=\dfrac{4(x-5)}{(x+5)-2}=\dfrac{4x-20}{x+3}$

f) $\dfrac{\dfrac{x}{x+2}+2}{\dfrac{x^2-1}{x-1}}=\dfrac{\dfrac{x}{x+2}+2}{\dfrac{(x+1)(x-1)}{x-1}}=\dfrac{\dfrac{x}{x+2}+2}{x+1}=\dfrac{\left(\dfrac{x}{x+2}+2\right)(x+2)}{(x+1)(x+2)}=\dfrac{\left(\dfrac{x}{x+2}\right)(x+2)+2(x+2)}{(x+1)(x+2)}$

$=\dfrac{x+2(x+2)}{(x+1)(x+2)}=\dfrac{x+2x+4}{(x+1)(x+2)}=\dfrac{3x+4}{(x+1)(x+2)}$

12.6 Solving Equations with Rational Expressions

Learning Objectives:

1. Distinguish between expressions and equations.
2. Solve equations with rational expressions.
3. Solve a formula for a specified variable.

Solved Examples:

1. Solve and check.

   a) $\dfrac{2}{5}y - \dfrac{1}{3}y = 5$

   b) $\dfrac{3y+6}{5} = 1 + \dfrac{3}{4}y$

   Solution

   Be sure to check your work by substituting the solution into the original equation.

   a) $\dfrac{2}{5}y - \dfrac{1}{3}y = 5$

   Multiply by the LCD, 15, to clear the fractions.

   $(15)\left(\dfrac{2}{5}y - \dfrac{1}{3}y\right) = 5(15)$

   $6y - 5y = 75$

   $y = 75$

   The solution set is $\{75\}$.

   b) $\dfrac{3y+6}{5} = 1 + \dfrac{3}{4}y$

   Multiply by the LCD, 20, to clear the fractions.

   $(20)\left(\dfrac{3y+6}{5}\right) = \left(1 + \dfrac{3}{4}y\right)(20)$

   $12y + 24 = 20 + 15y$

   $24 = 20 + 3y$ Subtract $12y$.

   $4 = 3y$ Subtract 20.

   $\dfrac{4}{3} = y$ Divide by 3.

   The solution set is $\left\{\dfrac{4}{3}\right\}$.

2. Solve and check.

   a) $\dfrac{15}{x} = 4 - \dfrac{1}{x}$

   b) $1 + \dfrac{1}{x} = \dfrac{12}{x^2}$

   c) $\dfrac{5-a}{a} + \dfrac{3}{4} = \dfrac{7}{a}$

   d) $\dfrac{2}{x} = \dfrac{x}{5x-12}$

   e) $\dfrac{3}{y+5} - \dfrac{5}{y-5} = \dfrac{6}{y^2-25}$

Solution

Remember to check your work by substituting each solution in the original equation.

a) $\dfrac{15}{x} = 4 - \dfrac{1}{x}$

Multiply by the LCD, x.

$$x\left(\dfrac{15}{x}\right) = \left(4 - \dfrac{1}{x}\right)x$$

$\qquad 15 = 4x - 1$

$\qquad 16 = 4x \qquad$ Add 1.

$\qquad 4 = x \qquad$ Divide by 4.

The solution set is $\{4\}$.

b) $1 + \dfrac{1}{x} = \dfrac{12}{x^2}$

Multiply by the LCD, x^2.

$$x^2\left(1 + \dfrac{1}{x}\right) = \left(\dfrac{12}{x^2}\right)x^2$$

$\qquad x^2 + x = 12$

$\qquad x^2 + x - 12 = 0 \qquad$ Standard form

$\qquad (x - 3)(x + 4) = 0 \qquad$ Factor.

Using the zero-factor property, we have

$\qquad x - 3 = 0 \quad$ or $\quad x + 4 = 0$

$\qquad\quad x = 3 \qquad\qquad x = -4$

Solution set: $\{3, -4\}$

c) $\dfrac{5 - a}{a} + \dfrac{3}{4} = \dfrac{7}{a}$

Multiply by the LCD, $4a$.

$$4a\left(\dfrac{5 - a}{a} + \dfrac{3}{4}\right) = \left(\dfrac{7}{a}\right)4a$$

$\qquad 4(5 - a) + 3a = 28$

$\qquad 20 - 4a + 3a = 28 \qquad$ Distributive property

$\qquad 20 - a = 28 \qquad$ Combine like terms.

$\qquad -a = 8 \qquad$ Subtract 20.

$\qquad a = -8 \qquad$ Multiply by -1.

Solution set: $\{-8\}$

d) $\dfrac{2}{x} = \dfrac{x}{5x - 12}$

Multiply by the LCD, $x(5x - 12)$.

$$x(5x - 12)\left(\dfrac{2}{x}\right) = \left(\dfrac{x}{5x - 12}\right)x(5x - 12)$$

$\qquad 2(5x - 12) = x^2$

$\qquad 10x - 24 = x^2 \qquad$ Distributive property

$\qquad x^2 - 10x + 24 = 0 \qquad$ Standard form

$\qquad (x - 4)(x - 6) = 0 \qquad$ Factor.

Using the zero-factor property, we have

$\qquad x - 4 = 0 \quad$ or $\quad x - 6 = 0$

$\qquad\quad x = 4 \qquad\qquad x = 6$

Solution set: $\{4, 6\}$

e) $\dfrac{3}{y + 5} - \dfrac{5}{y - 5} = \dfrac{6}{y^2 - 25}$

Multiply by the LCD, $y^2 - 25$ or $(y - 5)(y + 5)$.

$$(y + 5)(y - 5)\left(\dfrac{3}{y + 5} - \dfrac{5}{y - 5}\right) = \left(\dfrac{6}{y^2 - 25}\right)(y^2 - 25)$$

$\qquad 3(y - 5) - 5(y + 5) = 6$

$\qquad 3y - 15 - 5y - 25 = 6 \qquad$ Distributive property

$\qquad -2y - 40 = 6 \qquad$ Combine like terms.

$\qquad -2y = 46 \qquad$ Add 40.

$\qquad y = -23 \qquad$ Divide by -2.

Solution set: $\{-23\}$

3. Solve and check. If there is no solution, so indicate.

a) $\dfrac{2}{x + 3} = 5 + \dfrac{2}{x + 3}$

b) $\dfrac{x}{2x + 2} = \dfrac{-2x}{4x + 4} + \dfrac{2x - 3}{x + 1}$

c) $\dfrac{6x}{x+5} - \dfrac{30}{x-5} = \dfrac{6x^2+150}{x^2-25}$

d) $\dfrac{-2}{x+3} = \dfrac{1}{x+6} - \dfrac{6}{x^2+9x+18}$

Solution

Remember to check your work by substituting each solution in the original equation.

a) $\dfrac{2}{x+3} = 5 + \dfrac{2}{x+3}$

Multiply by the LCD, $x + 3$.

$$(x+3)\left(\dfrac{2}{x+3}\right) = \left(5 + \dfrac{2}{x+3}\right)(x+3)$$

$$2 = 5(x+3)+2$$
$$2 = 5x+15+2 \quad \text{Multiply.}$$
$$2 = 5x+17 \qquad \text{Combine like}$$
$$\qquad\qquad\qquad \text{terms.}$$
$$-15 = 5x \qquad\quad \text{Subtract 17.}$$
$$-3 = x \qquad\qquad \text{Divide by 5.}$$

However, -3 cannot be a solution because it makes the denominators in the original equation equal zero. There is no solution. Solution set: $\varnothing$

b) $\dfrac{x}{2x+2} = \dfrac{-2x}{4x+4} + \dfrac{2x-3}{x+1}$

Multiply by the LCD, $4(x+1)$.

$$4(x+1)\left(\dfrac{x}{2x+2}\right) = \left(\dfrac{-2x}{4x+4} + \dfrac{2x-3}{x+1}\right)4(x+1)$$

$$4(x+1)\left(\dfrac{x}{2(x+1)}\right) = \left(\dfrac{-2x}{4(x+1)} + \dfrac{2x-3}{x+1}\right)4(x+1)$$

$$2x = -2x+4(2x-3)$$
$$2x = -2x+8x-12 \qquad \text{Multiply.}$$
$$2x = 6x-12 \qquad\qquad \text{Combine like terms.}$$
$$-4x = -12 \qquad\qquad\quad \text{Subtract } 6x.$$
$$x = 3 \qquad\qquad\qquad \text{Divide by } -4.$$

Solution set: $\{3\}$

c) $\dfrac{6x}{x+5} - \dfrac{30}{x-5} = \dfrac{6x^2+150}{x^2-25}$

Multiply by the LCD, $x^2 - 25$ or $(x-5)(x+5)$.

$$(x-5)(x+5)\left(\dfrac{6x}{x+5} - \dfrac{30}{x-5}\right) = \left(\dfrac{6x^2+150}{x^2-25}\right)(x^2-25)$$

$$6x(x-5)-30(x+5) = 6x^2+150$$
$$6x^2-30x-30x-150 = 6x^2+150 \qquad \text{Multiply.}$$
$$6x^2-60x-150 = 6x^2+150 \qquad \text{Combine like terms.}$$
$$-60x = 300 \qquad\qquad\qquad \text{Subtract } 6x^2.$$
$$x = -5 \qquad\qquad\qquad\quad \text{Divide by } -60.$$

However, -5 cannot be a solution because it makes denominators in the original equation equal zero. There is no solution. Solution set: $\varnothing$

d)
$$\frac{-2}{x+3} = \frac{1}{x+6} - \frac{6}{x^2+9x+18}$$
$$\frac{-2}{x+3} = \frac{1}{x+6} - \frac{6}{(x+3)(x+6)}$$

Multiply by the LCD, $(x+3)(x+6)$.

$$(x+3)(x+6)\left(\frac{-2}{x+3}\right) = \left[\frac{1}{x+6} - \frac{6}{(x+3)(x+6)}\right](x+3)(x+6)$$

$$-2(x+6) = (x+3) - 6$$

$-2x - 12 = x - 3$	Multiply; combine like terms.
$-12 = 3x - 3$	Add $2x$.
$-9 = 3x$	Add 3.
$-3 = x$	Divide by 3.

However, -3 cannot be a solution because it makes denominators in the original equation equal zero. There is no solution. Solution set: $\varnothing$

4. Solve each equation for the specified variable.

a) $P = \dfrac{R-C}{n}$ for C

b) $\dfrac{1}{x} + \dfrac{2}{y} - \dfrac{1}{z} = 5$ for z

Solution

a) $P = \dfrac{R-C}{n}$

Multiply by the LCD, n.

$$Pn = \left(\frac{R-C}{n}\right)n$$
$$Pn = R - C$$

$Pn - R = -C$	Subtract R.
$R - Pn = C$	Multiply by -1.

b) $\dfrac{1}{x} + \dfrac{2}{y} - \dfrac{1}{z} = 5$

Multiply by the LCD, xyz.

$$xyz\left(\frac{1}{x} + \frac{2}{y} - \frac{1}{z}\right) = 5xyz$$
$$yz + 2xz - xy = 5xyz$$

$yz + 2xz - 5xyz - xy = 0$	Subtract $5xyz$.
$yz + 2xz - 5xyz = xy$	Add xy.
$z(y + 2x - 5xy) = xy$	Factor out z.

$$z = \frac{xy}{y + 2x - 5xy}$$

Divide by $y + 2x - 5xy$.

12.7 Applications of Rational Expressions

Learning Objectives:

1. Solving a problem about an unknown number.
2. Solve problems about distance, rate, and time.
3. Solve problems about work.

Solved Examples:

1. One third of a number is 2 more than one-fifth of the same number. What is the number?

   Solution

   Let n = the number. Then the equation is $\frac{1}{3}n = 2 + \frac{1}{5}n$.

   $$\frac{1}{3}n = 2 + \frac{1}{5}n$$

   $$15\left(\frac{1}{3}n\right) = \left(2 + \frac{1}{5}n\right)(15) \quad \text{Multiply by the LCD, 15.}$$

   $$\begin{array}{ll} 5n = 30 + 3n & \text{Multiply.} \\ 2n = 30 & \text{Subtract } 3n. \\ n = 15 & \text{Divide by 2.} \end{array}$$

   The number is 15.

2. The denominator of a fraction is $2\frac{1}{2}$ times the numerator. If the numerator is doubled, and 2 is subtracted from the denominator, the resulting fraction is equivalent to 1. What is the original fraction?

   Solution

   Let n = the numerator of the fraction. Then $\left(2\frac{1}{2}\right)n = \frac{5}{2}n$ = the denominator of the fraction. The equation

   is $\dfrac{2n}{\frac{5}{2}n - 2} = 1$.

   $$\frac{2n}{\frac{5}{2}n - 2} = 1$$

   $$\left(\tfrac{5}{2}n - 2\right)\left(\frac{2n}{\frac{5}{2}n - 2}\right) = 1\left(\tfrac{5}{2}n - 2\right) \quad \text{Multiply by } \tfrac{5}{2}n - 2.$$

   $$\begin{array}{ll} 2n = \tfrac{5}{2}n - 2 & \text{Multiply.} \\ 2(2n) = \left(\tfrac{5}{2}n - 2\right)2 & \text{Multiply by 2.} \\ 4n = 5n - 4 & \text{Multiply.} \\ -n = -4 & \text{Subtract } 5n. \\ n = 4 & \text{Multiply by } -1. \\ \tfrac{5}{2}n = \tfrac{5}{2}(4) = 10 & \text{Determine the denominator.} \end{array}$$

   The original fraction is $\dfrac{4}{10}$.

3. A cyclist bikes at a constant speed for 22 miles. He then returns home at the same speed but takes a different route. His return trip takes one hour longer and is 27 miles. Find his speed.

   Solution

   Let s = his speed. Using the formula $d = rt$, we have

	Rate	Time	Distance
Trip out	s	$\dfrac{22}{s}$	22
Return trip	s	$\dfrac{27}{s}$	27

   The equation is $\dfrac{27}{s} = 1 + \dfrac{22}{s}$.

   $$\frac{27}{s} = 1 + \frac{22}{s}$$

   $$s\left(\frac{27}{s}\right) = \left(1 + \frac{22}{s}\right)s \qquad \text{Multiply by the LCD, } s.$$

   $$27 = s + 22 \qquad \text{Multiply.}$$

   $$5 = s \qquad \text{Subtract 22.}$$

   The cyclist rode at 5 miles per hour.

4. Julie and Eric row their boat at a constant speed of 6 miles per hour in still water. It takes them as long to go 40 miles upstream as 80 miles downstream. Find the speed of the current.

   Solution

   Let r = the rate of the current. Using the formula $d = rt$, we have

	Rate	Time	Distance
Downstream	$6 + r$	$\dfrac{80}{6+r}$	80
Upstream	$6 - r$	$\dfrac{40}{6-r}$	40

   The equation is $\dfrac{80}{6+r} = \dfrac{40}{6-r}$.

   $$\frac{80}{6+r} = \frac{40}{6-r}$$

   $$(6+r)(6-r)\left(\frac{80}{6+r}\right) = \left(\frac{40}{6-r}\right)(6+r)(6-r) \quad \text{Multiply by the LCD, } (6+r)(6-r).$$

   $$80(6-r) = 40(6+r) \qquad \text{Multiply.}$$

   $$480 - 80r = 240 + 40r \qquad \text{Multiply.}$$

   $$480 = 240 + 120r \qquad \text{Add } 80r.$$

   $$240 = 120r \qquad \text{Subtract 240.}$$

   $$2 = r \qquad \text{Divide by 120.}$$

   The current's speed is 2 miles per hour.

5. A painter can finish painting a house in 7 hours. His assistant takes 9 hours to finish the same job. How long would it take them to complete the job if they were working together?

Solution

Let t = the time they work together. Then

	Rate working alone	Time working together	Fractional Part of the Job Done
Painter	$\dfrac{1}{7}$	t	$\dfrac{1}{7}t$
Assistant	$\dfrac{1}{9}$	t	$\dfrac{1}{9}t$

The equation is $\dfrac{1}{7}t + \dfrac{1}{9}t = 1$.

$$\frac{1}{7}t + \frac{1}{9}t = 1$$

$$63\left(\frac{1}{7}t + \frac{1}{9}t\right) = 1(63) \qquad \text{Multiply by the LCD, 63.}$$

$$9t + 7t = 63 \qquad \text{Multiply.}$$

$$16t = 63 \qquad \text{Combine like terms.}$$

$$t = \frac{63}{16} \qquad \text{Divide by 16.}$$

$$= 3.9375$$

$$.9375(60) = 56.25 \qquad \text{Convert the fractional part of an hour to minutes.}$$

It will take them 3 hours, 56 minutes working together.

12.8 Variation

Learning Objectives:

1. Solve problems about direct variation.
2. Solve problems about inverse variation.
3. Solve applications involving direct and inverse variation.
4. Determine whether the situation is direct or inverse variation.
5. Solve similar triangle problems.

Solved Examples:

1. If x varies directly as y, and $x = 2$ when $y = 5$, find x when $y = 10$.

   Solution

   y varies directly as x if there exists a constant k such that $y = kx$. Substitute the given values into this equation and solve for k:

   $5 = 2k$

   $\dfrac{5}{2} = k$

   Now substitute $k = \dfrac{5}{2}$ and $y = 10$ into $y = kx$ and solve for x.

   $$10 = \frac{5}{2}x$$
   $$\frac{2}{5} \cdot 10 = \frac{2}{5} \cdot \frac{5}{2}x$$
   $$4 = x$$

   Thus, $x = 4$ when $y = 10$.

2. If r varies inversely as s, and $r = 14$ when $s = 4$, find r when $s = 7$.

   Solution

   r varies inversely as s if there exists a constant k such that $r = \dfrac{k}{s}$. Substitute the given values into this equation and solve for k:

   $14 = \dfrac{k}{4}$

   $56 = k$ Multiply by 4.

   Now substitute $k = 56$ and $s = 7$ into $r = \dfrac{k}{s}$ and solve for r.

   $r = \dfrac{56}{7} = 8$

   Thus, $r = 8$ when $s = 7$.

3. The interest on a money market account varies directly as the rate of interest. If the interest is $450 when the interest rate is 6%, find the interest when the rate is $3\frac{1}{2}$%.

    Solution

    i varies directly as r if there exists a constant k such that $i = kr$. Substitute the given values into this equation and solve for k:

    $450 = 0.06k$
    $7500 = k$

    Now substitute $k = 7500$ and $r = 3\frac{1}{2}\% = 0.035$ into $i = kr$ and solve for i.

    $i = 7500 \cdot 0.035 = 262.5$

    Thus, the interest is $262.50 when the interest rate is $3\frac{1}{2}$%.

4. If the fuel efficiency of your car, in miles per gallon (mpg), varies inversely with your average driving speed, in miles per hour (mph), and you get 30 miles per gallon when driving at an average speed of 50 mph, what will your fuel efficiency be when you drive at an average speed of 60 mph?

    Solution

    f varies inversely as s if there exists a constant k such that $f = \dfrac{k}{s}$. Substitute the given values into this equation and solve for k:

    $30 = \dfrac{k}{50}$
    $1500 = k$

    Now substitute $k = 1500$ and $s = 60$ into $f = \dfrac{k}{s}$ and solve for f.

    $f = \dfrac{1500}{60} = 25$

    Thus, fuel efficiency is 25 miles per gallon when speed is 60 miles per hour.

5. Find the missing length in the similar triangles.

    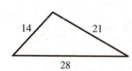

    Solution

    Two triangles are similar if they have the same shape (but not necessarily the same size). Similar triangles have side lengths that are proportional—that is, vary directly.

    $\dfrac{x}{28} = \dfrac{18}{14}$

    $28\left(\dfrac{x}{28}\right) = 28\left(\dfrac{18}{14}\right)$

    $x = 36$

    Thus, the missing length is 36.

13.1 Solving Systems of Linear Equations by Graphing

Learning Objectives:

1. Decide whether an ordered pair is a solution of a system.
2. Solve linear systems by graphing.
3. Describe systems of equations.
4. Solve application problems.

Solved Examples:

1. Determine whether the ordered pair is a solution of the system of linear equations.

   a) $(4,-1)$; $2x - y = 3$
   $$x + y = 3$$

   b) $(-2,5)$; $2x + 3y = 11$
   $$5x - 2y = -20$$

   Solution

   a) Substitute 4 for x and -1 for y in each equation.

   $$\begin{array}{c|c} 2x - y = 3 & x + y = 3 \\ ? & ? \\ 2(4)-(-1)=3 & 4+(-1)=3 \\ 9 \neq 3 & 3 = 3 \end{array}$$

   No, $(4,-1)$ is not a solution of the system.

   b) Substitute -2 for x and 5 for y in each equation.

   $$\begin{array}{c|c} 2x + 3y = 11 & 5x - 2y = -20 \\ ? & ? \\ 2(-2)+3(5)=11 & 5(-2)-2(5)=-20 \\ 11 = 11 & -20 = -20 \end{array}$$

   Yes, $(-2,5)$ is a solution of the system.

2. Complete each statement

   a) Two straight lines with different slopes have _____ intersection point(s).

   b) Two straight lines with the same slope but different y-intercept values have _____ intersection point(s).

   c) Two straight lines with the same slope and same y-intercept value have _____ intersection point(s).

   Solution

   a) 1

   b) 0

   c) infinitely many

3. Solve by graphing. Each system has a unique solution (**consistent** system).

   a) $x - y = 2$
   $x + y = 4$

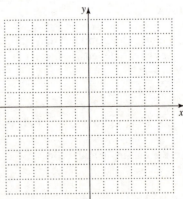

b) $-2x + 2y = 6$
$4x + 2y = 18$

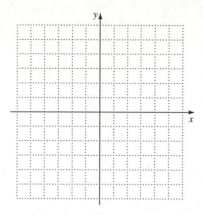

c) $-3x + y - 4 = 0$
$4x + y + 3 = 0$

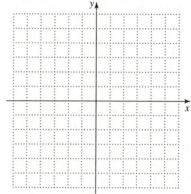

d) $\dfrac{1}{3}x - y = 1;\ \ x = 3$

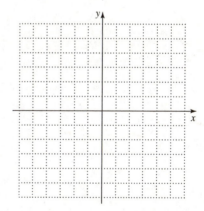

Solution

a)

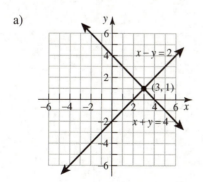

b)

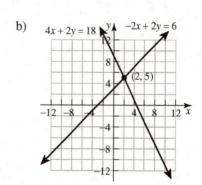

c)

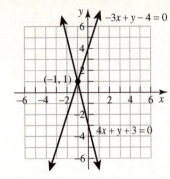

$-3x + y - 4 = 0$

$(-1, 1)$

$4x + y + 3 = 0$

d)

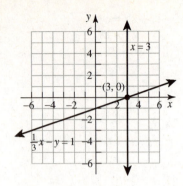

$x = 3$

$(3, 0)$

$\frac{1}{3}x - y = 1$

4. Solve by graphing. Each system has either no solution (*inconsistent* system) or an infinite number of solutions (*dependent* system)

a) $x + y = 3$
    $2x + 2y = 6$

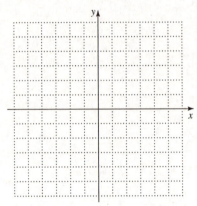

b) $2x + y = 5$
    $2x + y = 8$

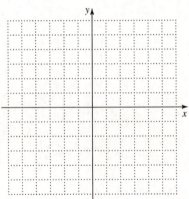

c) $y - 5x = 2$
    $5y = 25x + 10$

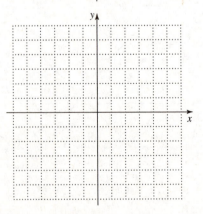

d) $x = -y$
       $y + x = 6$

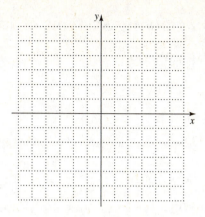

Solution

a)

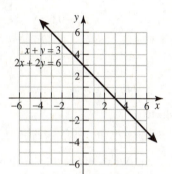

b)

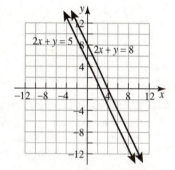

c)

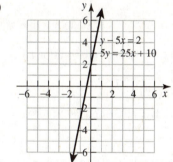

d)

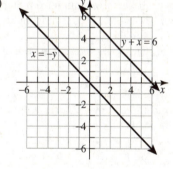

13.2 Solving Systems of Linear Equations by Substitution

Learning Objectives:

1. Solve systems using substitution.

Solved Examples:

1. Find the solution to each system of equations by the substitution method. Check your answers.

a) $x + y = 6$
$ x = y - 2$

b) $x - 6y = -18$
$ 2x - 7y = -16$

c) $x + 5y = 6$
$ -2x + 6y = -12$

d) $8x + 5y = -3$
$ -2x + 3y = 5$

e) $6x + 5y = 25$
$ 4x - 3y = -15$

f) $4x - 3y = 30 + x$
$ 4x = -(y + 2) + 3x$

g) $-\dfrac{x}{3} + \dfrac{3y}{4} = \dfrac{1}{2}$
$ \dfrac{x}{6} + \dfrac{y}{8} = \dfrac{3}{4}$

h) $-2(2x - y - 11) = y - 1$
$ 3(3x + y) = 4(2x + y) + 17$

i) $-0.5x + 0.2y = 0.5$
$ 0.8x - 0.1y = 0.3$

Solution

To solve a linear system by substitution, use the following steps.

Step 1: Solve one equation for either variable. If one of the variables has coefficient 1 or choose it, since it usually makes the substitution easier.

Step 2: Substitute for that variable in the other equation. The result should be an equation with just one variable.

Step 3: Solve the equation from Step 2.

Step 4: Substitute the result from Step 3 into the equation from Step 1 to find the value of the other variable.

Step 5: Check the solution in both of the original equations. Then write the solution set.

a) $x + y = 6$
$ x = y - 2$

Substitute $x = y - 2$ for x in the first equation, then solve for y.

$(y - 2) + y = 6$
$ 2y - 2 = 6$
$ 2y = 8$
$ y = 4$

Now substitute 4 for y in the second equation and solve for x.

$x = 4 - 2 = 2$

Check:

$\begin{array}{c|c} x + y = 6 & x = y - 2 \\ ? & ? \\ 2 + 4 = 6 & 2 = 4 - 2 \\ 6 = 6 & 2 = 2 \end{array}$

Solution set: $\{(2, 4)\}$

b) $x - 6y = -18$
$ 2x - 7y = -16$

Solve the first equation for x.

$x - 6y = -18$
$ x = 6y - 18$

Now substitute $6y - 18$ for x in the second equation, then solve for y.

$2(6y - 18) - 7y = -16$
$12y - 36 - 7y = -16$
$ 5y - 36 = -16$
$ 5y = 20$
$ y = 4$

Substitute 4 for y into the first equation and solve for x.

$x - 6(4) = -18$
$ x - 24 = -18$
$ x = 6$

Check:

$\begin{array}{c|c} x - 6y = -18 & 2x - 7y = -16 \\ ? & ? \\ 6 - 6(4) = -18 & 2(6) - 7(4) = -16 \\ -18 = -18 & -16 = -16 \end{array}$

Solution set: $\{(6, 4)\}$

c) $x + 5y = 6$
$-2x + 6y = -12$

Solve the first equation for x.
$x + 5y = 6$
$x = 6 - 5y$

Now substitute $6 - 5y$ for x in the second equation, then solve for y.
$-2(6 - 5y) + 6y = -12$
$-12 + 10y + 6y = -12$
$-12 + 16y = -12$
$16y = 0$
$y = 0$

Now substitute 0 for y in the first equation and solve for x.
$x + 5(0) = 6$
$x + 0 = 6$
$x = 6$

Check:

$x + 5y = 6$	$-2x + 6y = -12$
$6 + 5(0) \overset{?}{=} 6$	$-2(6) + 6(0) \overset{?}{=} -12$
$6 = 6$	$-12 = -12$

Solution set: $\{(6, 0)\}$

d) $8x + 5y = -3$
$-2x + 3y = 5$

Solve the second equation for x.
$-2x + 3y = 5$
$-2x = -3y + 5$
$x = \dfrac{3}{2}y - \dfrac{5}{2}$

Substitute $\dfrac{3}{2}y - \dfrac{5}{2}$ for x in the first equation, then solve for y.
$8\left(\dfrac{3}{2}y - \dfrac{5}{2}\right) + 5y = -3$
$12y - 20 + 5y = -3$
$17y - 20 = -3$
$17y = 17$
$y = 1$

Now substitute 1 for y in the second equation and solve for x.
$-2x + 3(1) = 5$
$-2x + 3 = 5$
$-2x = 2$
$x = -1$

Check:

$8x + 5y = -3$	$-2x + 3y = 5$
$8(-1) + 5(1) \overset{?}{=} -3$	$-2(-1) + 3(1) \overset{?}{=} 5$
$-3 = -3$	$5 = 5$

Solution set: $\{(-1, 1)\}$

e) $6x + 5y = 25$
$4x - 3y = -15$

Solve the second equation for x.
$4x - 3y = -15$
$4x = 3y - 15$
$x = \dfrac{3}{4}y - \dfrac{15}{4}$

Now substitute $\dfrac{3}{4}y - \dfrac{15}{4}$ for x in the first equation and solve for x.
$6\left(\dfrac{3}{4}y - \dfrac{15}{4}\right) + 5y = 25$
$\dfrac{9}{2}y - \dfrac{45}{2} + 5y = 25$
$9y - 45 + 10y = 50$
$19y - 45 = 50$
$19y = 95$
$y = 5$

Now substitute 5 for y in the second equation and solve for x.
$4x - 3(5) = -15$
$4x - 15 = -15$
$4x = 0$
$x = 0$

Check:

$6x + 5y = 25$	$4x - 3y = -15$
$8(0) + 5(5) \overset{?}{=} 25$	$4(0) - 3(5) \overset{?}{=} -15$
$25 = 25$	$-15 = -15$

Solution set: $\{(0, 5)\}$

f) $4x - 3y = 30 + x$
   $4x = -(y + 2) + 3x$

Solve the second equation for x.

$4x = -(y + 2) + 3x$
$x = -(y + 2) = -y - 2$

Substitute $-y - 2$ for x in the first equation, then solve for x.

$4(-y - 2) - 3y = 30 + (-y - 2)$
$-4y - 8 - 3y = 30 - y - 2$
$-7y - 8 = 28 - y$
$-6y - 8 = 28$
$-6y = 36$
$y = -6$

Now substitute -6 for y in the second equation and solve for x.

$4x = -(-6 + 2) + 3x$
$4x = -(-4) + 3x$
$4x = 4 + 3x$
$x = 4$

Check:

$4x - 3y = 30 + x$
$4(4) - 3(-6) \overset{?}{=} 30 + 4$
$16 + 18 \overset{?}{=} 34$
$34 = 34$

$4x = -(y + 2) + 3x$
$4(4) \overset{?}{=} -(-6 + 2) + 3(4)$
$16 \overset{?}{=} -(-4) + 12$
$16 = 16$

Solution set: $\{(4, -6)\}$

g) $-\dfrac{x}{3} + \dfrac{3y}{4} = \dfrac{1}{2}$
   $\dfrac{x}{6} + \dfrac{y}{8} = \dfrac{3}{4}$

Solve the second equation for x.

$\dfrac{x}{6} + \dfrac{y}{8} = \dfrac{3}{4}$

$24\left(\dfrac{x}{6} + \dfrac{y}{8}\right) = 24\left(\dfrac{3}{4}\right)$

$4x + 3y = 18$
$4x = -3y + 18$
$x = -\dfrac{3}{4}y + \dfrac{9}{2}$

Substitute $-\dfrac{3}{4}y + \dfrac{9}{2}$ for x in the first equation, then solve for y.

$-\dfrac{1}{3}\left(-\dfrac{3}{4}y + \dfrac{9}{2}\right) + \dfrac{3y}{4} = \dfrac{1}{2}$

$\dfrac{1}{4}y - \dfrac{3}{2} + \dfrac{3}{4}y = \dfrac{1}{2}$

$y - \dfrac{3}{2} = \dfrac{1}{2}$

$y = \dfrac{4}{2} = 2$

Now substitute 2 for y in the second equation and solve for x.

$\dfrac{x}{6} + \dfrac{2}{8} = \dfrac{3}{4}$

$\dfrac{x}{6} + \dfrac{1}{4} = \dfrac{3}{4}$

$\dfrac{x}{6} = \dfrac{2}{4} = \dfrac{1}{2}$

$x = 3$

Check:

$-\dfrac{x}{3} + \dfrac{3y}{4} = \dfrac{1}{2} \quad \bigg| \quad \dfrac{x}{6} + \dfrac{y}{8} = \dfrac{3}{4}$

$-\dfrac{3}{3} + \dfrac{3(2)}{4} \overset{?}{=} \dfrac{1}{2} \quad \bigg| \quad \dfrac{3}{6} + \dfrac{2}{8} \overset{?}{=} \dfrac{3}{4}$

$-1 + \dfrac{6}{4} \overset{?}{=} \dfrac{1}{2} \quad \bigg| \quad \dfrac{1}{2} + \dfrac{1}{4} \overset{?}{=} \dfrac{3}{4}$

$\dfrac{1}{2} = \dfrac{1}{2} \quad \bigg| \quad \dfrac{3}{4} = \dfrac{3}{4}$

Solution set: $\{(3, 2)\}$

h) $-2(2x-y-11)=y-1$

$\qquad 3(3x+y)=4(2x+y)+17$

Solve the first equation for y.

$-2(2x-y-11)=y-1$

$-4x+2y+22=y-1$

$-4x+22=-y-1$

$-4x+23=-y$

$4x-23=y$

Now substitute $4x-23$ for y in the second equation and solve for x.

$3\left[3x+(4x-23)\right]=4\left[2x+(4x-23)\right]+17$

$3(7x-23)=4(6x-23)+17$

$21x-69=24x-92+17$

$21x-69=24x-75$

$21x+6=24x$

$6=3x$

$2=x$

Substitute 2 for x in the first equation and solve for y.

$-2\left[2(2)-y-11\right]=y-1$

$-2(4-y-11)=y-1$

$-8+2y+22=y-1$

$2y+14=y-1$

$y+14=-1$

$y=-15$

Check:

$-2(2x-y-11)=y-1$

$-2\left[2(2)-(-15)-11\right]\overset{?}{=}-15-1$

$-2(8)\overset{?}{=}-16$

$-16=-16$

$3(3x+y)=4(2x+y)+17$

$3\left[3(2)+(-15)\right]\overset{?}{=}4\left[2(2)+(-15)\right]+17$

$3(-9)\overset{?}{=}4(-11)+17$

$-27=-27$

Solution set: $\{(2,-15)\}$

i) $-0.5x+0.2y=0.5$

$\qquad 0.8x-0.1y=0.3$

Solve the first equation for y.

$-0.5x+0.2y=0.5$

$-5x+2y=5$

$2y=5x+5$

$y=2.5x+2.5$

Now substitute $2.5x+2.5$ for y in the second equation and solve for x.

$0.8x-0.1(2.5x+2.5)=0.3$

$0.8x-0.25x-0.25=0.3$

$80x-25x-25=30$

$55x-25=30$

$55x=55$

$x=1$

Substitute 1 for x in the first equation and solve for y.

$-0.5(1)+0.2y=0.5$

$-0.5+0.2y=0.5$

$0.2y=1$

$y=5$

Check:

$-0.5x+0.2y=0.5$	$0.8x-0.1y=0.3$
$-0.5(1)+0.2(5)\overset{?}{=}0.5$	$0.8(1)-0.1(5)\overset{?}{=}0.3$
$-0.5+1\overset{?}{=}0.5$	$0.8-0.5\overset{?}{=}0.3$
$0.5=0.5$	$0.3=0.3$

Solution set: $\{(1,5)\}$

3. A construction equipment rental company in Worcester charges $1400 for initial delivery and a rental fee of $1000 per week. A company in Boston charges $400 for initial delivery and $1200 per week. Create a cost equation for each company where y is the total cost of renting the construction equipment and x is the number of weeks the equipment is rented. Write a system of equations.

Solution

$y=1000x+1400 \quad$ (Worcester)

$y=1200x+400 \quad$ (Boston)

13.3 Solving Systems of Linear Equations by Elimination

Learning Objectives:

1. Solve systems using elimination.

Solved Examples:

1. Find the solution using the elimination method.

a) $x + y = -12$
   $x - y = 2$

b) $x + 8y = -15$
   $7x + 8y = 39$

c) $x + 7y = -53$
   $-5x + 6y = -22$

d) $9x + 8y = 72$
   $-7x - 4y = -56$

e) $9x - 7y = 44$
   $6x + 2y = -4$

f) $9x + 7y = -14$
   $-7x + 2y = -4$

g) $\dfrac{1}{2}x + \dfrac{1}{2}y = -3$
   $x - y = 8$

h) $2x - 2y = -10$
   $4x + \dfrac{4}{5}y = -\dfrac{28}{5}$

i) $\dfrac{3x}{8} - \dfrac{3y}{5} = \dfrac{33}{80}$
   $\dfrac{4x}{7} + \dfrac{4y}{5} = \dfrac{37}{35}$

j) $-0.1x + 0.6y = 1.3$
   $0.8x + 0.6y = -5$

k) $-0.07x + 0.01y = -0.19$
   $0.01x + 0.01y = -0.03$

l) $-0.2x - 6y = -49.8$
   $-0.3x + 0.6y = 2.1$

Solution

To solve a linear system by elimination, use the following steps.

Step 1: Write both equations in standard form
Step 2: Transform so that the coefficients of one pair of variable terms are opposites. Multiply one or both equations by appropriate numbers so that the sum of the coefficients of either the *x*- or *y*-terms is 0.
Step 3: Add the new equations to eliminate a variable. The sum should be an equation with just one variable.
Step 4: Solve the equation from Step 3 for the remaining variable.
Step 5: Substitute the result from Step 4 into either of the original equations and solve for the other variable.
Step 6: Check the solution in both of the original equations. Then write the solution set.

a) $x + y = -12$
   $x - y = 2$

Add the equations to eliminate *y*. Then solve for *x*.

$$x + y = -12$$
$$\underline{x - y = 2}$$
$$2x = -10$$
$$x = -5$$

Substitute –5 for *x* in the first equation, then solve for *y*.

$$-5 + y = -12$$
$$y = -7$$

Check by substituting –5 for *x* and –7 for *y* in both equations.
Solution set: $\{(-5, -7)\}$

b) $x + 8y = -15$
   $7x + 8y = 39$

Multiply the first equation by –7, then add the result to the second equation to eliminate *x*. Solve for *y*.

$$-7(x + 8y) = -15(-7)$$
$$-7x - 56y = 105$$
$$\underline{7x + 8y = 39}$$
$$-48y = 144$$
$$y = -3$$

Substitute –3 for *y* in the first equation, then solve for *x*.

$$x + 8(-3) = -15$$
$$x - 24 = -15$$
$$x = 9$$

Check by substituting 9 for *x* and –3 for *y* in both equations.
Solution set: $\{(9, -3)\}$

c) $x + 7y = -53$
$-5x + 6y = -22$

Multiply the first equation by 5, then add the result to the second equation to eliminate x. Solve for y.

$5(x + 7y) = -53(5)$
$5x + 35y = -265$
$\underline{-5x + 6y = -22}$
$41y = -287$
$y = -7$

Substitute -7 for y in the first equation, then solve for x.

$x + 7(-7) = -53$
$x - 49 = -53$
$x = -4$

Check by substituting -4 for x and -7 for y in both equations.
Solution set: $\{(-4, -7)\}$

d) $9x + 8y = 72$
$-7x - 4y = -56$

Multiply the second equation by 2, then add the result to the first equation to eliminate y.

$2(-7x - 4y) = -56(2)$
$-14x - 8y = -112$
$\underline{9x + 8y = 72}$
$-5x = -40$
$x = 8$

Substitute 8 for x in the first equation, then solve for y.

$9(8) + 8y = 72$
$72 + 8y = 72$
$8y = 0$
$y = 0$

Check by substituting 8 for x and 0 for y in both equations.
Solution set: $\{(8, 0)\}$

e) $9x - 7y = 44$
$6x + 2y = -4$

Multiply the first equation by 2 and the second equation by 7, then add the resulting equations to eliminate y.

$2(9x - 7y) = 44(2) \Rightarrow 18x - 14y = 88$
$7(6x + 2y) = -4(7) \Rightarrow 42x + 14y = -28$

$18x - 14y = 88$
$\underline{42x + 14y = -28}$
$60x = 60$
$x = 1$

Substitute 1 for x in the second equation, then solve for y.

$6(1) + 2y = -4$
$6 + 2y = -4$
$2y = -10$
$y = -5$

Check by substituting 1 for x and -5 for y in both equations.
Solution set: $\{(1, -5)\}$

f) $9x + 7y = -14$
$-7x + 2y = -4$

Multiply the first equation by 7 and the second equation by 9, then add the resulting equations to eliminate x.

$7(9x + 7y) = -14(7) \Rightarrow 63x + 49y = -98$
$9(-7x + 2y) = -4(9) \Rightarrow -63x + 18y = -36$

$63x + 49y = -98$
$\underline{-63x + 18y = -36}$
$67y = -134$
$y = -2$

Substitute -2 for y in the first equation, then solve for x.

$9x + 7(-2) = -14$
$9x - 14 = -14$
$9x = 0$
$x = 0$

Check by substituting 0 for x and -2 for y in both equations.
Solution set: $\{(0, -2)\}$

g) $\dfrac{1}{2}x + \dfrac{1}{2}y = -3$
$x - y = 8$

Multiply equation 1 by 2, then add the result to the second equation to eliminate y.

$2\left(\dfrac{1}{2}x + \dfrac{1}{2}y\right) = -3(2)$
$x + y = -6$
$\underline{x - y = 8}$
$2x = 2$
$x = 1$

Substitute 1 for x in the second equation and solve for y.

$1 - y = 8$
$-y = 7$
$y = -7$

Check by substituting 1 for x and -7 for y in both equations.
Solution set: $\{(1, -7)\}$

h) $2x - 2y = -10$

$4x + \dfrac{4}{5}y = -\dfrac{28}{5}$

First, multiply the second equation by the LCD, 5, to clear the fractions.

$5\left(4x + \dfrac{4}{5}y\right) = \left(-\dfrac{28}{5}\right)5 \Rightarrow 20x + 4y = -28$ (3)

Now, multiply the first equation by 2, then add the result to equation (3) to eliminate y.

$2(2x - 2y) = (-10)2$

$\begin{array}{r} 4x - 4y = -20 \\ \underline{20x + 4y = -28} \\ 24x = -48 \\ x = -2 \end{array}$

Substitute -2 for x in the first equation and solve for y.

$\begin{array}{r} 2(-2) - 2y = -10 \\ -4 - 2y = -10 \\ -2y = -6 \\ y = 3 \end{array}$

Check by substituting -2 for x and 3 for y in both equations.

Solution set: $\{(1, -7)\}$

i) $\dfrac{3x}{8} - \dfrac{3y}{5} = \dfrac{33}{80}$

$\dfrac{4x}{7} + \dfrac{4y}{5} = \dfrac{37}{35}$

First, clear the fractions in each equation by multiplying by the LCD.

$80\left(\dfrac{3x}{8} - \dfrac{3y}{5}\right) = \left(\dfrac{33}{80}\right)80 \Rightarrow 30x - 48y = 33$ (3)

$35\left(\dfrac{4x}{7} + \dfrac{4y}{5}\right) = \left(\dfrac{37}{35}\right)35 \Rightarrow 20x + 28y = 37$ (4)

Now, multiply equation (3) by 2 and equation (4) by -3, then add the resulting equations to eliminate x.

$2(30x - 48y) = 33(2) \Rightarrow 60x - 96y = 66$

$-3(20x + 28y) = 37(-3) \Rightarrow -60x - 84y = -111$

$\begin{array}{r} 60x - 96y = 66 \\ \underline{-60x - 84y = -111} \\ -180y = -45 \\ y = \dfrac{45}{180} = \dfrac{1}{4} \end{array}$

Substitute $\dfrac{1}{4}$ for y into the first equation and solve for x.

$\dfrac{3x}{8} - \dfrac{3}{5}\left(\dfrac{1}{4}\right) = \dfrac{33}{80}$

$\dfrac{3x}{8} - \dfrac{3}{20} = \dfrac{33}{80}$

$80\left(\dfrac{3x}{8} - \dfrac{3}{20}\right) = \left(\dfrac{33}{80}\right)80$

$30x - 12 = 33$

$30x = 45$

$x = \dfrac{45}{30} = \dfrac{3}{2}$

Check by substituting $\dfrac{3}{2}$ for x and $\dfrac{1}{4}$ for y in both equations.

Solution set: $\left\{\left(\dfrac{3}{2}, \dfrac{1}{4}\right)\right\}$

j) $-0.1x + 0.6y = 1.3$

$0.8x + 0.6y = -5$

Multiply the second equation by -1, then add the result to the first equation to eliminate y.

$-1(0.8x + 0.6y) = (-5)(-1)$

$\begin{array}{r} -0.8x - 0.6y = 5 \\ \underline{-0.1x + 0.6y = 1.3} \\ -0.9x = 6.3 \\ x = -7 \end{array}$

Substitute -7 for x in the first equation and solve for y.

$-0.1(-7) + 0.6y = 1.3$

$0.7 + 0.6y = 1.3$

$0.6y = 0.6$

$y = 1$

Check by substituting -7 for x and 1 for y in both equations.

Solution set: $\{(-7, 1)\}$

k) $-0.07x + 0.01y = -0.19$
    $0.01x + 0.01y = -0.03$

Multiply the first equation by -1, then add the result to the second equation to eliminate y.

$$(-1)(-0.07x + 0.01y) = (-0.19)(-1)$$
$$0.07x - 0.01y = 0.19$$
$$\underline{0.01x + 0.01y = -0.03}$$
$$0.08x = 0.16$$
$$x = 2$$

Substitute 2 for x in the second equation and solve for y.

$$0.01(2) + 0.01y = -0.03$$
$$0.02 + 0.01y = -0.03$$
$$0.01y = -0.05$$
$$y = 5$$

Check by substituting 2 for x and 5 for y in both equations.
Solution set: $\{(2, 5)\}$

l) $-0.2x + 0.4y = -49.8$
    $-0.3x + 0.6y = 2.1$

Multiply the first equation by 3 and the second equation by -2, then add the resulting equations to eliminate x.

$$3(-0.2x + 0.4y) = (-49.8)3 \Rightarrow$$
$$-0.6x + 1.2y = -149.4$$
$$(-2)(-0.3x + 0.6y) = (2.1)(-2) \Rightarrow$$
$$0.6x - 1.2y = -4.2$$

$$-0.6x + 1.2y = -149.4$$
$$\underline{0.6x - 1.2y = -4.2}$$
$$0 = -153.6$$

This is a false statement, so there is no solution. The equations are independent and the system is inconsistent.
Solution set: $\varnothing$

4. When using the elimination method to solve a special system and a true statement results, the system is _____ ,but when a false statement results, the system is _____ .

    Solution

    Dependent; inconsistent

13.4 Applications of Linear Systems

Learning Objectives:

1. Solve applications of linear systems.

Solved Examples:

1. Two numbers total –6, and their difference is 12. Find the two numbers.

   Solution

   Let x = one number and let y = the other number. The equations are $x + y = -6$ and $x - y = 12$.
   Solve the system using elimination.

   $$\begin{array}{ll} x + y = -6 & 3 + y = -6 \\ \underline{x - y = 12} & y = -9 \\ 2x = 6 & \\ x = 3 & \end{array}$$

   The numbers are 3 and –9. Check by substituting the answers in the words of the original problem.

2. Dillon purchased tickets to an air show for 4 adults and 2 children. The total cost was $86. The cost of a child's ticket was $5 less than the cost of an adult's ticket. Find the price of an adult's ticket and a child's ticket.

   Solution

   Let a = the cost of the adult's ticket and let c = the cost of the child's ticket. The equations are
   $4a + 2c = 86$ and $c = a - 5$. Solve the system using substitution.

   $$\begin{array}{ll} 4a + 2(a - 5) = 86 & c = 16 - 5 = 11 \\ 4a + 2a - 10 = 86 & \\ 6a - 10 = 86 & \\ 6a = 96 & \\ a = 16 & \end{array}$$

   The adult's ticket cost $16 and the child's ticket costs $11. Check by substituting the answers in the words of the original problem.

3. Jason always throws loose change into a pencil holder on his desk and takes it out every two weeks. This time it is all nickels and dimes. There are 7 times as many dimes as nickels, and the value of the dimes is $3.90 more than the value of the nickels. How many nickels and dimes does Jason have?

   Solution

   Let n = the number of nickels and let d = the number of dimes. The equations are $d = 7n$ and
   $0.10d = 3.90 + 0.05n.$ Solve the system by substitution.

   $$\begin{array}{ll} 0.10(7n) = 3.90 + 0.05n & d = 7(6) = 42 \\ 0.70n = 3.90 + 0.05n & \\ 0.65n = 3.90 & \\ n = 6 & \end{array}$$

   There are 6 nickels and 42 dimes. Check by substituting the answers in the words of the original problem.

4. Kelly is a partner in an internet-based seed and garden supply business. The company offers a blend of exotic wildflower seeds for $95 per pound and a blend of common wildflower seeds for $20 per pound. Kelly is creating 40 pounds of a medium-price product by mixing together the more expensive blend with the less expensive blend. If the new blend sells for $76.25 per pound, how many pounds of each are needed?

Solution

Let x = the number of pounds of exotic seeds and let y = the number of pounds of common seeds. Summarize the information given in the problem in a table.

	Pounds of seeds	Price per pound	Total
Exotic seeds	x	$95	$95x$
Common seeds	y	$20	$20y$
Mixture	40	$76.25	$40(76.25) = 3050$

The equations are $x + y = 40$ and $95x + 20y = 3050$. Solve the system by substitution. $x = 40 - y$

$$95(40 - y) + 20y = 3050 \qquad x + 10 = 40$$
$$3800 - 95y + 20y = 3050 \qquad x = 30$$
$$3800 - 75y = 3050$$
$$-75y = -750$$
$$y = 10$$

The new blend consists of 30 pounds of the exotic seeds and 10 pounds of the common seeds. Check by substituting the answers in the words of the original problem.

5. A 520-mile trip from one city to another takes 4 hours when a plane is flying with the wind. The return trip against the wind takes 5 hours. Find the rate of the plane in still air and the rate of the wind.

Solution

Let x = the rate of the plane in still air and let y = the rate of the wind. Summarize the information given in the problem in a table.

	Rate	Time	Distance
With the wind	$x + y$	4	520
Against the wind	$x - y$	5	520

The equations are $4(x + y) = 520$ and $5(x - y) = 520$. Solve the system by elimination.

$$4(x + y) = 520 \Rightarrow x + y = 130$$
$$5(x - y) = 520 \Rightarrow x - y = 104$$

$$\begin{aligned} x + y &= 130 \\ x - y &= 104 \\ \hline 2x &= 234 \\ x &= 117 \end{aligned} \qquad \begin{aligned} 4(117 + y) &= 520 \\ 468 + 4y &= 520 \\ 4y &= 68 \\ y &= 17 \end{aligned}$$

The plane's speed in still air is 117 miles per hour and the wind speed is 17 miles per hour. Check by substituting the answers in the words of the original problem.

13.5 Solving Systems of Linear Inequalities

Learning Objectives:

Graph systems of linear inequalities.

Solved Examples:

1. Match the correct system of linear equations with the graph.

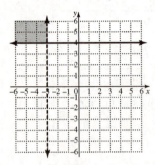

    a) $y \leq -3$ b) $y \geq 4$ c) $x \leq -3$

            $x > 4$ $x < -3$ $y > 4$

Solution

Choice (b)

2. Match the correct system of linear equations with the graph.

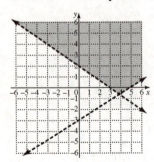

    a) $3x - 5y > 15$ b) $2x + 3y > 6$ c) $2x + 3y > 6$

            $2x + 3y < 6$ $3x - 5y > 15$ $3x - 5y < 15$

Solution

Choice (c)

3. Match the correct system of linear equations with the graph.

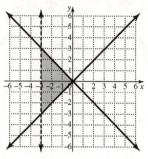

a) $y \geq x$
    $y \leq -x$
    $x > -3$

b) $y \leq x$
    $y \leq -x$
    $x > -3$

c) $y \geq x$
    $y \geq -x$
    $x < -3$

Solution

Choice (a)

14.1 Evaluating Roots

Learning Objectives:

1. Find square roots.
2. Find squares of radical expressions.
3. Find rational and approximate irrational square roots.
4. Use the Pythagorean formula.
5. Find higher roots.

Solved Examples:

1. Find the two square roots of each number.

   a) 9   b) 25   c) 36   d) 49   e) 100

   Solution

   a) 3, –3   b) 5, –5   c) 6, –6   d) 7, –7   e) 10, –10

2. Find the square root, if it exists. Do not use a calculator or a table of square roots.

   a) $\sqrt{4}$ b) $-\sqrt{4}$ c) $\sqrt{-4}$ d) $\sqrt{25}$

   e) $\sqrt{0.25}$ f) $-\sqrt{121}$ g) $-\sqrt{\dfrac{49}{36}}$ h) $\sqrt{\dfrac{81}{400}}$

   Solution

   a) $\sqrt{4} = 2$ b) $-\sqrt{4} = -2$ c) $\sqrt{-4}$ is a real number   d) $\sqrt{25} = 5$

   e) $\sqrt{0.25} = 0.5$ f) $-\sqrt{121} = -11$ g) $-\sqrt{\dfrac{49}{36}} = -\dfrac{7}{2}$ h) $\sqrt{\dfrac{81}{400}} = \dfrac{9}{20}$

3. Use a calculator to approximate to the nearest thousandth.

   a) $\sqrt{5}$ b) $-\sqrt{7}$ c) $\sqrt{59}$ d) $-\sqrt{951}$

   Solution

   a) $\sqrt{5} \approx 2.236$ b) $-\sqrt{7} \approx -2.646$ c) $\sqrt{59} \approx 7.681$ d) $-\sqrt{951} \approx -30.838$

4. Draw a diagram and use the Pythagorean theorem to solve. Round to the nearest tenth.

   a) A 12-foot ladder is leaning against a house with the base of the ladder 4 feet from the house. How high up the house does the ladder reach?

   b) One end of a wire is attached to the top of a 25 foot pole, and the other end is anchored into the ground 20 feet from the base of the pole. Find the length of the wire.

Solution

Use the Pythagorean formula to solve each problem.

a)

$$x^2 + 4^2 = 12^2$$
$$x^2 + 16 = 144$$
$$x^2 = 128$$
$$x \approx 11.3$$

The ladder reaches about 11.3 ft up the side of the house.

b)

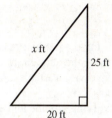

$$25^2 + 20^2 = x^2$$
$$625 + 400 = x^2$$
$$1025 = x^2$$
$$32.0 \approx x$$

The wire is about 32.0 feet long.

4. Find each root.

    a) $\sqrt[3]{8}$ b) $-\sqrt[3]{8}$ c) $\sqrt[3]{-8}$

    d) $\sqrt[4]{16}$ e) $-\sqrt[4]{16}$ f) $\sqrt[4]{-16}$

Solution

    a) $\sqrt[3]{8} = 2$ b) $-\sqrt[3]{8} = -2$ c) $\sqrt[3]{-8} = -2$

    d) $\sqrt[4]{16} = 2$ e) $-\sqrt[4]{16} = -2$ f) $\sqrt[4]{-16}$ is not a real number

14.2 Multiplying, Dividing, and Simplifying Radicals

Learning Objectives:

1. Multiply radicals.
2. Simplify radicals using the product rule.
3. Simplify radicals using the quotient rule.
4. Simplify radicals involving variables.
5. Simplify higher roots.
6. Solve problems involving volume and area.

Solved Examples:

1. Use the product rule and the quotient rule to find each product or quotient.

     a) $\sqrt{5} \cdot \sqrt{7}$ b) $\sqrt{6} \cdot \sqrt{7}$ c) $\sqrt{11} \cdot \sqrt{t}$ d) $\dfrac{\sqrt{252}}{\sqrt{7}}$

Solution

Product rule: For nonnegative real numbers a and b, $\sqrt{a} \cdot \sqrt{b} = \sqrt{ab}$ and $\sqrt{a} \cdot \sqrt{b} = \sqrt{ab} = \sqrt{a} \cdot \sqrt{b}$.

Quotient rule: If a and b are nonnegative real numbers and $b \neq 0$, then $\sqrt{\dfrac{a}{b}} = \dfrac{\sqrt{a}}{\sqrt{b}}$ and $\sqrt{\dfrac{a}{b}} = \dfrac{\sqrt{a}}{\sqrt{b}} = \sqrt{\dfrac{a}{b}}$.

     a) $\sqrt{5} \cdot \sqrt{7} = \sqrt{35}$ b) $\sqrt{6} \cdot \sqrt{7} = \sqrt{42}$

     c) $\sqrt{11} \cdot \sqrt{t} = \sqrt{11t}$ d) $\dfrac{\sqrt{252}}{\sqrt{7}} = \sqrt{\dfrac{252}{7}} = \sqrt{36} = 6$

2. Simplify. Assume that all variables represent positive numbers.

     a) $\sqrt{3^2}$ b) $\sqrt{6^2}$ c) $\sqrt{2^4}$ d) $\sqrt{13^2}$

     e) $\sqrt{x^2}$ f) $\sqrt{x^4}$ g) $\sqrt{x^6}$ h) $\sqrt{x^{14}}$

     i) $\sqrt{25x^2}$ j) $\sqrt{49x^6y^2}$ k) $\sqrt{100x^{34}}$ l) $\sqrt{x^{44}y^{50}}$

Solution

     a) $\sqrt{3^2} = 3$ b) $\sqrt{6^2} = 6$ c) $\sqrt{2^4} = 2$ d) $\sqrt{13^2} = 13$

     e) $\sqrt{x^2} = x$ f) $\sqrt{x^4} = x^2$ g) $\sqrt{x^6} = x^3$ h) $\sqrt{x^{14}} = x^7$

     i) $\sqrt{25x^2} = 5x$ j) $\sqrt{49x^6y^2} = 7x^3y$ k) $\sqrt{100x^{34}} = 10x^{17}$ l) $\sqrt{x^{44}y^{50}} = x^{22}y^{25}$

3. Simplify. Assume that all variables represent positive numbers.

a) $\sqrt{8}$ b) $-\sqrt{18}$ c) $\sqrt{72}$ d) $\sqrt{147}$

e) $\sqrt{x^3}$ f) $\sqrt{x^5}$ g) $\sqrt{x^{13}}$ h) $\sqrt{x^{14}}$

i) $\sqrt{80x^3}$ j) $\sqrt{25x^5y^2}$ k) $\sqrt{128x^6y^8}$ l) $\sqrt{63x^9y^3w^7}$

Solution

a) $\sqrt{8} = \sqrt{4\cdot 2} = \sqrt{4}\cdot\sqrt{2} = 2\sqrt{2}$

b) $-\sqrt{18} = -\sqrt{9\cdot 2} = -\sqrt{9}\cdot\sqrt{2} = -3\sqrt{2}$

c) $\sqrt{72} = \sqrt{36\cdot 2} = \sqrt{36}\cdot\sqrt{2} = 6\sqrt{2}$

d) $\sqrt{147} = \sqrt{49\cdot 3} = \sqrt{49}\cdot\sqrt{3} = 7\sqrt{3}$

e) $\sqrt{x^3} = \sqrt{x^2\cdot x} = \sqrt{x^2}\cdot\sqrt{x} = x\sqrt{x}$

f) $\sqrt{x^5} = \sqrt{x^4\cdot x} = \sqrt{x^4}\cdot\sqrt{x} = x^2\sqrt{x}$

g) $\sqrt{x^{13}} = \sqrt{x^{12}\cdot x} = \sqrt{x^{12}}\cdot\sqrt{x} = x^6\sqrt{x}$

h) $\sqrt{x^{14}} = x^7$

i) $\sqrt{80x^3} = \sqrt{16\cdot 5\cdot x^2\cdot x}$
$= \sqrt{16x^2}\cdot\sqrt{5x} = 4x\sqrt{5x}$

j) $\sqrt{25x^5y^2} = \sqrt{25x^4xy^2} = \sqrt{25x^4y^2}\cdot\sqrt{x}$
$= 5x^2y\sqrt{x}$

k) $\sqrt{128x^6y^8} = \sqrt{64\cdot 2x^6y^8} = \sqrt{64x^6y^8}\cdot\sqrt{2}$
$= 8x^3y^4\sqrt{2}$

l) $\sqrt{63x^9y^3w^7} = \sqrt{9\cdot 7x^8xy^2yw^6w}$
$= \sqrt{9x^8y^2w^6}\cdot\sqrt{7xyw}$
$= 3x^4yw^3\sqrt{7xyw}$

4. Simplify.

a) $\sqrt[3]{24}$ b) $\sqrt[3]{54}$ c) $\sqrt[3]{40x^4y^7}$ d) $\sqrt[3]{-\dfrac{1}{512}}$

Solution

a) $\sqrt[3]{24} = \sqrt[3]{8\cdot 3} = \sqrt[3]{8}\cdot\sqrt[3]{3} = 2\sqrt[3]{3}$

b) $\sqrt[3]{54} = \sqrt[3]{27\cdot 2} = \sqrt[3]{27}\cdot\sqrt[3]{2} = 3\sqrt[3]{2}$

c) $\sqrt[3]{40x^4y^7} = \sqrt[3]{8\cdot 5x^3xy^6y} = \sqrt[3]{8x^3y^6}\cdot\sqrt[3]{5xy}$
$= 2xy^2\sqrt[3]{5xy}$

d) $\sqrt[3]{-\dfrac{1}{512}} = -\dfrac{1}{8}$

14.3 Adding and Subtracting Radicals

Learning Objectives:

1. Add and subtract radicals.
2. Simplify radical sums and differences.
3. Simplify more complicated radical expressions.

Solved Examples:

1. Combine, if possible. Do not use a calculator.

    a) $3\sqrt{7} + 2\sqrt{7} - \sqrt{7}$ b) $4\sqrt{3} - 6\sqrt{2} + 7\sqrt{3} + 3\sqrt{2}$ c) $\sqrt{x} - 3\sqrt{x}$

    d) $-2.3\sqrt{2a} + 4.1\sqrt{2a}$ e) $3\sqrt{w} - 2\sqrt{v} - \sqrt{w}$

    Solution

    a) $3\sqrt{7} + 2\sqrt{7} - \sqrt{7} = 4\sqrt{7}$

    b) $4\sqrt{3} - 6\sqrt{2} + 7\sqrt{3} + 3\sqrt{2}$
    $= \left(4\sqrt{3} + 7\sqrt{3}\right) + \left(-6\sqrt{2} + 3\sqrt{2}\right)$
    $= 11\sqrt{3} - 3\sqrt{2}$

    c) $\sqrt{x} - 3\sqrt{x} = -2\sqrt{x}$

    d) $-2.3\sqrt{2a} + 4.1\sqrt{2a} = 1.8\sqrt{2a}$

    e) $3\sqrt{w} - 2\sqrt{v} - \sqrt{w} = \left(3\sqrt{w} - \sqrt{w}\right) - 2\sqrt{v}$
    $= 2\sqrt{w} - 2\sqrt{v}$

2. Combine, if possible. Do not use a calculator.

    a) $\sqrt{12} + \sqrt{27}$ b) $5\sqrt{7} + 3\sqrt{63}$ c) $\sqrt{50} - 6\sqrt{98} + 8\sqrt{72}$

    d) $\sqrt{36} + 4\sqrt{72} - 3\sqrt{12}$ e) $3\sqrt{48} - 2\sqrt{8} + \sqrt{50}$ f) $9\sqrt{5y} - 2\sqrt{20y}$

    g) $x\sqrt{x} + 3\sqrt{x^3}$ h) $-4\sqrt{27y^3} + 5y\sqrt{12y}$ i) $6x\sqrt{50x} - 4\sqrt{18x^3}$

    Solution

    a) $\sqrt{12} + \sqrt{27} = \sqrt{4 \cdot 3} + \sqrt{9 \cdot 3} = 2\sqrt{3} + 3\sqrt{3}$
    $= 5\sqrt{3}$

    b) $5\sqrt{7} + 3\sqrt{63} = 5\sqrt{7} + 3\sqrt{9 \cdot 7} = 5\sqrt{7} + 3(3)\sqrt{7}$
    $= 5\sqrt{7} + 9\sqrt{7} = 14\sqrt{7}$

    c) $\sqrt{50} - 6\sqrt{98} + 8\sqrt{72}$
    $= \sqrt{25 \cdot 2} - 6\sqrt{49 \cdot 2} + 8\sqrt{36 \cdot 2}$
    $= 5\sqrt{2} - 6(7)\sqrt{2} + 8(6)\sqrt{2}$
    $= 5\sqrt{2} - 42\sqrt{2} + 48\sqrt{2} = 11\sqrt{2}$

    d) $\sqrt{36} + 4\sqrt{72} - 3\sqrt{12} = 6 + 4\sqrt{36 \cdot 2} - 3\sqrt{4 \cdot 3}$
    $= 6 + 4(6)\sqrt{2} - 3(2)\sqrt{3}$
    $= 6 + 24\sqrt{2} - 6\sqrt{3}$

    e) $3\sqrt{48} - 2\sqrt{8} + \sqrt{50}$
    $= 3\sqrt{16 \cdot 3} - 2\sqrt{4 \cdot 2} + \sqrt{25 \cdot 2}$
    $= 3(4)\sqrt{3} - 2(2)\sqrt{2} + 5\sqrt{2}$
    $= 12\sqrt{3} - 4\sqrt{2} + 5\sqrt{2} = 12\sqrt{3} + \sqrt{2}$

    f) $9\sqrt{5y} - 2\sqrt{20y} = 9\sqrt{5y} - 2\sqrt{4 \cdot 5y}$
    $= 9\sqrt{5y} - 2(2)\sqrt{5y}$
    $= 9\sqrt{5y} - 4\sqrt{5y} = 5\sqrt{5y}$

g) $x\sqrt{x} + 3\sqrt{x^3} = x\sqrt{x} + 3\sqrt{x^2 x} = x\sqrt{x} + 3x\sqrt{x}$
$\qquad\qquad = 4x\sqrt{x}$

h) $-4\sqrt{27y^3} + 5y\sqrt{12y}$
$\qquad = -4\sqrt{9 \cdot 3y^2 y} + 5y\sqrt{4 \cdot 3y}$
$\qquad = -4(3y)\sqrt{3y} + 5y(2)\sqrt{3y}$
$\qquad = -12y\sqrt{3y} + 10y\sqrt{3y} = -2y\sqrt{3y}$

i) $6x\sqrt{50x} - 4\sqrt{18x^3} = 6x\sqrt{25 \cdot 2x} - 4\sqrt{9 \cdot 2x^2 x}$
$\qquad\qquad\qquad = 6x(5)\sqrt{2x} - 4(3x)\sqrt{2x}$
$\qquad\qquad\qquad = 30x\sqrt{2x} - 12x\sqrt{2x}$
$\qquad\qquad\qquad = 18x\sqrt{2x}$

3. Solve the following applied problems. Do not use a calculator.

   a) A rectangular plot of land has a length of $\left(24\sqrt{2} + 6\sqrt{3}\right)$ yards and a width of $\left(2\sqrt{2} + 6\sqrt{3}\right)$ yards. Find the perimeter of the plot of land.

   b) To suspend a prop from the ceiling on the set of the school play, Ernie needs 10 pieces of wire. Each of five of the pieces of wire needs a length of $\sqrt{54}$ meters, while each of the other five needs a length of $\sqrt{27}$ meters. Find the total length of wire required.

Solution

a) $P = 2l + 2w$
$\quad = 2\left(24\sqrt{2} + 6\sqrt{3}\right) + 2\left(2\sqrt{2} + 6\sqrt{3}\right)$
$\quad = 48\sqrt{2} + 12\sqrt{3} + 4\sqrt{2} + 12\sqrt{3}$
$\quad = 52\sqrt{2} + 24\sqrt{3}$

The perimeter is $52\sqrt{2} + 24\sqrt{3}$ yards.

b) $5\sqrt{54} + 5\sqrt{27} = 5\sqrt{9 \cdot 6} + 5\sqrt{9 \cdot 3}$
$\qquad\qquad\qquad = 5(3)\sqrt{6} + 5(3)\sqrt{3}$
$\qquad\qquad\qquad = 15\sqrt{6} + 15\sqrt{3}$

Ernie needs $15\sqrt{6} + 15\sqrt{3}$ meters of wire.

14.4 Rationalizing the Denominator

Learning Objectives:

1. Rationalize denominators with square roots.
2. Write radicals in simplified form.
3. Rationalize denominators with cube roots.
4. Solve applications.

Solved Examples:

1. Simplify. Do not use a calculator.

   a) $\dfrac{\sqrt{20}}{\sqrt{5}}$
      b) $\dfrac{\sqrt{6}}{\sqrt{54}}$
      c) $\dfrac{\sqrt{12x^3}}{\sqrt{3x^2}}$
      d) $\dfrac{\sqrt{27x^7}}{\sqrt{3x}}$

Solution

a) $\dfrac{\sqrt{20}}{\sqrt{5}} = \sqrt{\dfrac{20}{5}}$
$= \sqrt{4} = 2$

b) $\dfrac{\sqrt{6}}{\sqrt{54}} = \sqrt{\dfrac{6}{54}} = \sqrt{\dfrac{1}{9}} = \dfrac{1}{3}$

c) $\dfrac{\sqrt{12x^3}}{\sqrt{3x^2}} = \sqrt{\dfrac{12x^3}{3x^2}} = \sqrt{4x} = 2\sqrt{x}$

d) $\dfrac{\sqrt{27x^7}}{\sqrt{3x}} = \sqrt{\dfrac{27x^7}{3x}} = \sqrt{9x^6} = 3x^3$

2. Rationalize the denominator. Simplify your answer.

   a) $\dfrac{3}{\sqrt{11}}$
      b) $\dfrac{\sqrt{6}}{\sqrt{x}}$
      c) $\sqrt{\dfrac{13}{17}}$

   d) $\dfrac{x\sqrt{x}}{\sqrt{5}}$
      e) $\dfrac{3x}{\sqrt{x^5}}$
      f) $\dfrac{\sqrt{32}}{\sqrt{2x^3}}$

Solution

a) $\dfrac{3}{\sqrt{11}} = \dfrac{3}{\sqrt{11}} \cdot \dfrac{\sqrt{11}}{\sqrt{11}} = \dfrac{3\sqrt{11}}{11}$

b) $\dfrac{\sqrt{6}}{\sqrt{x}} = \dfrac{\sqrt{6}}{\sqrt{x}} \cdot \dfrac{\sqrt{x}}{\sqrt{x}} = \dfrac{\sqrt{6x}}{x}$

c) $\sqrt{\dfrac{13}{17}} = \dfrac{\sqrt{13}}{\sqrt{17}} = \dfrac{\sqrt{13}}{\sqrt{17}} \cdot \dfrac{\sqrt{17}}{\sqrt{17}} = \dfrac{\sqrt{221}}{17}$

d) $\dfrac{x\sqrt{x}}{\sqrt{5}} = \dfrac{x\sqrt{x}}{\sqrt{5}} \cdot \dfrac{\sqrt{5}}{\sqrt{5}} = \dfrac{x\sqrt{5x}}{5}$

e) $\dfrac{3x}{\sqrt{x^5}} = \dfrac{3x}{\sqrt{x^4 x}} = \dfrac{3x}{x^2\sqrt{x}} = \dfrac{3x}{x^2\sqrt{x}} \cdot \dfrac{\sqrt{x}}{\sqrt{x}}$
$= \dfrac{3x}{x^2 \cdot x} = \dfrac{3}{x^2}$

f) $\dfrac{\sqrt{32}}{\sqrt{2x^3}} = \dfrac{\sqrt{16}}{\sqrt{x^2 x}} = \dfrac{4}{x\sqrt{x}} \cdot \dfrac{\sqrt{x}}{\sqrt{x}} = \dfrac{4\sqrt{x}}{x \cdot x} = \dfrac{4\sqrt{x}}{x^2}$

3. Rationalize the denominator. Simplify your answer.

a) $\dfrac{4}{\sqrt[3]{2}}$

b) $\sqrt[3]{\dfrac{2x}{9y}}$

c) $\sqrt[3]{\dfrac{8x^3y^6}{9z}}$

Solution

a) $\dfrac{4}{\sqrt[3]{2}} = \dfrac{4}{\sqrt[3]{2}} \cdot \dfrac{\sqrt[3]{2^2}}{\sqrt[3]{2^2}} = \dfrac{4\sqrt[3]{2^2}}{\sqrt[3]{2^3}} = \dfrac{4\sqrt[3]{4}}{2} = 2\sqrt[3]{4}$

b) $\sqrt[3]{\dfrac{2x}{9y}} = \dfrac{\sqrt[3]{2x}}{\sqrt[3]{9y}} \cdot \dfrac{\sqrt[3]{(9y)^2}}{\sqrt[3]{(9y)^2}} = \dfrac{\sqrt[3]{2x} \cdot \sqrt[3]{81y^2}}{\sqrt[3]{(9y)^3}}$

$\phantom{\sqrt[3]{\dfrac{2x}{9y}}} = \dfrac{\sqrt[3]{2x} \cdot \sqrt[3]{27 \cdot 3y^2}}{9y} = \dfrac{\sqrt[3]{2x} \cdot 3\sqrt[3]{3y^2}}{9y}$

$\phantom{\sqrt[3]{\dfrac{2x}{9y}}} = \dfrac{3\sqrt[3]{2x \cdot 3y^2}}{9y} = \dfrac{\sqrt[3]{6xy^2}}{y}$

c) $\sqrt[3]{\dfrac{8x^3y^6}{9z}} = \dfrac{\sqrt[3]{8x^3y^6}}{\sqrt[3]{9z}} = \dfrac{2xy^2}{\sqrt[3]{9z}} \cdot \dfrac{\sqrt[3]{(9z)^2}}{\sqrt[3]{(9z)^2}} = \dfrac{2xy^2\sqrt[3]{81z^2}}{\sqrt[3]{(9z)^3}} = \dfrac{2xy^2\sqrt[3]{27 \cdot 3z^2}}{9z}$

$\phantom{\sqrt[3]{\dfrac{8x^3y^6}{9z}}} = \dfrac{2xy^2(3)\sqrt[3]{3z^2}}{9z} = \dfrac{2xy^2\sqrt[3]{3z^2}}{3z}$

14.5 More Simplifying and Operations with Radicals

Learning Objectives:

1. Simplify expressions involving radicals.
2. Use conjugates to rationalize denominators of radical expressions.
3. Write radical expressions with quotients in lowest terms.

Solved Examples:

1. Multiply. Be sure to simplify any radicals in your answer. Do not use a calculator.

    a) $\sqrt{3}\sqrt{7}$ b) $\sqrt{3}\sqrt{33}$ c) $\sqrt{6}\sqrt{18}$

    d) $\sqrt{6x}\sqrt{60x}$ e) $\left(2\sqrt{10x}\right)\left(5\sqrt{5x}\right)$ f) $\left(-6\sqrt{ab}\right)\left(4\sqrt{b}\right)$

Solution

a) $\sqrt{3}\sqrt{7} = \sqrt{3\cdot7} = \sqrt{21}$ b) $\sqrt{3}\sqrt{33} = \sqrt{3\cdot33} = \sqrt{99} = \sqrt{9\cdot11} = 3\sqrt{11}$

c) $\sqrt{6}\sqrt{18} = \sqrt{6\cdot18} = \sqrt{6\cdot6\cdot3} = 6\sqrt{3}$ d) $\sqrt{6x}\sqrt{60x} = \sqrt{6x\cdot60x} = \sqrt{6x\cdot6\cdot10x}$
$$= \sqrt{6x\cdot6x\cdot10} = 6x\sqrt{10}$$

e) $\left(2\sqrt{10x}\right)\left(5\sqrt{5x}\right) = 2(5)\sqrt{10x\cdot5x}$ f) $\left(-6\sqrt{ab}\right)\left(4\sqrt{b}\right) = (-6)(4)\sqrt{ab\cdot b} = -24b\sqrt{a}$
$$= 10\sqrt{5x\cdot2\cdot5x}$$
$$= 10(5x)\sqrt{2} = 50x\sqrt{2}$$

2. Multiply. Be sure to simplify any radicals in your answer. Do not use a calculator.

    a) $\sqrt{5}\left(\sqrt{3}+\sqrt{7}\right)$ b) $\sqrt{6}\left(4\sqrt{12}-3\sqrt{3}\right)$ c) $-5\sqrt{b}\left(6\sqrt{a}+5\sqrt{b}\right)$

    d) $\sqrt{10}\left(2\sqrt{5}-2\sqrt{10}+6\sqrt{2}\right)$ e) $2\sqrt{x}\left(\sqrt{y}+3\sqrt{xy}-5\sqrt{x}\right)$

Solution

a) $\sqrt{5}\left(\sqrt{3}+\sqrt{7}\right) = 5\sqrt{3}+5\sqrt{7}$ b) $\sqrt{6}\left(4\sqrt{12}-3\sqrt{3}\right) = 4\sqrt{12\cdot6}-3\sqrt{3\cdot6}$
$$= 4\sqrt{2\cdot6\cdot6}-3\sqrt{3\cdot3\cdot2}$$
$$= 4(6)\sqrt{2}-3(3)\sqrt{2}$$
$$= 24\sqrt{2}-9\sqrt{2} = 15\sqrt{2}$$

c) $-5\sqrt{b}\left(6\sqrt{a}+5\sqrt{b}\right)$ d) $\sqrt{10}\left(2\sqrt{5}-2\sqrt{10}+6\sqrt{2}\right)$
$$= (-5)(6)\sqrt{ab}-5(5)\sqrt{b\cdot b} \qquad\qquad = 2\sqrt{5\cdot10}-2\sqrt{10\cdot10}+6\sqrt{2\cdot10}$$
$$= -30\sqrt{ab}-25b \qquad\qquad\qquad\quad = 2\sqrt{5\cdot5\cdot2}-2\sqrt{10\cdot10}+6\sqrt{2\cdot2\cdot5}$$
$$= 2(5)\sqrt{2}-2(10)+6(2)\sqrt{5}$$
$$= 10\sqrt{2}-20+12\sqrt{5}$$

e) $2\sqrt{x}\left(\sqrt{y}+3\sqrt{xy}-5\sqrt{x}\right)$
$$= 2\sqrt{xy}+3(2)\sqrt{x\cdot xy}-5(2)\sqrt{x\cdot x}$$
$$= 2\sqrt{xy}+6x\sqrt{y}-10x$$

3. Multiply. Be sure to simplify any radicals in your answer. Do not use a calculator.

   a) $\left(2\sqrt{5}+\sqrt{3}\right)\left(\sqrt{5}+\sqrt{3}\right)$ b) $\left(2\sqrt{2}+5\sqrt{10}\right)\left(2\sqrt{2}-\sqrt{10}\right)$ c) $\left(3\sqrt{5}-2\right)^2$

   d) $\left(4\sqrt{3}+6\sqrt{5}\right)^2$ e) $\left(2a\sqrt{3}-3\sqrt{2}\right)\left(2a\sqrt{3}+3\sqrt{2}\right)$

Solution

   a) Use FOIL: $\left(2\sqrt{5}+\sqrt{3}\right)\left(\sqrt{5}+\sqrt{3}\right)=2\sqrt{5\cdot5}+2\sqrt{5\cdot3}+\sqrt{3\cdot5}+\sqrt{3\cdot3}=10+3\sqrt{15}+3$

   b) Use FOIL:

   $$\left(2\sqrt{2}+5\sqrt{10}\right)\left(2\sqrt{2}-\sqrt{10}\right)=2(2)\sqrt{2\cdot2}-2\sqrt{2\cdot10}+5(2)\sqrt{2\cdot10}-5\sqrt{10\cdot10}$$
   $$=4(2)+8\sqrt{20}-5(10)=8+8\sqrt{4\cdot5}-50=-42+8(2)\sqrt{5}=-42+16\sqrt{5}$$

   c) Use $(a-b)^2=a^2-2ab+b^2$

   $$\left(3\sqrt{5}-2\right)^2=\left(3\sqrt{5}\right)^2-2\left(3\sqrt{5}\right)(2)+2^2=9(5)-12\sqrt{5}+4=49-12\sqrt{5}$$

   d) Use $(a+b)^2=a^2+2ab+b^2$

   $$\left(4\sqrt{3}+6\sqrt{5}\right)^2=\left(4\sqrt{3}\right)^2+2\left(4\sqrt{3}\right)\left(6\sqrt{5}\right)+\left(6\sqrt{5}\right)^2=16(3)+48\sqrt{15}+36(5)$$
   $$=48+48\sqrt{15}+180=228+48\sqrt{15}$$

   e) Use $(a+b)(a-b)=a^2-b^2$

   $$\left(2a\sqrt{3}-3\sqrt{2}\right)\left(2a\sqrt{3}+3\sqrt{2}\right)=\left(2a\sqrt{3}\right)^2-\left(3\sqrt{2}\right)^2=4a^2(3)-9(2)=12a^2-18$$

4. Rationalize the denominator. Simplify your answer.

   a) $\dfrac{3}{\sqrt{3}+1}$ b) $\dfrac{7}{\sqrt{13}-\sqrt{6}}$ c) $\dfrac{\sqrt{8}}{\sqrt{5}+\sqrt{8}}$

   d) $\dfrac{3x}{2\sqrt{5}+2\sqrt{6}}$ e) $\dfrac{\sqrt{x}}{\sqrt{3}+\sqrt{2}}$ f) $\dfrac{5\sqrt{3}+2\sqrt{5}}{\sqrt{5}-\sqrt{3}}$

Solution

   a) $\dfrac{3}{\sqrt{3}+1}=\dfrac{3}{\sqrt{3}+1}\cdot\dfrac{\sqrt{3}-1}{\sqrt{3}-1}=\dfrac{3\left(\sqrt{3}-1\right)}{\left(\sqrt{3}\right)^2-1^2}=\dfrac{3\sqrt{3}-3}{3-1}=\dfrac{3\sqrt{3}-3}{2}$

   b) $\dfrac{7}{\sqrt{13}-\sqrt{6}}=\dfrac{7}{\sqrt{13}-\sqrt{6}}\cdot\dfrac{\sqrt{13}+\sqrt{6}}{\sqrt{13}+\sqrt{6}}=\dfrac{7\left(\sqrt{13}+\sqrt{6}\right)}{\left(\sqrt{13}\right)^2-\left(\sqrt{6}\right)^2}=\dfrac{7\left(\sqrt{13}+\sqrt{6}\right)}{13-6}=\dfrac{7\left(\sqrt{13}+\sqrt{6}\right)}{7}=\sqrt{13}+\sqrt{6}$

   c) $\dfrac{\sqrt{8}}{\sqrt{5}+\sqrt{8}}=\dfrac{\sqrt{8}}{\sqrt{5}+\sqrt{8}}\cdot\dfrac{\sqrt{5}-\sqrt{8}}{\sqrt{5}-\sqrt{8}}=\dfrac{\sqrt{8}\left(\sqrt{5}-\sqrt{8}\right)}{\left(\sqrt{5}\right)^2-\left(\sqrt{8}\right)^2}=\dfrac{\sqrt{8\cdot5}-\sqrt{8\cdot8}}{5-8}=\dfrac{\sqrt{40}-8}{-3}$

   $$=\dfrac{\sqrt{4\cdot10}-8}{-3}=-\dfrac{2\sqrt{10}-8}{3}=\dfrac{8-2\sqrt{10}}{3}$$

d) $\dfrac{3x}{2\sqrt{5}+2\sqrt{6}} = \dfrac{3x}{2\sqrt{5}+2\sqrt{6}} \cdot \dfrac{2\sqrt{5}-2\sqrt{6}}{2\sqrt{5}-2\sqrt{6}} = \dfrac{3x\left(2\sqrt{5}-2\sqrt{6}\right)}{\left(2\sqrt{5}\right)^2 - \left(2\sqrt{6}\right)^2} = \dfrac{2(3x)\left(\sqrt{5}-\sqrt{6}\right)}{4\cdot 5 - 4\cdot 6}$

$= \dfrac{2\left(3x\sqrt{5}-3x\sqrt{6}\right)}{-4} = -\dfrac{3x\sqrt{5}-3x\sqrt{6}}{2} = \dfrac{3x\sqrt{6}-3x\sqrt{5}}{2}$

e) $\dfrac{\sqrt{x}}{\sqrt{3}+\sqrt{2}} = \dfrac{\sqrt{x}}{\sqrt{3}+\sqrt{2}} \cdot \dfrac{\sqrt{3}-\sqrt{2}}{\sqrt{3}-\sqrt{2}} = \dfrac{\sqrt{x}\left(\sqrt{3}-\sqrt{2}\right)}{\left(\sqrt{3}\right)^2 - \left(\sqrt{2}\right)^2} = \dfrac{\sqrt{3x}-\sqrt{2x}}{3-2} = \sqrt{3x}-\sqrt{2x}$

f) $\dfrac{5\sqrt{3}+2\sqrt{5}}{\sqrt{5}-\sqrt{3}} = \dfrac{5\sqrt{3}+2\sqrt{5}}{\sqrt{5}-\sqrt{3}} \cdot \dfrac{\sqrt{5}+\sqrt{3}}{\sqrt{5}+\sqrt{3}} = \dfrac{5\sqrt{3}\sqrt{5}+5\sqrt{3}\sqrt{3}+2\sqrt{5}\sqrt{5}+2\sqrt{5}\sqrt{3}}{\left(\sqrt{5}\right)^2 - \left(\sqrt{3}\right)^2}$

$= \dfrac{5\sqrt{15}+5(3)+2(5)+2\sqrt{15}}{5-3} = \dfrac{25+7\sqrt{15}}{2}$

14.6 Solving Equations with Radicals

Learning Objectives:

1. Solve radical equations.
2. Solve equations by squaring a binomial.
3. Solve applications.

Solved Examples:

1. A right triangle has legs a and b and hypotenuse c. Find the length of the missing side. Leave any irrational answers in radical form.

   a) $a = 2, b = 3$. Find c.

   b) $a = \sqrt{3}$, $b = \sqrt{5}$. Find c.

   c) $a = 9, c = 15$. Find b.

   d) $b = \sqrt{3}$, $c = \sqrt{8}$. Find a.

   Solution

   Use the Pythagorean formula $a^2 + b^2 = c^2$.

   a) $2^2 + 3^2 = c^2$
   $4 + 9 = c^2$
   $13 = c^2$
   $\sqrt{13} = c$

   b) $\left(\sqrt{3}\right)^2 + \left(\sqrt{5}\right)^2 = c^2$
   $3 + 5 = c^2$
   $8 = c^2$
   $8 = 2\sqrt{2} = c$

   c) $9^2 + b^2 = 15^2$
   $81 + b^2 = 225$
   $b^2 = 144$
   $b = \sqrt{144} = 12$

   d) $a^2 + \left(\sqrt{3}\right)^2 = \left(\sqrt{8}\right)^2$
   $a^2 + 3 = 8$
   $a^2 = 5$
   $a = \sqrt{5}$

2. Solve for the variable. Check your solutions.

   a) $\sqrt{x} = 7$

   b) $\sqrt{y+3} = 6$

   c) $\sqrt{10x-9} = 9$

   d) $\sqrt{5x} - 3 = 7$

   e) $\sqrt{8x-2} = \sqrt{x+4}$

   f) $\sqrt{5x-6} = x$

   g) $\sqrt{y+9} = y+3$

   h) $\sqrt{12y-1} - 5y = y$

   i) $\sqrt{x} + 3 = \sqrt{x+21}$

   Solution

   a) $\sqrt{x} = 7$ Check:
   $\left(\sqrt{x}\right)^2 = 7^2$ $\sqrt{x} = 7$
   $x = 49$ $\sqrt{49} \overset{?}{=} 7$
      $7 = 7$ ✓
   Solution set: $\{49\}$

   b) $\sqrt{y+3} = 6$ Check:
   $\left(\sqrt{y+3}\right)^2 = 6^2$ $\sqrt{y+3} = 6$
   $y + 3 = 36$ $\sqrt{33+3} \overset{?}{=} 6$
   $y = 33$ $\sqrt{36} \overset{?}{=} 6$
      $6 = 6$ ✓
   Solution set: $\{33\}$

c) $\sqrt{10x-9}=9$ Check:

$\left(\sqrt{10x-9}\right)^2 = 9^2$ $\sqrt{10x-9}=9$

$10x-9=81$ $\sqrt{10(9)-9}\overset{?}{=}9$

$10x=90$

$x=9$ $\sqrt{81}\overset{?}{=}9$

$9=9$ ✓

Solution set: {9}

d) $\sqrt{5x}-3=7$ Check:

$\sqrt{5x}=10$ $\sqrt{5x}-3=7$

$\left(\sqrt{5x}\right)^2 = 10^2$ $\sqrt{5(20)}-3\overset{?}{=}7$

$5x=100$

$x=20$ $\sqrt{100}-3\overset{?}{=}7$

$10-3\overset{?}{=}7$

$7=7$ ✓

Solution set: {20}

e) $\sqrt{8x-2}=\sqrt{x+4}$

$\left(\sqrt{8x-2}\right)^2 = \left(\sqrt{x+4}\right)^2$

$8x-2=x+4$

$8x=x+6$

$7x=6$

$x=\dfrac{6}{7}$

Check:

$\sqrt{8x-2}=\sqrt{x+4}$

$\sqrt{8\left(\dfrac{6}{7}\right)-2}\overset{?}{=}\sqrt{\dfrac{6}{7}+4}$

$\sqrt{\dfrac{48}{7}-2}\overset{?}{=}\sqrt{\dfrac{34}{7}}$

$\sqrt{\dfrac{34}{7}}=\sqrt{\dfrac{34}{7}}$ ✓

Solution set: $\left\{\dfrac{6}{7}\right\}$

f) $\sqrt{5x-6}=x$

$\left(\sqrt{5x-6}\right)^2 = x^2$

$5x-6=x^2$

$0=x^2-5x-6$

$0=(x-2)(x-3)$

$x-2=0 \mid x-3=0$

$x=2 \mid x=3$

Check:

$\sqrt{5x-6}=x$ $\sqrt{5x-6}=x$

$\sqrt{5(2)-6}\overset{?}{=}2$ $\sqrt{5(3)-6}\overset{?}{=}3$

$\sqrt{10-6}\overset{?}{=}2$ $\sqrt{15-6}\overset{?}{=}3$

$\sqrt{4}\overset{?}{=}2$ $\sqrt{9}\overset{?}{=}3$

$2=2$ ✓ $3=3$ ✓

Solution set: {2, 3}

g) $\sqrt{y+9}=y+3$

$\left(\sqrt{y+9}\right)^2 = (y+3)^2$

$y+9=y^2+6y+9$

$0=y^2+5y$

$0=y(y+5)$

$y=0 \mid y+5=0$

$\mid y=-5$

Check:

$\sqrt{y+9}=y+3$ $\sqrt{y+9}=y+3$

$\sqrt{0+9}\overset{?}{=}0+3$ $\sqrt{-5+9}\overset{?}{=}-5+3$

$\sqrt{9}\overset{?}{=}3$ $\sqrt{4}\overset{?}{=}-2$

$3=3$ ✓ $2\neq 2$

Solution set: {0}

h) $\sqrt{12y-1} - 5y = y$

$\sqrt{12y-1} = 6y$

$\left(\sqrt{12y-1}\right)^2 = (6y)^2$

$12y - 1 = 36y^2$

$0 = 36y^2 - 12y + 1$

$0 = (6y-1)^2$

$0 = 6y - 1$

$1 = 6y$

$\dfrac{1}{6} = y$

Check:

$\sqrt{12y-1} - 5y = y$

$\sqrt{12\left(\dfrac{1}{6}\right)-1} - 5\left(\dfrac{1}{6}\right) \overset{?}{=} \dfrac{1}{6}$

$\sqrt{2-1} - \dfrac{5}{6} \overset{?}{=} \dfrac{1}{6}$

$1 - \dfrac{5}{6} \overset{?}{=} \dfrac{1}{6}$

$\dfrac{1}{6} = \dfrac{1}{6}$ ✓

Solution set: $\left\{\dfrac{1}{6}\right\}$

i) $\sqrt{x} + 3 = \sqrt{x+21}$

$\left(\sqrt{x}+3\right)^2 = \left(\sqrt{x+21}\right)^2$

$x + 6\sqrt{x} + 9 = x + 21$

$6\sqrt{x} = 12$

$\sqrt{x} = 2$

$x = 4$

Solution set: {4}

Check:

$\sqrt{x} + 3 = \sqrt{x+21}$

$\sqrt{4} + 3 \overset{?}{=} \sqrt{4+21}$

$2 + 3 \overset{?}{=} \sqrt{25}$

$5 = 5$ ✓

3. Three times the square root of 7 equals the square root of the sum of some number and 5. What is the number?

Solution

Let x = the number. The equation is $3\sqrt{7} = \sqrt{x+5}$.

$3\sqrt{7} = \sqrt{x+5}$

$\left(3\sqrt{7}\right)^2 = \left(\sqrt{x+5}\right)^2$

$63 = x + 5$

$58 = x$

Check:

$3\sqrt{7} = \sqrt{58+5}$

$3\sqrt{7} \overset{?}{=} \sqrt{63}$

$3\sqrt{7} \overset{?}{=} \sqrt{9\cdot7}$

$3\sqrt{7} = 3\sqrt{7}$ ✓

The number is 58.

15.1 Solving Equations by the Square Root Property

Learning Objectives:

1. Solve equations of the form $x^2 = k$.
2. Solve equations of the form $(a+b)^2 = k$.
3. Use formulas involving squared variables.

Solved Examples:

1. Solve using the square root property.

   a) $x^2 = 9$

   b) $x^2 = 20$

   c) $x^2 - 75 = 0$

   d) $4x^2 = 16$

   e) $3x^2 + 4 = 64$

   f) $(x-5)^2 = 25$

   g) $(x+3)^2 = 11$

   h) $x^2 + 36 = 0$

   i) $(5x-3)^2 = 48$

   j) $(4x+1)^2 = 36$

   k) $k^2 = 22.09$

   l) $(m+3)^2 + 25 = 0$

   m) $\left(y - \dfrac{2}{3}\right)^2 = 4$

   n) $(2x-2)^2 = 25$

   Solution

   Be sure to check each solution in the original equation. We show the checks only for those examples with extraneous solutions.

   a) $x^2 = 9$
   $x = \pm\sqrt{9} = \pm 3$
   Solution set: $\{\pm 3\}$

   b) $x^2 = 20$
   $x = \pm\sqrt{20} = \pm 2\sqrt{5}$
   Solution set: $\left\{\pm 2\sqrt{5}\right\}$

   c) $x^2 - 75 = 0$
   $x^2 = 75$
   $x = \pm\sqrt{75} = \pm 5\sqrt{3}$
   Solution set: $\left\{\pm 5\sqrt{3}\right\}$

   d) $4x^2 = 16$
   $x^2 = 4$
   $x = \pm\sqrt{4} = \pm 2$
   Solution set: $\{\pm 2\}$

   e) $3x^2 + 4 = 64$
   $3x^2 = 60$
   $x^2 = 20$
   $x = \pm\sqrt{20} = \pm 2\sqrt{5}$
   Solution set: $\left\{\pm 2\sqrt{5}\right\}$

   f) $(x-5)^2 = 25$
   $\sqrt{(x-5)^2} = \pm\sqrt{25}$
   $x - 5 = \pm 5$
   $x - 5 = 5 \mid x - 5 = -5$
   $x = 10 \mid x = 0$
   Solution set: $\{0, 10\}$

   g) $(x+3)^2 = 11$
   $\sqrt{(x+3)^2} = \pm\sqrt{11}$
   $x + 3 = \pm\sqrt{11}$
   $x = 3 \pm \sqrt{11}$
   Solution set: $\left\{3 - \sqrt{11}, 3 + \sqrt{11}\right\}$

   h) $x^2 + 36 = 0$
   $x^2 = -36$
   Because -36 is a negative number and because the square of a real number cannot be negative, there is no real number solution of this equation. Solution set: $\varnothing$

i) $(5x-3)^2 = 48$

$\sqrt{(5x-3)^2} = \pm\sqrt{48}$

$5x-3 = \pm4\sqrt{3}$

$5x-3 = -4\sqrt{3} \quad | \quad 5x-3 = 4\sqrt{3}$

$5x = 3-4\sqrt{3} \qquad 5x = 3+4\sqrt{3}$

$x = \dfrac{3-4\sqrt{3}}{5} \quad | \quad x = \dfrac{3+4\sqrt{3}}{5}$

Solution set: $\left\{ \dfrac{3-4\sqrt{3}}{5}, \dfrac{3+4\sqrt{3}}{5} \right\}$

j) $(4x+1)^2 = 36$

$\sqrt{(4x+1)^2} = \pm\sqrt{36}$

$4x+1 = \pm6$

$4x+1 = -6 \quad | \quad 4x+1 = 6$

$4x = -7 \qquad 4x = 5$

$x = -\dfrac{7}{4} \quad | \quad x = \dfrac{5}{4}$

Solution set: $\left\{ -\dfrac{7}{4}, \dfrac{5}{4} \right\}$

k) $k^2 = 22.09$

$k = \pm\sqrt{22.09} = \pm4.7$

Solution set: $\{\pm4.7\}$

l) $(m+3)^2 + 25 = 0$

$(m+3)^2 = -25$

Because -25 is a negative number and because the square of a real number cannot be negative, there is no real number solution of this equation. Solution set: $\varnothing$

m) $\left(y - \dfrac{2}{3}\right)^2 = 4$

$\sqrt{\left(y - \dfrac{2}{3}\right)^2} = \pm\sqrt{4}$

$y - \dfrac{2}{3} = \pm2$

$y = \dfrac{2}{3} \pm 2 = -\dfrac{4}{3}, \dfrac{8}{3}$

Solution set: $\left\{ -\dfrac{4}{3}, \dfrac{8}{3} \right\}$

n) $(2x-2)^2 = 25$

$\sqrt{(2x-2)^2} = \pm\sqrt{25}$

$2x-2 = \pm5$

$2x-2 = -5 \quad | \quad 2x-2 = 5$

$2x = -3 \qquad 2x = 7$

$x = -\dfrac{3}{2} \quad | \quad x = \dfrac{7}{2}$

Solution set: $\left\{ -\dfrac{3}{2}, \dfrac{7}{2} \right\}$

2. The value, P, of an investment of A dollars after 2 years is given by the formula $P = A(1+r)^2$, where r is the annual interest rate earned by the investment. If an initial investment of \$3500 grew to a value of \$4082.40 in 2 years, what was the annual percentage rate?

Solution

$P = A(1+r)^2$

$4082.40 = 3500(1+r)^2$

$1.664 = (1+r)^2$

$1.08 = 1+r \qquad$ Disregard the negative root since rates cannot be negative.

$0.08 = 8\% = r$

The rate was 8%

15.2 Solving Quadratic Equations by Completing the Square

Learning Objectives:

1. Complete the square.
2. Solve by completing the square when the coefficient of the squared term is 1.
3. Solve by completing the square when the coefficient of the squared term is not 1.
4. Simplify before solving.
5. Solve applied problems.

Solved Examples:

1. Solve by completing the square.

   a) $x^2 + 4x = -3$ b) $x^2 - 2x = 35$ c) $x^2 + 11x + 30 = 0$ d) $x^2 - 16x = 0$

   e) $x^2 + 18x + 67 = 0$ f) $6x^2 - 36x = 0$ g) $2x^2 - 5x = 3$ h) $2x^2 + 11x = -12$

   i) $2x^2 + 5x - 3 = 0$ j) $q^2 + 3q - 1 = 0$ k) $x^2 + 5x + 7 = 0$

   l) $(d-1)(d-2) = 3d - 5$ m) $4m + 2 = (m-1)(m+3)$

Solution

To solve a quadratic equation by completing the square, follow these steps.

Step 1: Be sure the second-degree term has coefficient 1. If the coefficient of the second-degree term is 1, proceed to Step 2. If it is not 1, but some other nonzero number a, divide each side of the equation by a.

Step 2: Write in correct form. Make sure that all terms with variables are on one side of the equals sign and that all constant terms are on the other side.

Step 3: Complete the square. Take half the coefficient of the first degree term, and square it. Add the square to each side of the equation. Factor the variable side, and simplify on the other side.

Step 4: Solve the equation by using the square root property.

Be sure to check each solution in the original equation. We show the checks only for those examples with extraneous solutions.

a)
$$x^2 + 4x = -3$$
$$x^2 + 4x + \left(\frac{4}{2}\right)^2 = -3 + \left(\frac{4}{2}\right)^2$$
$$x^2 + 4x + 4 = -3 + 4$$
$$x^2 + 4x + 4 = 1$$
$$(x + 2)^2 = 1$$
$$x + 2 = \pm 1$$
$$x = -2 \pm 1 = -3, -1$$
Solution set: $\{-3, -1\}$

b)
$$x^2 - 2x = 35$$
$$x^2 - 2x + \left(\frac{2}{2}\right)^2 = 35 + \left(\frac{2}{2}\right)^2$$
$$x^2 - 2x + 1 = 36$$
$$(x - 1)^2 = 36$$
$$x - 1 = \pm 6$$
$$x = 1 \pm 6 = -5, 7$$
Solution set: $\{-5, 7\}$

c) $\quad x^2 + 11x + 30 = 0$

$$x^2 + 11x = -30$$

$$x^2 + 11x + \left(\frac{11}{2}\right)^2 = -30 + \left(\frac{11}{2}\right)^2$$

$$\left(x + \frac{11}{2}\right)^2 = \frac{1}{4}$$

$$x + \frac{11}{2} = \pm\sqrt{\frac{1}{4}} = \pm\frac{1}{2}$$

$$x = -\frac{11}{2} \pm \frac{1}{2} = -6, -5$$

Solution set: $\{-6, -5\}$

d) $\quad x^2 - 16x = 0$

$$x^2 - 16x + \left(\frac{16}{2}\right)^2 = 0 + \left(\frac{16}{2}\right)^2$$

$$x^2 - 16x + 64 = 64$$

$$(x - 8)^2 = 64$$

$$\sqrt{(x-8)^2} = \pm\sqrt{64}$$

$$x - 8 = \pm 8$$

$$x = 8 \pm 8 = 0, 16$$

Solution set: $\{0, 16\}$

e) $\quad x^2 + 18x + 67 = 0$

$$x^2 + 18x = -67$$

$$x^2 + 18x + \left(\frac{18}{2}\right)^2 = -67 + \left(\frac{18}{2}\right)^2$$

$$x^2 + 18x + 81 = -67 + 81$$

$$(x + 9)^2 = 14$$

$$\sqrt{(x+9)^2} = \pm\sqrt{14}$$

$$x + 9 = \pm\sqrt{14}$$

$$x = -9 \pm \sqrt{14}$$

Solution set: $\left\{-9 - \sqrt{14}, -9 + \sqrt{14}\right\}$

f) $\quad 6x^2 - 36x = 0$

$$x^2 - 6x = 0 \qquad \text{Divide by 6}$$

$$x^2 - 6x + \left(\frac{6}{2}\right)^2 = 0 + \left(\frac{6}{2}\right)^2$$

$$x^2 - 6x + 9 = 9$$

$$(x - 3)^2 = 9$$

$$\sqrt{(x-3)^2} = \pm\sqrt{9}$$

$$x - 3 = \pm 3$$

$$x = 3 \pm 3 = 0, 6$$

Solution set: $\{0, 6\}$

g) $\quad 2x^2 - 5x = 3$

$$x^2 - \frac{5}{2}x = \frac{3}{2} \qquad \text{Divide by 2.}$$

$$x^2 - \frac{5}{2}x + \left[\frac{1}{2}\left(\frac{5}{2}\right)\right]^2 = \frac{3}{2} + \left[\frac{1}{2}\left(\frac{5}{2}\right)\right]^2$$

$$x^2 - \frac{5}{2}x + \frac{25}{16} = \frac{49}{16}$$

$$\left(x - \frac{5}{4}\right)^2 = \frac{49}{16}$$

$$\sqrt{\left(x - \frac{5}{4}\right)^2} = \pm\sqrt{\frac{49}{16}}$$

$$x - \frac{5}{4} = \pm\frac{7}{4}$$

$$x = \frac{5}{4} \pm \frac{7}{4} = -\frac{1}{2}, 3$$

Solution set: $\left\{-\frac{1}{2}, 3\right\}$

h) $\quad 2x^2 + 11x = -12$

$$x^2 + \frac{11}{2}x = -6 \qquad \text{Divide by 2.}$$

$$x^2 + \frac{11}{2}x + \left[\frac{1}{2}\left(\frac{11}{2}\right)\right]^2 = -6 + \left[\frac{1}{2}\left(\frac{11}{2}\right)\right]^2$$

$$x^2 + \frac{11}{2}x + \frac{121}{16} = -6 + \frac{121}{16}$$

$$\left(x + \frac{11}{4}\right)^2 = \frac{25}{16}$$

$$\sqrt{\left(x + \frac{11}{4}\right)^2} = \pm\sqrt{\frac{25}{16}}$$

$$x + \frac{11}{4} = \pm\frac{5}{4}$$

$$x = -\frac{11}{4} \pm \frac{5}{4}$$

$$= -4, -\frac{3}{2}$$

Solution set: $\left\{-4, -\frac{3}{2}\right\}$

i)
$$2x^2 + 5x - 3 = 0$$
$$2x^2 + 5x = 3$$
$$x^2 + \frac{5}{2}x = \frac{3}{2}$$
$$x^2 + \frac{5}{2}x + \left[\frac{1}{2}\left(\frac{5}{2}\right)\right]^2 = \frac{3}{2} + \left[\frac{1}{2}\left(\frac{5}{2}\right)\right]^2$$
$$x^2 + \frac{5}{2}x + \frac{25}{16} = \frac{3}{2} + \frac{25}{16}$$
$$\left(x + \frac{5}{4}\right)^2 = \frac{49}{16}$$
$$\sqrt{\left(x + \frac{5}{4}\right)^2} = \pm\sqrt{\frac{49}{16}}$$
$$x + \frac{5}{4} = \pm\frac{7}{4}$$
$$x = -\frac{5}{4} \pm \frac{7}{4} = -3, \frac{1}{2}$$

Solution set: $\left\{-3, \frac{1}{2}\right\}$

j)
$$q^2 + 3q - 1 = 0$$
$$q^2 + 3q = 1$$
$$q^2 + 3q + \left(\frac{3}{2}\right)^2 = 1 + \left(\frac{3}{2}\right)^2$$
$$\left(q + \frac{3}{2}\right)^2 = \frac{13}{4}$$
$$\sqrt{\left(q + \frac{3}{2}\right)^2} = \pm\sqrt{\frac{13}{4}}$$
$$q + \frac{3}{2} = \pm\frac{\sqrt{13}}{2}$$
$$q = -\frac{3}{2} \pm \frac{\sqrt{13}}{2} = \frac{-3 \pm \sqrt{13}}{2}$$

Solution set: $\left\{\dfrac{-3 - \sqrt{13}}{2}, \dfrac{-3 + \sqrt{13}}{2}\right\}$

k)
$$x^2 + 5x + 7 = 0$$
$$x^2 + 5x = -7$$
$$x^2 + 5x + \left(\frac{5}{2}\right)^2 = -7 + \left(\frac{5}{2}\right)^2$$
$$\left(x + \frac{5}{2}\right)^2 = -\frac{3}{4}$$

Because $-\dfrac{3}{4}$ is a negative number and because the square of a real number cannot be negative, there is no real number solution of this equation. Solution set: $\varnothing$

l)
$$(d - 1)(d - 2) = 3d - 5$$
$$d^2 - 3d + 2 = 3d - 5$$
$$d^2 - 6d + 2 = -5$$
$$d^2 - 6d = -7$$
$$d^2 - 6d + \left(\frac{6}{2}\right)^2 = -7 + \left(\frac{6}{2}\right)^2$$
$$d^2 - 6d + 9 = 2$$
$$(d - 3)^2 = 2$$
$$\sqrt{(d - 3)^2} = \pm\sqrt{2}$$
$$d - 3 = \pm\sqrt{2}$$
$$d = 3 \pm \sqrt{2}$$

Solution set: $\left\{3 - \sqrt{2},\ 3 + \sqrt{2}\right\}$

m)
$$4m + 2 = (m - 1)(m + 3)$$
$$4m + 2 = m^2 + 2m - 3$$
$$2 = m^2 - 2m - 3$$
$$5 = m^2 - 2m$$
$$5 + \left(\frac{2}{2}\right)^2 = m^2 - 2m + \left(\frac{2}{2}\right)^2$$
$$6 = m^2 - 2m + 1$$
$$6 = (m - 1)^2$$
$$\pm\sqrt{6} = \sqrt{(m - 1)^2}$$
$$\pm\sqrt{6} = m - 1$$
$$1 \pm \sqrt{6} = m$$

Solution set: $\left\{1 - \sqrt{6}, 1 + \sqrt{6}\right\}$

2. A pellet is fired into the air with an initial velocity of 48 ft/sec. The equation, $h = 48t - 16t^2$, gives the height of the pellet, h, in feet above the ground after t seconds. At what times will the pellet be 32 feet above the ground?

Solution

$$48t - 16t^2 = h$$
$$48t - 16t^2 = 32$$
$$t^2 - 3t = -2 \qquad \text{Divide by } -16.$$
$$t^2 - 3t + \left(\frac{3}{2}\right)^2 = -2 + \left(\frac{3}{2}\right)^2$$
$$\left(t - \frac{3}{2}\right)^2 = \frac{1}{4}$$
$$\sqrt{\left(t - \frac{3}{2}\right)^2} = \pm\sqrt{\frac{1}{4}}$$
$$t - \frac{3}{2} = \pm\frac{1}{2}$$
$$t = \frac{3}{2} \pm \frac{1}{2} = 1, \, 2$$

The pellet be 32 feet above the ground at 1 second and at 2 seconds after being fired.

15.3 Solving Quadratic Equations by the Quadratic Formula

Learning Objectives:

1. Identify the values a, b, and c.
2. Use the quadratic formula to solve equations.
3. Solve applied problems.

Solved Examples:

1. Solve using the quadratic formula. If there are no real roots, say so.

a) $x^2 + 5x + 6 = 0$

b) $x^2 + 4x - 7 = 0$

c) $3x^2 - 9x = -2$

d) $3x^2 - 4x + 8 = 0$

e) $5x^2 = -10x - 3$

f) $x^2 + \dfrac{4}{5}x = -\dfrac{1}{5}$

Solution

The quadratic formula is $x = \dfrac{-b \pm \sqrt{b^2 - 4ac}}{2a}$.

a) $x^2 + 5x + 6 = 0$
$a = 1, b = 5, c = 6$

$$x = \frac{-5 \pm \sqrt{5^2 - 4(1)(6)}}{2(1)} = \frac{-5 \pm \sqrt{1}}{2}$$

$$= \frac{-5 \pm 1}{2} = -3, -2$$

Solution set: $\{-3, -2\}$

b) $x^2 + 4x - 7 = 0$
$a = 1, b = 4, c = -7$

$$x = \frac{-4 \pm \sqrt{4^2 - 4(1)(-7)}}{2(1)} = \frac{-4 \pm \sqrt{44}}{2}$$

$$= \frac{-4 \pm 2\sqrt{11}}{2} = -2 \pm \sqrt{11}$$

Solution set: $\left\{-2 - \sqrt{11}, -2 + \sqrt{11}\right\}$

c) $3x^2 - 9x = -2$
$3x^2 - 9x + 22 = 0$ Standard form
$a = 3, b = -9, c = 2$

$$x = \frac{-(-9) \pm \sqrt{(-9)^2 - 4(3)(2)}}{2(3)} = \frac{9 \pm \sqrt{57}}{6}$$

Solution set: $\left\{\dfrac{9 - \sqrt{57}}{6}, \dfrac{9 + \sqrt{57}}{6}\right\}$

d) $3x^2 - 4x + 8 = 0$
$a = 3, b = -4, c = 8$

$$x = \frac{-(-4) \pm \sqrt{(-4)^2 - 4(3)(8)}}{2(3)} = \frac{4 \pm \sqrt{-80}}{6}$$

Because $\sqrt{-80}$ is not a real number, the solution set is $\varnothing$.

e) $5x^2 = -10x - 3$
$5x^2 + 10x + 3 = 0$ Standard form
$a = 5, b = 10, c = 3$

$$x = \frac{-10 \pm \sqrt{10^2 - 4(5)(3)}}{2(5)} = \frac{-10 \pm \sqrt{40}}{10}$$

$$= \frac{-10 \pm 2\sqrt{10}}{10} = \frac{-5 \pm \sqrt{10}}{5}$$

Solution set: $\left\{\dfrac{-5 - \sqrt{10}}{5}, \dfrac{-5 + \sqrt{10}}{5}\right\}$

f) $x^2 + \dfrac{4}{5}x = -\dfrac{1}{5}$

2. Use the quadratic formula to find the roots.

    a) $9x^2 - 6x + 1 = 0$ b) $25x^2 + 20x + 4 = 0$

    Solution

    a) $9x^2 - 6x + 1 = 0$
       $a = 9, b = -6, c = 1$

    $$x = \frac{-(-6) \pm \sqrt{(-6)^2 - 4(9)(1)}}{2(9)} = \frac{6 \pm \sqrt{0}}{18} = \frac{1}{3}$$

    Solution set: $\left\{ \dfrac{1}{3} \right\}$

    b) $25x^2 + 20x + 4 = 0$
       $a = 25, b = 20, c = 4$

    $$x = \frac{-20 \pm \sqrt{20^2 - 4(25)(4)}}{2(25)} = \frac{-20 \pm \sqrt{0}}{50} = -\frac{2}{5}$$

    Solution set: $\left\{ -\dfrac{2}{5} \right\}$

3. Solve for the variable. Choose the method you feel will work best.

    a) $8x^2 - 15x - 2 = 0$ b) $(2x + 9)^2 = 36$ c) $3x^2 + 10x = -6$

    d) $4x^2 + 6x + 1 = 0$ e) $x^2 + 4x = 0$ f) $(x + 4)(x - 3) = 10$

    g) $x^2 + x + 3 = 0$ h) $x^2 = -16x - 100$ i) $3 + x(x + 4) = 8$

    Solution

    a) $8x^2 - 15x - 2 = 0$
       Use the quadratic formula.
       $a = 8, b = -15, c = -2$

    $$x = \frac{-(-15) \pm \sqrt{(-15)^2 - 4(8)(-2)}}{2(8)}$$

    $$= \frac{15 \pm \sqrt{289}}{16} = \frac{15 \pm 17}{16} = -\frac{1}{8}, 2$$

    Solution set: $\left\{ -\dfrac{1}{8}, 2 \right\}$

    b) $(2x + 9)^2 = 36$
       Use the square root property.
       $(2x + 9)^2 = 36$
       $\sqrt{(2x + 9)^2} = \pm\sqrt{36}$
       $2x + 9 = \pm 6$

$2x + 9 = -6$	$2x + 9 = 6$
$2x = -15$	$2x = -3$
$x = -\dfrac{15}{2}$	$x = -\dfrac{3}{2}$

       Solution set: $\left\{ -\dfrac{15}{2}, -\dfrac{3}{2} \right\}$

    c) $3x^2 + 10x = -6$
       Use the quadratic formula.
       $$3x^2 + 10x = -6$$
       $$3x^2 + 10x + 6 = 0$$
       $a = 3, b = 10, c = 6$

    $$x = \frac{-10 \pm \sqrt{10^2 - 4(3)(6)}}{2(3)} = \frac{-10 \pm \sqrt{28}}{6}$$

    $$= \frac{-10 \pm 2\sqrt{7}}{6} = \frac{-5 \pm \sqrt{7}}{3}$$

    Solution set: $\left\{ \dfrac{-5 - \sqrt{7}}{3}, \dfrac{-5 + \sqrt{7}}{3} \right\}$

    d) $4x^2 + 6x + 1 = 0$
       Use the quadratic formula.
       $a = 4, b = 6, c = 1$

    $$x = \frac{-6 \pm \sqrt{6^2 - 4(4)(1)}}{2(4)} = \frac{-6 \pm \sqrt{20}}{8}$$

    $$= \frac{-6 \pm 2\sqrt{5}}{8} = \frac{-3 \pm \sqrt{5}}{4}$$

    Solution set: $\left\{ \dfrac{-3 - \sqrt{5}}{4}, \dfrac{-3 + \sqrt{5}}{4} \right\}$

e) $x^2 + 4x = 0$

Solve by factoring.

$x^2 + 4x = 0$

$x(x + 4) = 0$

$x = 0 \quad | \quad x + 4 = 0$

$ x = -4$

Solution set: $\{-4, 0\}$

f) $(x + 4)(x - 3) = 10$

$x^2 + x - 12 = 10$

Solve by completing the square.

$x^2 + x = 22$

$x^2 + x + \left(\dfrac{1}{2}\right)^2 = 22 + \left(\dfrac{1}{2}\right)^2$

$\left(x + \dfrac{1}{2}\right)^2 = \dfrac{89}{4}$

$\sqrt{\left(x + \dfrac{1}{2}\right)^2} = \pm\sqrt{\dfrac{89}{4}}$

$x + \dfrac{1}{2} = \pm\dfrac{\sqrt{89}}{2}$

$x = \dfrac{1}{2} \pm \dfrac{\sqrt{89}}{2} = \dfrac{1 \pm \sqrt{89}}{2}$

Solution set: $\left\{\dfrac{1 - \sqrt{89}}{2}, \dfrac{1 + \sqrt{89}}{2}\right\}$

g) $x^2 + x + 3 = 0$

Use the quadratic formula.

$a = 1, b = 1, c = 3$

$x = \dfrac{-1 \pm \sqrt{1^2 - 4(1)(3)}}{2(1)} = \dfrac{-1 \pm \sqrt{-12}}{1}$

Because $\sqrt{-12}$ is not a real number, the solution set is $\varnothing$.

h) $x^2 = -16x - 100$

Use the quadratic formula.

$x^2 = -16x - 100$

$x^2 + 16x + 100 = 0$

$a = 1, b = 16, c = 100$

$x = \dfrac{-16 \pm \sqrt{16^2 - 4(1)(100)}}{2(1)} = \dfrac{-16 \pm \sqrt{-144}}{2}$

Because $\sqrt{-144}$ is not a real number, the solution set is $\varnothing$.

i) $3 + x(x + 4) = 8$

$3 + x^2 + 4x = 8$

$x^2 + 4x - 5 = 0$

Solve by factoring.

$x^2 + 4x - 5 = 0$

$(x + 5)(x - 1) = 0$

$x + 5 = 0 \quad | \quad x - 1 = 0$

$x = -5 x = 1$

Solution set: $\{-5, 1\}$

15.4 Graphing Quadratic Equations and Inequalities

Learning Objectives:

1. Graph quadratic equations.
2. Graph a quadratic inequality.

Solved Examples:

1. Sketch the parabolas $y = x^2$ and $y = -x^2$. Discuss the following points.

   a) The vertex is the lowest point (if parabola opens up) or highest point (if parabola opens down). Label the vertex on each parabola.

   b) Imagine moving the parabolas in any direction. How many x-intercepts can a parabola have?

   c) How many y-intercepts can a parabola have?

   Solution

   a)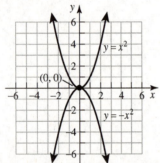

   The vertex of each parabola is $(0, 0)$

   b) A parabola can have 0, 1, or 2 x-intercepts.

   c) A parabola can have only 1 y-intercept.

2. Find five ordered pairs for the equation. Then graph it.

   a) $y = x^2 + 1$

   b) $y = x^2 - 2$

   c) $y = \dfrac{1}{2}x^2$

   d) $y = -\dfrac{1}{4}x^2$

   e) $y = -2x^2 + 2$

   f) $y = (x+3)^2$

   g) $y = (x-2)^2$

   h) $y = \dfrac{1}{4}(x+1)^2$

   Solution

   a)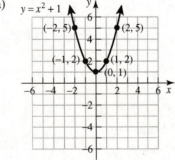

   b)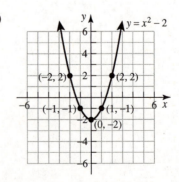

SE-282

c)

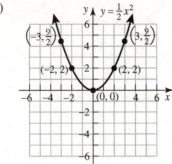

d)

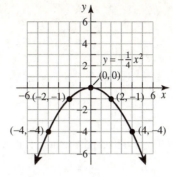

e)

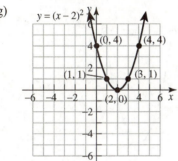

f)

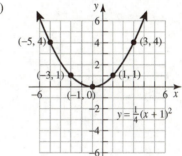

g)

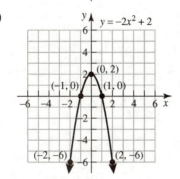

h)

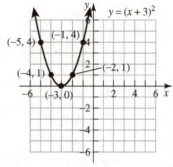

3. Determine the vertex and the x-intercepts. Then sketch the graph.

a) $y = x^2 - 1$

b) $y = x^2 + 4x$

c) $y = x^2 + 2x - 3$

d) $y = -x^2 + 2x + 3$

e) $y = x^2 + 5x + 4$

Solution

Find the x-coordinate of the vertex using $x = -\dfrac{b}{2a}$. Find the x-intercepts by substituting 0 for y and solving the equation.

a) $\quad y = x^2 - 1$

$x = -\dfrac{0}{2(1)} = 0; \; y = 0^2 - 1 = -1$

The vertex is at $(0, -1)$.

$x^2 - 1 = 0$

$\quad x^2 = 1$

$\quad x = \pm 1$

The x-intercepts are at $(-1, 0)$ and $(1, 0)$.

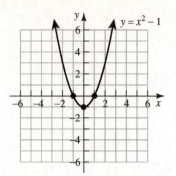

b) $\quad y = x^2 + 4x$

$x = -\dfrac{4}{2(1)} = -4; \; y = (-4)^2 + 4(-4) = 0$

The vertex is at $(-2, 0)$.

$x^2 + 4x = 0$

$x(x + 4) = 0$

$x = 0 \;\Big|\; x + 4 = 0$

$\qquad \quad x = -4$

The x-intercepts are at $(-4, 0)$ and $(0, 0)$.

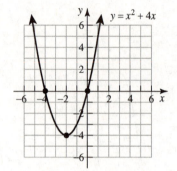

c) $\quad y = x^2 + 2x - 3$

$x = -\dfrac{2}{2(1)} = -1; \; y = (-1)^2 + 2(-1) - 3 = -4$

The vertex is at $(-1, -4)$.

$0 = x^2 + 2x - 3$

$0 = (x + 3)(x - 1)$

$x + 3 = 0 \;\Big|\; x - 1 = 0$

$\quad x = -3 \;\Big|\; \quad x = 1$

The x-intercepts are at $(-3, 0)$ and $(1, 0)$.

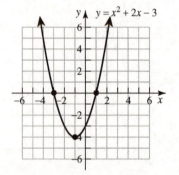

d) $\quad y = -x^2 + 2x + 3$

$x = -\dfrac{2}{2(-1)} = 1; \; y = -(1)^2 + 2(1) + 3 = 4$

The vertex is at $(1, 4)$.

$0 = -x^2 + 2x + 3$

$0 = x^2 - 2x - 3$

$0 = (x - 3)(x + 1)$

$x - 3 = 0 \;\Big|\; x + 1 = 0$

$\quad x = 3 \;\Big|\; \quad x = -1$

The x-intercepts are at $(-1, 0)$ and $(3, 0)$.

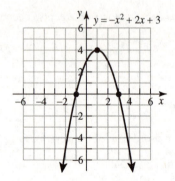

e)
$$y = x^2 + 5x + 4$$

$$x = -\frac{5}{2(1)} = -\frac{5}{2}$$

$$y = \left(-\frac{5}{2}\right)^2 + 5\left(-\frac{5}{2}\right) + 4 = -\frac{9}{4}$$

The vertex is at $\left(-\frac{5}{2}, -\frac{9}{4}\right)$.

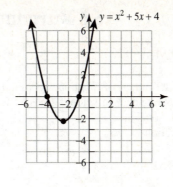

$$0 = x^2 + 5x + 4$$
$$0 = (x+4)(x+1)$$

$$\begin{array}{c|c} x+4 = 0 & x+1 = 0 \\ x = -4 & x = -1 \end{array}$$

The x-intercepts are at (−4, 0) and (−1, 0).

15.5 Introduction to Functions

Learning Objectives:

1. Understand definitions of functions and relations.
2. Decide whether and an equation defines a function.
3. Use function notation.
4. Solve applications.

Solved Examples:

1. Complete the table for the function defined by $f(x) = x - 3$

	x	$x - 3$	$f(x)$	(x, y)
a	-3			
b	-2			
c	-1			
d	0			
e	1			

Solution

	x	$x - 3$	$f(x)$	(x, y)
a)	-3	$-3 - 3 = -6$	-6	$(-3, -6)$
b)	-2	$-2 - 3 = -5$	-5	$(-2, -5)$
c)	-1	$-1 - 3 = -4$	-4	$(-1, -4)$
d)	0	$0 - 3 = -3$	-3	$(0, -3)$
e)	1	$1 - 3 = -2$	-2	$(1, -2)$

2. Determine whether each relation represents a function.

   a) $\{(1,3),(2,5),(4,1)\}$ b) $\{(-1,3),(1,3),(2,-5)\}$ c) $\{(7,-1),(3,-1),(7,4)\}$

   Solution

   a) Yes, the relation is a function. b) Yes, the relation is a function.

   c) No, the relation is not a function because the first component 7 appears in two ordered pairs and corresponds to two different second components.

3. Decide whether each relation represents a function.

   a)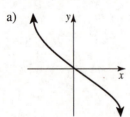

   b)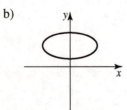

   c) $y = 7x - 3$

   d) $x = y^2 + 5$

Solution

Use the vertical line test to determine if a graph is a function. If a vertical line intersects a graph in more than one point, the graph is not the graph of a function.

a) Yes, the relation is a function.

b) No, the relation is not a function.

c) This is a linear equation. Thus, it is a function.

d) This is not a function. For each value of $x \geq 5$, there are two values of y. For example if $x = 10$, $y = \pm\sqrt{5}$.

4. For each function, find the indicated function value.

a) $f(x) = -2x + 6$; $f(-1)$

b) $f(x) = 3x^2 + x - 5$; $f(0)$

c) $f(x) = 2|x| - 1$; $f(-3)$

Solution

a) $f(-1) = -2(-1) + 6 = 8$

b) $f(0) = 3(0)^2 + 0 - 5 = -5$

c) $f(-3) = 2|-3| - 1 = 5$

Chapter 1 WHOLE NUMBERS

1.1   Reading and Writing Whole Numbers

Indicate whether each number is a whole number or not a whole number

1. 48

1._____

2. $7\frac{1}{2}$

2._____

3. 357

3._____

4. 3.14159

4._____

Fill in the digit for the given place value in each of the following whole numbers.

5. 9841   thousands   tens

5._____

6. 25,016   ten-thousands   hundreds

6._____

7. 86,331   ten-thousands   ones

7._____

8. 5,813,207   millions   thousands

8._____

9. 2,800,439,012   billions   millions

9._____

Fill in the number for the given period in each of the following whole numbers.

10. 29,176   thousands   ones

10._____

11. 75,229,301   millions   thousands   ones

11._____

12. 70,000,603,214   billions   millions   12._____

                         thousands   ones

13. 300,459,200,005   billions   millions   13._____

                          thousands   ones

Rewrite the following numbers in words.

14. 8714   14._____

15. 39,015   15._____

16. 834,768   16._____

17. 2,015,102   17._____

18. 96,543,228   18._____

Rewrite each of the following numbers using digits.

19. Four thousand, one hundred twenty-seven   19._____

20. Twenty-nine thousand, five hundred sixteen   20._____

21. Six hundred eight-five million, two hundred fifty-nine   21._____

22. Three hundred million, seventy-five thousand, two   22._____

Rewrite the number from the following sentence using digits.

23. A bottle of a certain vaccine will give seven thousand, two hundred ten injections.

23._____

24. Every year, nine hundred seventy-two thousand, four hundred thirty people visit a certain historical area.

24._____

25. A supermarket has fifteen thousand three hundred thirteen different items for sale.

25._____

26. The population of a large city is six million, two hundred five thousand.

26._____

Use the table to find each of the following and write the number in digits.

Weight of Exerciser	123 lbs	130 lbs	143 lbs
Calories burned in 30 minutes			
Cycling	168	177	195
Running	324	342	375
Jumping Rope	273	288	315
Walking	162	171	189

Source: Fitness magazine

27. The number of calories burned by a 130-pound adult in 30 minutes of cycling.

27._____

28. The activity which will burn at least 300 calories when performed by a 123-pound adult.

28._____

29. The number of calories burned by a 143-pound adult in 30 minutes of jumping rope.

29._____

30. The weight of an adult who will burn 189 calories in 30 minutes of walking.

30._____

Chapter 1 WHOLE NUMBERS

1.2   Adding Whole Numbers

Add.

   1. $3 + 9$ 1._____

   2. $8 + 5$ 2._____

   3. $2 + 8$ 3._____

   4. $9 + 8$ 4._____

Add.

   5.   7   5._____
            2
            5
            3
          $+ 8$

   6.   9   6._____
            7
            2
            5
          $+ 6$

7.   3
        4
        9
        2
        7
     + 6

7. _____

8.   6
        4
        9
        8
        4
     + 3

8. _____

Add.

9.   42
    + 57

9. _____

10.   421
    + 567

10. _____

11.   86,305
    + 12,672

11. _____

12.   45,158
       20,340
    +   2401

12. _____

Add.

13.   83
      + 29

13. _____

14.   563
      + 478

14. _____

15.   7439
      + 8376

15. _____

16.   7033
        809
        2532
      +   41

16. _____

Using the map below, find the shortest distance between the following cities.

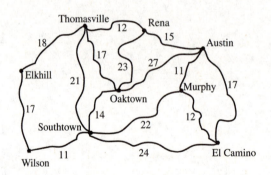

17. Murphy and Thomasville

17. _____

18. Wilson and Austin

18. _____

19. El Camino and Thomasville

19. _____

Solve the following application problems, using addition.

20. Kevin Levy has 52 nickels, 37 dimes, and 119 quarters. How many coins does he have altogether?

20._____

21. The theater sold 276 adult tickets, and 349 child tickets. How many tickets were sold altogether?

21._____

22. At a charity bazaar, a church has a total of 1873 books for sale, while a lodge has 3358 books for sale. How many books are for sale?

22._____

Find the perimeter or total distance around each of the following figures.

23.

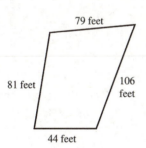

23._____

24.

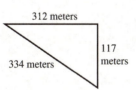

24._____

25.

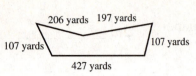

206 yards   197 yards

107 yards   107 yards

427 yards

25._____

Check the following additions. If an answer is incorrect, give the correct answer.

26.   67
           48
         + 83
         ─────
          198

26._____

27.   73
           9815
            390
         + 7002
         ──────
         16,270

27._____

28.   723
           681
            29
           412
         + 103
         ─────
          1947

28._____

Name: Date:
Instructor: Section:

29.

$$\begin{array}{r} 3028 \\ 335 \\ 2914 \\ 688 \\ +\,1647 \\ \hline 8612 \end{array}$$

29._______________

30.

$$\begin{array}{r} 72 \\ 38 \\ 5735 \\ 764 \\ +\ \ 16 \\ \hline 6625 \end{array}$$

30._______________

Chapter 1 WHOLE NUMBERS

1.3   Subtracting Whole Numbers

Write two subtraction problems for each addition problem.

    1. $6 + 9 = 15$ 1._____

    2. $17 + 9 = 26$ 2._____

    3. $149 + 38 = 187$ 3._____

    4. $478 + 239 = 717$ 4._____

Write an addition problem for each subtraction problem.

    5. $1211 - 426 = 785$ 5._____

    6. $204 - 87 = 117$ 6._____

    7. $5917 - 2196 = 3721$ 7._____

    8. $5094 - 113 = 4981$ 8._____

Identify the minuend, subtrahend, and difference in each of the following subtraction problems.

    9. $5 - 3 = 2$ 9._____

    10. $35 - 9 = 24$ 10._____

11. $98 - 36 = 62$

11._____

12. $236 - 142 = 94\ 30$

12._____

Subtract.

13. $\begin{array}{r} 95 \\ -\ 64 \\ \hline \end{array}$

13._____

14. $\begin{array}{r} 76 \\ -\ 55 \\ \hline \end{array}$

14._____

15. $\begin{array}{r} 5573 \\ -\ 422 \\ \hline \end{array}$

15._____

16. $\begin{array}{r} 8539 \\ -\ 2527 \\ \hline \end{array}$

16._____

Check the following subtractions. If an answer is not correct, give the correct answer.

17. $\begin{array}{r} 192 \\ -\ 39 \\ \hline 167 \end{array}$

17._____

18. $\begin{array}{r} 4847 \\ -\ 3768 \\ \hline 1121 \end{array}$

18._____

19.   5763
        − 2783
        ‾‾‾‾‾‾
         3980

19._____

20.   31,146
        − 7312
        ‾‾‾‾‾‾
        23,834

20._____

21. 82,004
        − 3917
        ‾‾‾‾‾‾
        79,193

21._____

Subtract.

22.   42
         − 35

22._____

23.   87
         − 48

23._____

24.   927
         − 729

24._____

25.   4687
        − 2769

25._____

26.   86,372
        $-$ 29,485

26._____

Solve the following application problems.

27. A Girl Scout has 52 boxes of cookies to sell. If she sells 27 boxes, how many boxes will she have left?

27._____

28. Nathaniel Best has $553 in his checking account. He writes a check for $134. How much is then left in the account.

28._____

29. One bid for painting a house was $2134. A second bid was $1954. How much would be saved using the second bid?

29._____

30. On Friday, 11, 594 people visited Eastridge Amusement Park, while 14,352 people visited park on Saturday. How many more people visited the park on Saturday?

30._____

Chapter 1 WHOLE NUMBERS

1.4   Multiplying Whole Numbers

Identify the factors and the product in each multiplication problem.

1. $5 \times 3 = 15$

1._____

2. $5(2) = 10$

2._____

3. $13 \cdot 3 = 39$

3._____

4. $108 = 9 \times 12$

4._____

Multiply.

5. $4 \times 4 \times 2$

5._____

6. $7 \cdot 4 \cdot 0$

6._____

7. $(6)(4)(8)$

7._____

8. $2 \cdot 6 \cdot 10$

8._____

Multiply.

9. 5×8

9._____

10. $5 \cdot 12$

10._____

11.   54
$\times\ 4$

11._____

12.   163
$\times\ \ 5$

12._____

13.   405
$\times\ \ 7$

13._____

14.   38,471
$\times\ \ \ \ \ 3$

14._____

Multiply.

15. 82×10

15._____

16. $(852)(30)$

16._____

17. 42×200

17._____

18. 3005×2000

18._____

19. $387\cdot20,000$

19._____

20. 47,000 · 6000

Multiply.

21.   46
      × 21

21._____

22.   47
      × 32

22._____

23.   644
      × 19

23._____

24.   409
      × 27

24._____

25.   8341
      × 59

25._____

26.   8621
      × 131

26._____

Solve the following application problems.

27. A fabric store has 16 bolts of silk. Each bolt contains 35 yards of silk. How many yards of silk does the fabric store have in all?

27._____

28. Heinen's Supermarket received a shipment of 28 cartons of canned vegetables. There were 24 cans in each carton. How many cans were there altogether?

28._____

Find the total cost of each of the following.

29. 24 soccer balls at $16 per ball

29._____

30. 47 watches at $29 per watch

30._____

Chapter 1 WHOLE NUMBERS

1.5   Dividing Whole Numbers

Write each division problem using two other symbols.

1. $15 \div 3 = 5$

1._____

2. $\dfrac{50}{25} = 2$

2._____

3. $16\overset{2}{\overline{)32}}$

3._____

Identify the dividend, divisor, and quotient.

4. $63 \div 7 = 9$

4._____

5. $5\overset{6}{\overline{)30}}$

5._____

6. $\dfrac{44}{11} = 4$

6._____

Divide, whenever possible.

7. $0 \div 15$

7._____

8. $\dfrac{0}{6}$

8._____

9. $12\overline{)0}$

9._____

Divide, whenever possible.

10. $\dfrac{7}{0}$

_____

11. $0\overline{)72}$

11._____

12. $9 \div 0$

12._____

Divide.

13. $18 \div 18$

13._____

14. $17\overline{)17}$

14._____

15. $\dfrac{8}{8}$

15._____

Divide.

16. $17 \div 1$

16._____

17. $\dfrac{128}{1}$

17._____

18. $1\overline{)38}$

18._____

Divide.

19. $6\overline{)72}$

19._____

20. $\dfrac{575}{5}$

20._____

21. $\dfrac{651}{9}$

21._____

22. $843 \div 7$

22._____

Check each answer. If an answer is incorrect, give the correct answer.

23. $5\overline{)135}$ with quotient 28

23._____

24. $2915 \div 8 = 364$

24._____

25. $46,650 \div 7 = 6664 \text{ R}2$

25._____

26. $\dfrac{34,176}{7} = 4882 \text{ R } 2$

26._____

$$
\begin{array}{r}
3389 \\
9\overline{)30{,}508}\ \ R\ 8
\end{array}
$$

27.

27._____

Determine if the following numbers are divisible by 2, 3, 5, or 10.

28. 50

28._____

29. 897

29._____

30. 6205

30._____

Chapter 1 WHOLE NUMBERS

1.6   Long Division

Divide using long division. Check each answer.

1. $32\overline{)2624}$ 1._____

2. $29\overline{)9396}$ 2._____

3. $42\overline{)3234}$ 3._____

4. $53\overline{)5406}$ 4._____

5. $89\overline{)7649}$ 5._____

6. $94\overline{)29,047}$ **6.** _____

7. $86\overline{)8,473,758}$ **7.** _____

8. $205\overline{)6,680,335}$ **8.** _____

9. $657\overline{)429,700}$ **9.** _____

10. $732\overline{)4,268,292}$ **10.** _____

Divide.

11. $30\overline{)270}$ **11.** _____

12. $80\overline{)560}$

12. _____

13. $700\overline{)4900}$

13. _____

14. $2000\overline{)12,000}$

14. _____

15. $50\overline{)800}$

15. _____

16. $400\overline{)6000}$

16. _____

17. $230\overline{)16,100}$

17. _____

18. $210\overline{)16,800}$

18. _____

19. $500\overline{)42,000}$

19. _____

20. $1200\overline{)960{,}000}$

20._____

Check each answer. If an answer is incorrect, give the correct answer.

21. $19\overline{)3299}$ R 12 (quotient 173)

21._____

22. $37\overline{)3235}$ R 16 (quotient 87)

22._____

23. $89\overline{)5790}$ R 5 (quotient 65)

23._____

24. $74\overline{)25{,}621}$ R 18 (quotient 346)

24._____

25. $103\overline{)4658}$ R 22 (quotient 44)

25._____

26. $205\overline{)47,538}$ $\overset{231}{}$ R 183

26._____

27. $318\overline{)94,207}$ $\overset{297}{}$ R 79

27._____

28. $428\overline{)196,883}$ $\overset{400}{}$ R 30

28._____

29. $537\overline{)431,042}$ $\overset{802}{}$ R 368

29._____

30. $614\overline{)152,923}$ $\overset{249}{}$ R 37

30._____

Chapter 1 WHOLE NUMBERS

1.7   Rounding Whole Numbers

Locate the place to which the number is rounded by underlining the appropriate digit.

1. 853   Nearest ten   1._____

2. 1037   Nearest hundred   2._____

3. 645,371   Nearest ten-thousand   3._____

4. 39,943,712   Nearest million   4._____

5. 257,301   Nearest ten   5._____

6. 2,781,421   Nearest thousand   6._____

Round as shown.

7. 7863 to the nearest hundred   7._____

8. 1382 to the nearest ten   8._____

9. 814 to the nearest ten   9._____

10. 32,576 to the nearest ten   10._____

11. 53,595 to the nearest hundred   11._____

12. 8398 to the nearest hundred

12._____

13. 16,668 to the nearest hundred

13._____

14. 3842 to the nearest thousand

14._____

Estimate the following answers by rounding to the nearest ten. Then find the exact answers.

15.   37
        24
        58
    + 91

15._____

16.   19
        87
        35
    + 20

16._____

17.   69
    − 42

17._____

18.   88
    − 52

18._____

Estimate the following answers by rounding to the nearest hundred. Then find the exact answer.

19.   276
     312
     174
  + 936

19. _____

20.   419
     188
     324
  + 194

20. _____

21.   971
  − 382

21. _____

22.   815
  − 678

22. _____

23.   912
  × 784

23. _____

24.   876
  × 141

24. _____

Estimate the following answers by using front end rounding. Then find the exact answer.

25.   571
            42
            215
         + 2452

25. _____

26.   731
          31
          709
        + 78

26. _____

27.   872
        − 39

27. _____

28.   313
        − 49

28. _____

29.   980
        × 37

29. _____

30.   437
        × 29

30. _____

Chapter 1 WHOLE NUMBERS

1.8   Exponents, Roots, and Order of Operations

Identify the exponent and the base. Simplify each expression

1. 7^2

                                                1._____

2. 4^2

                                                  2._____

3. 9^2

                                                  3._____

4. 3^3

                                                  4._____

5. 1^5

                                                  5._____

6. 10^4

                                                  6._____

7. 2^7

                                                  7._____

Find each square root.

8. $\sqrt{4}$

                                                  8._____

9. $\sqrt{9}$

                                                  9._____

10. $\sqrt{16}$

                                                 10._____

11. $\sqrt{121}$ **11.** _____

12. $\sqrt{64}$ **12.** _____

13. $\sqrt{100}$ **13.** _____

14. $\sqrt{49}$ **14.** _____

Complete each blank.

15. $18^2 =$ _____ so $\sqrt{} = 18$ **15.** _____

16. $14^2 =$ _____ so $\sqrt{196} =$ _____   **16.** _____

17. $50^2 =$ _____ so $\sqrt{} = 50$ **17.** _____

18. $23^2 =$ _____ so $\sqrt{} = 23$ **18.** _____

19. $15^2 =$ _____ so $\sqrt{} = 15$ **19.** _____

20. $36^2 =$ _____ so $\sqrt{1296} =$ _____   **20.** _____

21. $52^2 =$ _____ so $\sqrt{} = 52$ **21.** _____

Work each problem by using the order of operations.

22. $2^4 + 3 \cdot 4 - 5$

22._____

23. $6 \cdot 5 - 5 \div 0$

23._____

24. $7 - 25 \div 5$

24._____

25. $6 \cdot 3^2 + 0 \div 6$

25._____

26. $8 \cdot 5 - 12 \div (2 \cdot 3 - 6)$

26._____

27. $4 \cdot 3 + 8 \cdot 5 - 7$

27._____

28. $2^3 \cdot 3^2 + 5(3) \div 5$

28._____

29. $42 \div 6 + 3 \cdot \sqrt{49}$

29._____

30. $2 \cdot \sqrt{121} - 2 \div \sqrt{4} + (14 - 2 \cdot 7) \div 4$

30._____

Chapter 1 WHOLE NUMBERS

1.9   Reading Pictographs, Bar Graphs, and Line Graphs

Use the pictograph to answer the questions.

State Sales Tax

Georgia	$ $ $ $
Utah	$ $ $ $ $
Idaho	$ $ $ $ $
Texas	$ $ $ $ $ [
Minnesota	$ $ $ $ $ $

$ = 1% sales tax

Source: Federation of Tax Administrators

1. Which state shown in the pictograph charges the least sales tax?

    1._____

2. Which state has a sales tax of 5%?

    2._____

3. According to the pictograph, which state has the greatest sales tax?

    3._____

4. Which state has a sales tax of 4%?

    4._____

5. By how much does Minnesota's sales tax exceed Georgia's sales tax?

    5._____

Name: Date:
Instructor: Section:

Use the pictograph to answer the questions.

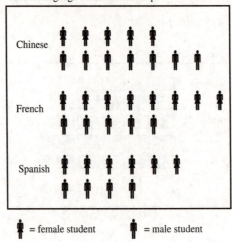

School Language Clubs Membership

† = female student   † = male student

6. Which language club has the largest number of members?

6._____

7. How many more female students are there than male students in the French club?

7._____

8. What is the total number of members in the Chinese club?

8._____

9. Which language club has the least number of members?

9._____

10. How many more female students are in the French or Spanish club than male students?

10._____

Use the bar graph, which shows the amount of blood donated by the different departments of a company, to answer the questions.

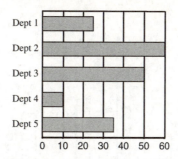

11. How many pints of blood were donated by Department 3?

11._____

12. How many more pints of blood were donated by Department 5 than Department 4?

12._____

13. Which Department donated the fewest pints of blood?

13._____

14. How many pints of blood were donated by Department 1?

14._____

15. Department 2 donated as much blood as which two departments combined?

15._____

Use the double bar graph, which shows the enrollment by gender in each class at a small college, to answer the questions.

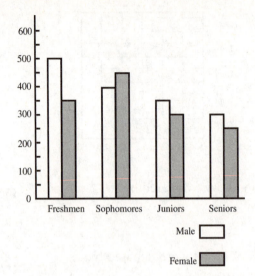

16. How many more male freshmen are there than female seniors?

16._____

17. Find the total number of students enrolled.

17._____

18. How many more sophomores are there than juniors?

18._____

19. Which class has the greatest difference between male students and female students?

19._____

20. Which class has more female students than male students?

20._____

Use the line graph to answer the questions.

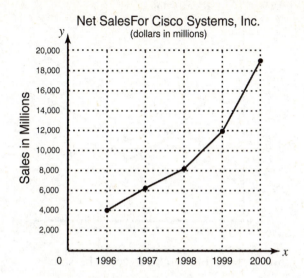

21. What trend or patterns is shown in the graph?

21._____

22. Approximately what were the net sales in 1996?

22._____

23. Which year had the largest increase over the previous year?

23._____

24. Which year had the highest net sales?

24._____

25. Between which two years was the increase smallest?

25._____

Use the comparison line graph, which shows the annual sales for two different stores for each of the past few years, to answer the questions.

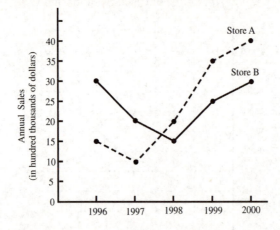

26. In which years did the sales of store A exceed the sales of store B?

26._____

27. Which year showed the least difference between the sales of store A and the sales of store B?

27._____

28. Which year showed the greatest difference between the sales of store A and the sales of store B?

28._____

29. What was the difference in annual sales between store A and store B in 1996?

29._____

30. Between which two years was the increase for both stores largest?

30._____

Chapter 1 WHOLE NUMBERS

1.10   Solving Application Problems

Write the operation that is indicated by the following words.

1. plus   1._____

2. decreased by   2._____

3. twice   3._____

4. sum of   4._____

5. product   5._____

6. less than   6._____

7. quotient   7._____

8. difference   8._____

9. more than   9._____

Solve each of the following application problems.

10. Andy's Auto Supply just raised the price of a rebuilt   10._____
engine to $1985. If this is $163 more than the old price,
find the old price.

11. Tim Rhinehart, coordinator of Toys for Tots, has collected 2548 toys. If his group wants to give the same number of toys to each of 637 children, how many toys will each child receive?

11. _____

12. If 843 movie tickets are sold per day, how many tickets will be sold in a 5-day period.

12. _____

13. Dana Burgess owes $2840 plus $168 interest on his credit union loan. If he wishes to pay the loan in full, how much must he pay?

13. _____

14. To qualify for a real estate loan at Uptown Bank, a borrower must have a monthly income of at least 4 times the monthly payment. What minimum monthly income must a borrower have to qualify for a monthly payment of $725?

14. _____

15. Naturalists report that 69 salmon are passing a fish ladder each hour. At this rate, how many salmon are passing the ladder in a 12-hour period?

15. _____

16. Lori Knight knows that her car gets 36 miles per gallon in town. How many miles can she travel on 26 gallons?

16. _____

17. A car loan of $12,672 is to be paid off in 36 months. What will the monthly payment be?

17._____

18. Diana Ditka spent $286 on tuition, $137 on books, and $32 on supplies. If this money is withdrawn form her checking account, which had a balance of $723, what is her new balance.

18._____

Estimate an answer for each problem by using front end rounding. Then find the exact answer.

19. A bus traveled 605 miles at 55 miles per hour. How long did the trip take?

19._____

20. The total cost for 23 baseball uniforms is $1817. Find the cost of each uniform.

20._____

21. A person borrows $47,000, and pays interest of $541. Find the total amount that must be repaid.

21._____

22. Amanda Raymond owes $5520 on a loan. Find her monthly payment if the loan is paid off in 48 months.

22._____

23. Two sisters share a legal bill of $1903. One sister pays
$954 toward the bill. How much must the other sister
pay?

23._____

24. Teisha Jordan can assemble 38 toasters on one hour.
How many hours would it take her to assemble 4066
toasters?

24._____

25. Tot total receipts at a concert were $191,800. Each
ticket cost $28. How many people attended the concert?

25._____

26. A new car costs $11,350 before a trade-in. The car can
be paid off in 36 monthly payments of $209 each after
the trade-in. Find the amount of the trade-in.

26._____

27. A biology class found 14 deer in one area, 158 in
another, and 417 in a third. How many deer did the class
find?

27._____

28. A Seiko quartz watch has a crystal that vibrates 32,768
times in one second. How many times will this crystal
vibrate in 30 seconds?

28._____

29. Edward Biondi has $3117 in his checking account. If he
   pays $340 for tires, $725 for equipment repairs, and
   $198 for fuel and oil, find the balance remaining in his
   account.

29._____

30. Cheryl Brown can type 12 forms per hours. How many
   forms can she type in 7 hours?

30._____

Chapter 2 FRACTIONS AND MIXED NUMBERS

2.1   Basics of Fractions

Write the fraction that represents the shaded area.

1.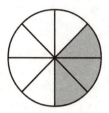

1._____

2.

2._____

3.

3._____

4.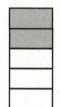

4._____

5.

5._____

6.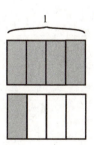

6._____

7.

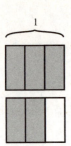

8.

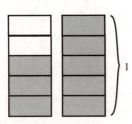

8._____

9.

9._____

10.

10._____

Identify the numerator and the denominator.

11. $\dfrac{4}{3}$

11._____

12. $\dfrac{1}{2}$

12._____

13. $\dfrac{2}{5}$

13._____

14. $\dfrac{9}{23}$

14._____

15. $\dfrac{8}{11}$

15._____

16. $\dfrac{11}{8}$

16._____

17. $\dfrac{112}{5}$

17._____

18. $\dfrac{19}{50}$

18._____

19. $\dfrac{7}{15}$

19._____

20. $\dfrac{19}{8}$

20._____

Write whether each fraction is proper or improper.

21. $\dfrac{9}{7}$

21._____

22. $\dfrac{5}{12}$

22._____

23. $\dfrac{7}{15}$

23._____

24. $\dfrac{17}{11}$

24._____

25. $\dfrac{4}{19}$

25._____

26. $\dfrac{1}{4}$

26._____

27. $\dfrac{11}{7}$

27._____

28. $\dfrac{18}{18}$

28._____

29. $\dfrac{5}{4}$

29._____

30. $\dfrac{10}{10}$

30._____

Chapter 2 FRACTIONS AND MIXED NUMBERS

2.2   Mixed Numbers

List the mixed numbers in each group.

1. $2\frac{1}{2}$, $\dfrac{3}{5}$, $1\frac{1}{6}$, $\dfrac{3}{4}$

    1._____

2. $\dfrac{3}{8}$, $5\frac{2}{3}$, $\dfrac{7}{4}$, $3\frac{1}{2}$

    2._____

3. $\dfrac{8}{7}$, $\dfrac{10}{10}$, $\dfrac{2}{3}$, $\dfrac{0}{5}$

    3._____

4. $\dfrac{9}{9}$, $3\frac{1}{2}$, $10\frac{1}{3}$, $\dfrac{8}{2}$, $\dfrac{7}{9}$

    4._____

Write each mixed number as an improper fraction.

5. $2\frac{7}{8}$

    5._____

6. $1\frac{5}{6}$

    6._____

7. $2\frac{4}{5}$

    7._____

8. $5\frac{4}{7}$

    8._____

9. $1\frac{3}{4}$

    9._____

10. $6\frac{1}{4}$

    10._____

11. $4\frac{2}{3}$

    11._____

12. $7\frac{1}{2}$

    12._____

13. $2\frac{7}{11}$

13._____

14. $5\frac{3}{7}$

14._____

15. $6\frac{2}{3}$

15._____

16. $8\frac{7}{9}$

16._____

17. $11\frac{1}{3}$

17._____

Write each improper fraction as a mixed number.

18. $\dfrac{11}{2}$

18._____

19. $\dfrac{8}{5}$

19._____

20. $\dfrac{9}{8}$

20._____

21. $\dfrac{33}{10}$

21._____

22. $\dfrac{14}{9}$

22._____

23. $\dfrac{20}{7}$

23._____

24. $\dfrac{29}{9}$

24._____

25. $\dfrac{26}{7}$

25._____

26. $\dfrac{21}{5}$

27. $\dfrac{41}{9}$

28. $\dfrac{25}{9}$

29. $\dfrac{29}{4}$

30. $\dfrac{58}{5}$

26. _____

27. _____

28. _____

29. _____

30. _____

Chapter 2 FRACTIONS AND MIXED NUMBERS

2.3   Factors

Find all the factors of each number.

1. 7

1._____

2. 12

2._____

3. 49

3._____

4. 15

4._____

5. 10

5._____

6. 36

6._____

7. 25

7._____

8. 24

8._____

9. 18

9._____

10. 30

10._____

Write whether each number is prime, composite, *or* neither.

11. 1   11._____

12. 5   12._____

13. 11   13._____

14. 15   14._____

15. 24   15._____

16. 45   16._____

17. 2   17._____

18. 31   18._____

19. 29   19._____

20. 38   20._____

Find the prime factorization of each number. Write the answer with exponents when repeated factors appear.

21. 12   21._____

22. 22   22._____

23. 15   23._____

24. 27   24._____

25. 28   25._____

26. 42   26._____

27. 32   **27.**_____

28. 24   **28.**_____

29. 63   **29.**_____

30. 100   **30.**_____

Chapter 2 FRACTIONS AND MIXED NUMBERS

2.4   Writing a Fraction in Lowest Terms

Write whether or not each fraction is in lowest terms.

1. $\dfrac{4}{12}$

2. $\dfrac{3}{7}$

3. $\dfrac{12}{18}$

4. $\dfrac{13}{17}$

5. $\dfrac{27}{30}$

6. $\dfrac{7}{19}$

7. $\dfrac{4}{9}$

8. $\dfrac{18}{21}$

1._____

2._____

3._____

4._____

5._____

6._____

7._____

8._____

Write each fraction in lowest terms.

9. $\dfrac{2}{8}$

10. $\dfrac{4}{12}$

11. $\dfrac{14}{49}$

9._____

10._____

11._____

12. $\dfrac{8}{36}$

12._____

13. $\dfrac{25}{30}$

13._____

14. $\dfrac{10}{18}$

14._____

15. $\dfrac{10}{35}$

15._____

Write each fraction in lowest terms using prime factors.

16. $\dfrac{63}{84}$

16._____

17. $\dfrac{28}{56}$

17._____

18. $\dfrac{180}{210}$

18._____

19. $\dfrac{72}{90}$

19._____

20. $\dfrac{36}{54}$

20._____

21. $\dfrac{71}{142}$

21._____

22. $\dfrac{75}{500}$

22._____

Decide whether each pair of fractions is equivalent *or* not equivalent.

23. $\dfrac{4}{12}$ and $\dfrac{1}{4}$

23._____

24. $\dfrac{2}{3}$ and $\dfrac{10}{15}$

24._____

25. $\dfrac{6}{21}$ and $\dfrac{3}{7}$

25._____

26. $\dfrac{8}{16}$ and $\dfrac{15}{20}$

26._____

27. $\dfrac{9}{12}$ and $\dfrac{6}{8}$

27._____

28. $\dfrac{8}{36}$ and $\dfrac{2}{9}$

28._____

29. $\dfrac{20}{24}$ and $\dfrac{15}{31}$

29._____

30. $\dfrac{3}{39}$ and $\dfrac{1}{3}$

30._____

Chapter 2 FRACTIONS AND MIXED NUMBERS

2.5   Multiplying Fractions

Multiply. Write the answer in lowest terms.

1. $\dfrac{3}{8} \cdot \dfrac{5}{9}$

2. $\dfrac{5}{9} \cdot \dfrac{7}{6}$

3. $\dfrac{1}{3} \cdot \dfrac{2}{5}$

4. $\dfrac{4}{7} \cdot \dfrac{3}{5}$

5. $\dfrac{5}{6} \cdot \dfrac{11}{4}$

6. $\dfrac{9}{10} \cdot \dfrac{3}{2}$

7. $\dfrac{7}{8} \cdot \dfrac{1}{5}$

8. $\dfrac{10}{7} \cdot \dfrac{4}{5}$

1._____

2._____

3._____

4._____

5._____

6._____

7._____

8._____

9. $\dfrac{1}{2} \cdot \dfrac{4}{3}$

9. _____

10. $\dfrac{2}{5} \cdot \dfrac{15}{16}$

10. _____

Use the multiplication shortcut to find each product. Write the answer in lowest terms.

11. $\dfrac{1}{6} \cdot \dfrac{9}{8}$

11. _____

12. $\dfrac{7}{6} \cdot \dfrac{3}{14}$

12. _____

13. $\dfrac{4}{9} \cdot \dfrac{15}{16}$

13. _____

14. $\dfrac{3}{5} \cdot \dfrac{25}{27}$

14. _____

15. $\dfrac{11}{4} \cdot \dfrac{8}{33}$

15. _____

16. $\dfrac{9}{10} \cdot \dfrac{2}{3}$

16. _____

17. $\dfrac{5}{6} \cdot \dfrac{4}{35}$

17. _____

18. $\dfrac{10}{7} \cdot \dfrac{63}{21}$

18._____

19. $\dfrac{4}{13} \cdot \dfrac{52}{64}$

19._____

20. $\dfrac{8}{11} \cdot \dfrac{55}{72}$

20._____

21. $\dfrac{3}{4} \cdot \dfrac{5}{9} \cdot \dfrac{2}{5}$

21._____

Multiply. Write the answer in lowest terms; change the answer to a whole or mixed number where possible.

22. $27 \cdot \dfrac{5}{9}$

22._____

23. $49 \cdot \dfrac{6}{7}$

23._____

24. $72 \cdot \dfrac{5}{8}$

24._____

25. $26 \cdot \dfrac{2}{13}$

25._____

26. $30 \cdot \dfrac{3}{5}$

26._____

27. $27 \cdot \dfrac{7}{54}$

27._____

28. $21 \cdot \dfrac{3}{7} \cdot \dfrac{7}{9}$

28._____

29. $8 \cdot \dfrac{7}{32} \cdot \dfrac{1}{2}$

29._____

30. $200 \cdot \dfrac{7}{50} \cdot \dfrac{5}{28}$

30._____

Chapter 2 FRACTIONS AND MIXED NUMBERS

2.6   Applications of Multiplication

Solve each application problem. Look for indicator words.

1. A bookstore sold 2800 books, $\frac{3}{5}$ of which were paperbacks. How many paperbacks were sold?

1._____

2. A store sells 3750 items, of which $\frac{2}{15}$ are classified as junk food. How many of the items are junk food?

2._____

3. Sara needs $2500 to go to school for one year. She earns $\frac{3}{5}$ of this amount in the summer. How much does she earn in the summer?

3._____

4. Lany paid $120 for textbooks this term. Of this amount, the bookstore kept $\frac{1}{4}$. How much did the bookstore keep?

4._____

5. Of the 570 employees of Grand Tire Service, $\frac{7}{30}$ have given to the United Fund. How many have given to the United Fund?

5._____

6. A school gives scholarships to $\frac{3}{25}$ of its 1900 freshmen. How many students receive scholarships?

6._____

7. Kim earns $1500 a month. If she uses $\frac{1}{3}$ of her income on housing, how much does she pay for housing?

7._____

8. Steve's home is $\frac{3}{5}$ of the way from Carolyn's home to Laplace College, a distance of 45 miles. How far is it from Carolyn's home to Steve's?

8._____

9. Ben sets puts $\frac{1}{12}$ of his weekly earnings in a retirement fund. If he makes \$1248 a week, how much does he put in his retirement fund each week?

9._____

10. The sophomore class at Lincoln High School has 312 students. If $\frac{7}{13}$ of the students are boys, how many boys are in the sophomore class?

10._____

11. The Donut Shack sells donuts, bagels, and muffins. During a typical week, they sell 1120 items, of which $\frac{2}{7}$ are muffins. How many muffins does the Donut Shack sell in a typical week?

11._____

12. During the month of February $\frac{5}{7}$ of the days had temperatures that were below normal. How many days had below normal temperatures? (assume that this is not a leap year)

12._____

13. Marcie is reading a 360 page book. How many pages has she read if she had completed $\frac{4}{9}$ of the book?

13._____

14. A local hospital is recruiting new blood donors. During the month of June, $\frac{3}{16}$ of the people who donated were first-time donors. If 112 people donated blood in June, how many were first time donors?

14._____

15. Major league baseball teams play 162 games during the regular season. If a team wins $\frac{15}{27}$ of its games, how many games does it win?

15._____

16. A lawnmower uses a gasoline/oil mixture in which $\frac{1}{30}$ of the mixture must be oil. If the tank holds 150 ounces of the mixture, how many ounces are oil?

16._____

17. During a local election, a candidate received $\frac{2}{3}$ of the votes. If 2532 people voted in the election, how many votes did the candidate receive?

17._____

18. George takes a train to work everyday. The distance from George's house to the train station is $\frac{1}{8}$ of the total distance of 48 miles to work. What is the distance from his home to the train station?

18._____

19. Mary must calculate the area of her home office for tax purposes. What is area of Mary's home office if it takes up $\frac{2}{15}$ of her entire 2340 square foot home?

19._____

20. Richard is participating in a 150 mile bike ride for charity. There are 10 equally-spaced stops including the finish line. How many miles has Richard completed after reaching the 7th stop?

20._____

21. Kelly noticed that her gas gauge moved from the $\frac{7}{8}$ mark to the $\frac{4}{8}$ mark during a recent trip. If her tank holds 16 gallons, how many gallons did she use during this trip?

21._____

22. A local college has 8700 undergraduate students and $\frac{7}{30}$ of these students commute to school. How many students commute to school?

22._____

The Hu family earned $36,000 last year. Use this fact to solve Problems 23 – 26.

23. They paid $\frac{1}{3}$ of their income for taxes. Find their tax amount.

23._____

24. They spend $\frac{2}{5}$ of their income for rent. Find the amount spent on rent.

24._____

25. They saved $\frac{1}{16}$ of their income. How much did they save?

25._____

26. They spend $\frac{1}{6}$ of their income on food. How much did they spend on food?

26._____

The Highview Condo Association collects $96,000 each year from its members. Use this fact to solve Problems 27 – 30.

27. The association spends $\frac{1}{6}$ of this money on landscaping. How much do they spend on landscaping?

27._____

28. The association spends $\frac{5}{12}$ of this money on routine maintenance. How much do they spend on routine maintenance?

28._____

29. The association spends $\frac{3}{8}$ on insurance. How much to they spend on insurance?

29._____

30. The association puts $\frac{3}{32}$ of this money in an emergency fund. How much money do they put in this fund?

30._____

Chapter 2 FRACTIONS AND MIXED NUMBERS

2.7   Dividing Fractions

Find the reciprocal of each fraction.

1. $\dfrac{3}{4}$

1._____

2. $\dfrac{9}{2}$

2._____

3. $\dfrac{1}{3}$

3._____

4. $\dfrac{6}{7}$

4._____

5. 10

5._____

6. $\dfrac{15}{4}$

6._____

7. $\dfrac{8}{5}$

7._____

Divide. Write the answer in lowest terms; change the answers to a whole or mixed number where possible.

8. $\dfrac{1}{9} \div \dfrac{1}{3}$

8._____

9. $\dfrac{4}{5} \div \dfrac{3}{8}$

9._____

10. $\dfrac{\frac{7}{10}}{\frac{14}{5}}$

10._____

11. $\dfrac{\frac{4}{9}}{\frac{16}{27}}$

11._____

12. $\dfrac{\frac{8}{15}}{\frac{10}{12}}$

12._____

13. $\dfrac{28}{5} \div \dfrac{42}{25}$

13._____

14. $9 \div \dfrac{3}{2}$

14._____

15. $15 \div \dfrac{2}{5}$

15._____

16. $\dfrac{5}{8} \div 15$

16._____

17. $\dfrac{\frac{6}{11}}{18}$

17._____

18. $\dfrac{\frac{11}{3}}{5}$

18._____

19. $4 \div \dfrac{12}{7}$

19._____

20. $5 \div \dfrac{13}{20}$

20._____

21. $\dfrac{36}{\frac{12}{25}}$

21._____

Solve each application problem by using division.

22. Abel has a piece of property with an area of $\frac{7}{8}$ acre. He wishes to divide it into four equal parts for his children. How many acres of land will each child get?

22._____

23. Amanda wants to make doll dresses to sell at a craft's fair. Each dress needs $\frac{1}{3}$ yard of material. She has 18 yards of material. Find the number of dresses that she can make.

23._____

24. It takes $\frac{4}{5}$ pound of salt to fill a large salt shaker. How many salt shakers can be filled with 32 pounds of salt?

24._____

25. Lynn has 2 gallons of lemonade. If each of her Brownies gets $\frac{1}{12}$ gallon of lemonade, how many Brownies does she have?

25._____

26. How many $\frac{1}{9}$-ounce medicine vials can be filled with 7 ounces of medicine?

26._____

27. Each guest at a party will eat $\frac{5}{16}$ pound of chips. How many guests can be served with 10 pounds of chips?

27._____

28. Samantha uses $\frac{2}{3}$ yard of ribbon to make a bow for each package she wraps at May's Department Store. How many bows can she make if she has 60 yards of ribbon?

28._____

29. Bill wishes to make hamburger patties that weight $\frac{5}{12}$ pound. How many hamburger patties can he make with 10 pounds of hamburger?

29._____

30. Glen has a small pickup truck that will carry $\frac{3}{4}$ cord of firewood. Find the number of trips needed to deliver 30 cords of wood.

30._____

Chapter 2 FRACTIONS AND MIXED NUMBERS

2.8   Multiplying and Dividing Mixed Numbers

First estimate the answer. Then multiply to find the exact answer. Write the answer as a mixed number or a whole number.

1. $5\frac{1}{3} \cdot 2\frac{1}{2}$

    1._____

2. $3\frac{1}{2} \cdot 4\frac{2}{7}$

    2._____

3. $5\frac{1}{4} \cdot 3\frac{1}{5}$

    3._____

4. $3\frac{1}{2} \cdot 1\frac{3}{7}$

    4._____

5. $4\frac{4}{9} \cdot 2\frac{2}{5}$

    5._____

6. $5\frac{2}{3} \cdot 7\frac{1}{8}$

    6._____

7. $2\frac{1}{6} \cdot 3\frac{3}{4}$

    7._____

8. $18 \cdot 2\frac{5}{9}$

    8._____

9. $6 \cdot 3\frac{1}{2}$

    9._____

10. $3\frac{2}{5} \cdot 15$

    10._____

11. $\dfrac{5}{6} \cdot 2\dfrac{1}{2} \cdot 2\dfrac{2}{5}$

12. $1\dfrac{1}{4} \cdot 1\dfrac{1}{3} \cdot 1\dfrac{1}{2}$

First estimate the answer. Then divide to find the exact answer. Write the answer as a mixed number or a whole number.

13. $5\dfrac{5}{6} \div 5\dfrac{1}{4}$

13._____

14. $3\dfrac{1}{8} \div 2\dfrac{3}{4}$

14._____

15. $4\dfrac{5}{8} \div 1\dfrac{1}{4}$

15._____

16. $4\dfrac{3}{8} \div 3\dfrac{1}{2}$

16._____

17. $5\dfrac{3}{5} \div 1\dfrac{1}{6}$

17._____

18. $6\dfrac{1}{4} \div 2\dfrac{1}{2}$

18._____

19. $5 \div 3\dfrac{3}{4}$

19._____

20. $14 \div 8\frac{2}{5}$

20._____

21. $7\frac{1}{3} \div 6$

21._____

22. $4\frac{2}{3} \div 2$

22._____

23. $7\frac{1}{2} \div \dfrac{2}{3}$

23._____

24. $2\frac{5}{8} \div 1\frac{3}{4}$

24._____

First estimate the answer. Then solve each application problem by using multiplication or division to find the exact answer.

25. Maria wants to make 20 dresses to sell at a bazaar. Each dress needs $3\frac{1}{4}$ yards of material. How many yards does she need?

25._____

26. Juan worked $38\frac{1}{4}$ hours at $9 per hour. How much did he make?

26._____

27. Each home in an area needs $41\frac{1}{3}$ yards of rain gutter. How much rain gutter would be needed for 6 homes?

27._____

28. A farmer applies fertilizer to this fields at a rate of $5\frac{5}{6}$ gallons per acre. How many acres can he fertilize with $65\frac{5}{6}$ gallons?

28. _____

29. Insect spray is mixed using $1\frac{3}{4}$ ounces of a chemical per gallon of water. How many ounces of the chemical are needed to mix with $28\frac{4}{5}$ gallons of water?

29. _____

30. How many $\frac{3}{4}$-pound peanut cans can be filled with 15 pounds of peanuts?

30. _____

Chapter 2 FRACTIONS AND MIXED NUMBERS

2.9   Adding and Subtracting Like Fractions

Write like *or* unlike *for each set of fractions.*

1. $\dfrac{9}{7}, \dfrac{2}{7}$ 1._____

2. $\dfrac{3}{5}, \dfrac{4}{10}$ 2._____

3. $\dfrac{2}{5}, \dfrac{3}{5}$ 3._____

4. $\dfrac{2}{3}, \dfrac{3}{2}$ 4._____

5. $\dfrac{2}{15}, \dfrac{3}{15}, \dfrac{1}{5}$ 5._____

6. $\dfrac{0}{3}, \dfrac{5}{3}, \dfrac{1}{3}$ 6._____

Add. Write the answer in lowest terms and as a mixed number when possible.

7. $\dfrac{1}{4} + \dfrac{3}{4}$ 7._____

8. $\dfrac{3}{7} + \dfrac{2}{7}$ 8._____

9. $\dfrac{11}{15} + \dfrac{1}{15}$ 9._____

10. $\dfrac{4}{3} + \dfrac{7}{3}$ 10._____

11. $\dfrac{11}{16} + \dfrac{7}{16}$ 11._____

12. $\dfrac{9}{8} + \dfrac{1}{8}$ 12._____

13. $\dfrac{1}{6} + \dfrac{5}{6}$ 13._____

14. $\dfrac{1}{5} + \dfrac{2}{5} + \dfrac{4}{5}$ 14._____

15. $\dfrac{6}{10} + \dfrac{4}{10} + \dfrac{3}{10}$ 15._____

Solve each application problem. Write the answer in lowest terms.

16. Last month the Yee family paid $\frac{2}{11}$ of a debt. This   16._____
month they paid an additional $\frac{5}{11}$ of the same debt. What
fraction of the debt has been paid?

17. Malika walked $\frac{3}{8}$ of a mile downhill and then $\frac{1}{8}$ of a   17._____
mile along a creek. How far did she walk altogether?

18. Brent painted $\frac{1}{6}$ of a house last week and another $\frac{3}{6}$ this week. How much of the house is painted?

18._____

Subtract. Write the answer in lowest terms and as a mixed number when possible.

19. $\dfrac{11}{13} - \dfrac{3}{13}$

19._____

20. $\dfrac{3}{10} - \dfrac{1}{10}$

20._____

21. $\dfrac{15}{7} - \dfrac{8}{7}$

21._____

22. $\dfrac{7}{8} - \dfrac{3}{8}$

22._____

23. $\dfrac{16}{15} - \dfrac{6}{15}$

23._____

24. $\dfrac{8}{25} - \dfrac{5}{25}$

24._____

25. $\dfrac{29}{35} - \dfrac{1}{35}$

25._____

26. $\dfrac{25}{28} - \dfrac{15}{28}$

26._____

27. $\dfrac{31}{36} - \dfrac{11}{36}$

27._____

Solve each application problem. Write the answer in lowest terms.

28. Bill must walk $\frac{9}{12}$ of a mile. He has already walked $\frac{1}{12}$ of a mile. How much farther must he walk?

28._____

29. The Thompsons still owe $\frac{8}{15}$ of a debt. If they pay $\frac{2}{15}$ of it this month, what fraction of the debt will they still owe?

29._____

30. Jeff planted $\frac{11}{18}$ of his garden in corn and potatoes. If $\frac{5}{18}$ of the garden is corn, how much of the garden is potatoes?

30._____

Chapter 2 FRACTIONS AND MIXED NUMBERS

2.10   Least Common Multiples

Find the least common multiple for each of the following by listing the common multiples of each number.

1. 7, 14

1._____

2. 12, 18

2._____

3. 21, 28

3._____

4. 20, 65

4._____

5. 40, 50

5._____

Find the least common multiple for each of the following by using multiples of the larger number.

6. 5, 12

6._____

7. 16, 20

7._____

8. 15, 25

8._____

9. 14, 35

9._____

10. 21, 28

10._____

Find the least common multiple for each of the following using prime factorization.

11. 14, 28

11._____

12. 28, 32

12._____

13. 10, 24, 32

13._____

14. 16, 20, 25

14._____

15. 7, 12, 21, 35

15._____

Find the least common multiple for each set of numbers by using an alternate method.

16. 9, 18

16._____

17. 22, 55

17._____

18. 26, 65

18._____

19. 4, 18, 27

19._____

20. 12, 30, 40

_____

21. 7, 20, 35

_____

Rewrite each fraction with the indicated denominator.

22. $\dfrac{1}{3} = \dfrac{}{12}$

22._____

23. $\dfrac{1}{9} = \dfrac{}{36}$

23._____

24. $\dfrac{4}{7} = \dfrac{}{35}$

24._____

25. $\dfrac{4}{9} = \dfrac{}{81}$

25._____

26. $\dfrac{1}{13} = \dfrac{}{78}$

26._____

27. $\dfrac{3}{8} = \dfrac{}{88}$

27._____

28. $\dfrac{3}{13} = \dfrac{}{52}$

28._____

29. $\dfrac{7}{12} = \dfrac{}{60}$

29._____

30. $\dfrac{21}{11} = \dfrac{}{55}$

30._____

Chapter 2 FRACTIONS AND MIXED NUMBERS

2.11   Adding and Subtracting Unlike Fractions

Add. Write the answer in lowest terms.

1. $\dfrac{2}{3} + \dfrac{1}{6}$ 1._____

2. $\dfrac{1}{3} + \dfrac{1}{2}$ 2._____

3. $\dfrac{1}{5} + \dfrac{5}{8}$ 3._____

4. $\dfrac{9}{13} + \dfrac{3}{26}$ 4._____

5. $\dfrac{3}{10} + \dfrac{7}{15}$ 5._____

6. $\dfrac{5}{8} + \dfrac{1}{4}$ 6._____

7. $\dfrac{3}{11} + \dfrac{2}{33}$ 7._____

8. $\dfrac{5}{12} + \dfrac{9}{16}$ 8._____

9. $\dfrac{3}{5} + \dfrac{2}{9}$ 9._____

10. $\dfrac{3}{8} + \dfrac{5}{12}$ 10._____

Solve each application problem.

11. A buyer for a grain company bought $\frac{3}{8}$ ton of wheat, $\frac{1}{6}$ 11._____
ton of rice, and $\frac{1}{4}$ ton of barley. How many tons of grain
were bought?

12. Michael Pippen paid $\frac{1}{9}$ of a debt in January, $\frac{1}{2}$ in   12._____
February, $\frac{1}{4}$ in March, and $\frac{1}{12}$ in April. What fraction of
the debt was paid in these four months?

Add. Write the answer in lowest terms.

13. $\dfrac{1}{2}$ 13._____

$+\dfrac{1}{3}$

14. $\dfrac{7}{12}$

$+\dfrac{3}{8}$

14. _____

15. $\dfrac{2}{15}$

$+\dfrac{7}{10}$

15. _____

16. $\dfrac{1}{6}$

$+\dfrac{2}{9}$

16. _____

17. $\dfrac{3}{7}$

$+\dfrac{4}{21}$

17. _____

18. $\dfrac{5}{22}$

$+\dfrac{7}{33}$

18. _____

19. $\dfrac{6}{13}$

$+\dfrac{15}{52}$

19. _____

20. $\dfrac{3}{14}$

      $+\dfrac{5}{21}$

20._____

Subtract. Write the answer in lowest terms.

21. $\dfrac{7}{8}-\dfrac{1}{2}$

21._____

22. $\dfrac{2}{3}-\dfrac{1}{6}$

22._____

23. $\dfrac{7}{12}-\dfrac{1}{4}$

23._____

24. $\dfrac{8}{15}-\dfrac{1}{5}$

24._____

25. $\dfrac{7}{8}$

      $-\dfrac{2}{3}$

25._____

26. $\dfrac{7}{8}$

      $-\dfrac{5}{6}$

26._____

27. $\dfrac{2}{3}$

$-\dfrac{3}{5}$

27._____

28. $\dfrac{7}{15}$

$-\dfrac{3}{10}$

28._____

Solve each application problem.

29. A company has $\frac{5}{8}$ acre of land. They sell $\frac{1}{3}$ acre. How much land is left?

29._____

30. Greg had $\frac{7}{12}$ of his savings goal to complete at the beginning of the month. During the month he saved another $\frac{1}{8}$ of the goal. How much of the goal is left to save?

30._____

Chapter 2 FRACTIONS AND MIXED NUMBERS

2.12   Adding and Subtracting Mixed Numbers

First estimate the answer. Then add or subtract to find the exact answer. Write each answer as a mixed number in lowest terms.

1. $5\frac{1}{7}$

   $+\ 4\frac{3}{7}$

   1._____

2. $7\frac{3}{4}$

   $+\ 4\frac{5}{8}$

   2._____

3. $17\frac{5}{8}$

   $12\frac{1}{4}$

   $+\ \ 5\frac{5}{6}$

   3._____

4. $126\frac{4}{5}$

   $28\frac{9}{10}$

   $+\ 13\frac{2}{15}$

   4._____

5. $5\frac{3}{5}$

   $-\ 2\frac{1}{10}$

   5._____

6. $12\frac{9}{16}$

   $-\ \ 2\frac{3}{8}$

   6._____

First estimate the answer. Then solve each application problem.

7. A mechanic had $8\frac{3}{4}$ gallons of transmission fluid. If he purchased $2\frac{1}{3}$ gallons of fluid, find the number of gallons he has altogether.

7._____

8. Nate bought $2\frac{3}{8}$ boxes of oranges and $2\frac{2}{3}$ boxes of lemons. How many boxes of fruit did he buy in all?

8._____

9. Paul worked $12\frac{3}{4}$ hours over the weekend. He worked $6\frac{3}{8}$ hours on Saturday. How many hours did he work on Sunday?

9._____

10. Marty Hirsch worked $6\frac{2}{5}$ hours on Monday, $7\frac{1}{2}$ hours on Tuesday, $8\frac{3}{4}$ hours on Wednesday, $7\frac{4}{5}$ hours on Thursday, and 8 hours on Friday. How many hours did he work altogether?

10._____

11. The Eastside Wholesale Vegetable Market sold $4\frac{3}{4}$ tons of broccoli, $8\frac{2}{3}$ tons of spinach, $2\frac{1}{2}$ tons of corn, and $1\frac{5}{12}$ tons of turnips last month. Find the total number of tons of these vegetables sold by the market last month.

11._____

First estimate the answer. Then subtract to find the exact answer. Write each answer as a mixed number in lowest terms.

12. $9\frac{1}{8}$

      $-\,7\frac{3}{8}$

12._____

13. $6\frac{1}{3}$

      $-\,5\frac{7}{12}$

13._____

14. $27\frac{2}{15}$

      $-\,18\frac{7}{10}$

14._____

15. $12\frac{5}{12}$

      $-\,11\frac{11}{16}$

15._____

16. $42\frac{7}{12}$

      $-\,29\frac{5}{8}$

16._____

17. 42

      $-\,19\frac{3}{4}$

17._____

Name: Date:

Instructor: Section:

First estimate the answer. Then solve each application problem.

18. Amy Atwood worked 40 hours during a certain week.
She worked $8\frac{1}{4}$ hours on Monday, $6\frac{3}{8}$ hours on
Tuesday, $7\frac{3}{4}$ hours on Wednesday, and $8\frac{3}{4}$ hours on
Thursday. How many hours did she work on Friday?

18. _____

19. Three sides of a parking lot are $35\frac{1}{4}$ yards, $42\frac{7}{8}$ yards,
and $32\frac{3}{4}$ yards. If the total distance around the lot is
$145\frac{1}{2}$ yards, find the length of the fourth side.

19. _____

20. A concrete truck is loaded with $11\frac{5}{8}$ cubic yards of
concrete. The driver unloads $1\frac{1}{6}$ cubic yards at the first
stop, and $2\frac{5}{12}$ cubic yards at the second stop. The
customer at the third stop gets 3 cubic yards. How much
concrete is left in the truck?

20. _____

21. Debbie Andersen bought 15 yards of material at a sale.
She made a shirt with $3\frac{1}{8}$ yards of the material, a dress
with $4\frac{7}{8}$ yards, and a jacket with $3\frac{3}{4}$ yards. How many
yards of material were left over?

21. _____

Find x in the following figures.

22.

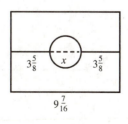

22. _____

WS-91

23.

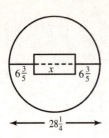

23._____

Add or subtract by changing mixed numbers to improper fractions. Write the answer as a mixed number.

24. $3\frac{3}{4}$
         $+1\frac{1}{2}$

24._____

25. $5\frac{1}{3}$
         $+2\frac{5}{6}$

25._____

26. $3\frac{7}{8}$
         $+1\frac{5}{12}$

26._____

27. $4\frac{3}{4}$
         $-2\frac{3}{8}$

27._____

28. $3\frac{1}{2}$
         $-1\frac{2}{3}$

28._____

29. $5\frac{5}{8}$
         $-2\frac{3}{4}$

29._____

30. $3\frac{2}{3}$
         $-1\frac{5}{6}$

30._____

Chapter 2 FRACTIONS AND MIXED NUMBERS

2.13   Order Relations and the Order of Operations

Write < or > to make a true statement.

1. $\dfrac{1}{2}$ ____ $\dfrac{5}{8}$

1._____

2. $\dfrac{2}{3}$ ____ $\dfrac{5}{6}$

2._____

3. $\dfrac{3}{8}$ ____ $\dfrac{5}{16}$

3._____

4. $\dfrac{7}{5}$ ____ $\dfrac{19}{15}$

4._____

5. $\dfrac{5}{12}$ ____ $\dfrac{3}{5}$

5._____

6. $\dfrac{11}{15}$ ____ $\dfrac{13}{20}$

6._____

7. $\dfrac{13}{24}$ ____ $\dfrac{23}{36}$

7._____

8. $\dfrac{23}{40}$ ____ $\dfrac{17}{30}$

8._____

9. $\dfrac{17}{25}$ ____ $\dfrac{9}{16}$

9._____

Evaluate each of the following.

10. $\left(\dfrac{1}{4}\right)^3$

10._____

11. $\left(\dfrac{1}{2}\right)^2$

11._____

12. $\left(\dfrac{2}{3}\right)^2$

13. $\left(\dfrac{1}{5}\right)^2$

13._____

14. $\left(\dfrac{5}{3}\right)^3$

14._____

15. $\left(\dfrac{1}{5}\right)^3$

15._____

16. $\left(\dfrac{1}{2}\right)^4$

16._____

17. $\left(\dfrac{3}{2}\right)^4$

17._____

18. $\left(\dfrac{3}{7}\right)^3$

18._____

19. $\left(\dfrac{3}{4}\right)^3$

19._____

Use the order of operations to simplify.

20. $\left(\dfrac{2}{3}\right)^2 \cdot 6$

20._____

21. $\left(\dfrac{4}{5}\right)^2 \cdot \dfrac{5}{12}$

21._____

22. $\left(\dfrac{3}{5}\right)^2 \cdot \left(\dfrac{2}{3}\right)^2$

22._____

23. $\left(\dfrac{4}{3}\right)^2 \cdot \left(\dfrac{1}{8}\right)^2$

23._____

24. $9 \cdot \left(\dfrac{3}{2}\right)^2 \cdot \left(\dfrac{1}{6}\right)^2$

24._____

25. $\dfrac{4}{3} - \dfrac{1}{2} + \dfrac{7}{12}$

25._____

26. $\dfrac{7}{8} - \dfrac{3}{4} + \dfrac{1}{2}$

26._____

27. $\dfrac{1}{3} \cdot \dfrac{3}{7} \cdot \dfrac{5}{4}$

27._____

28. $\dfrac{2}{5} \cdot \dfrac{15}{11} \cdot \dfrac{33}{8}$

28._____

29. $\dfrac{3}{14} \cdot \dfrac{7}{5} + \dfrac{1}{2} \cdot \dfrac{2}{5}$

29._____

30. $\dfrac{5}{8} - \dfrac{2}{3} \cdot \dfrac{3}{4}$

30._____

Chapter 3 DECIMALS

3.1   Reading and Writing Decimals

Write the portion of each square that is shaded as a fraction, as a decimal, and in words.

1.

1._____

2.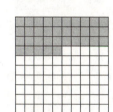

2._____

3.

3._____

4.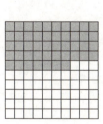

4._____

5.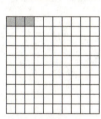

5._____

Identify the digit that has the given place value.

6. 43.507   tenths   hundredths

6._____

7. 5.632   tenths   hundredths

7._____

8. 0.769   hundredths   thousandths   8._____

9. 2.83714 thousandths   ten-thousandths   9._____

10. 42.692   tens   tenths   10._____

11. 302.9651   hundreds   Hundredths   11._____

Identify the place value of each digit in these decimals.

12. 0.73   7   3   12._____

13. 0.85   8   5   13._____

14. 0.782   7   8   2   14._____

15. 0.176   1   7   6   15._____

Tell how to read each decimal in words.

16. 0.08   16._____

17. 4.06   17._____

18. 0.0561   18._____

19. 2.304   19._____

20. 97.008   20._____

Write each decimal in numbers.

21. Five and four hundredths

21._____

22. Eleven and nine thousandths

22._____

23. Thirty-eight and fifty-two hundred-thousandths

23._____

24. Three hundred and twenty-three ten-thousandths

24._____

Write each decimal as a fraction or mixed number in lowest terms.

25. 0.8

25._____

26. 0.1

26._____

27. 3.6

27._____

28. 4.26

28._____

29. 0.95

29._____

30. 3.75

30._____

Chapter 3 DECIMALS

3.2   Rounding Decimals

Round each number to the place indicated.

1. 17.8937 to the nearest tenth

1._____

2. 489.84 to the nearest tenth

2._____

3. 785.4982 to the nearest thousandth

3._____

4. 43.51499 to the nearest ten-thousandth

4._____

5. 53.329 to the nearest hundredth

5._____

6. 75.399 to the nearest tenth

6._____

7. 486.496 to the nearest one

7._____

Round to the nearest hundredth and then to the nearest tenth. Remember to always round the original number.

8. 89.525

8._____

9. 21.769

9._____

10. 0.8948

10._____

11. 1.437

11._____

12. 0.0986

12._____

13. 114.038

13._____

14. 101.749

14._____

15. 78.695

15._____

16. 108.073

16._____

Round to the nearest dollar.

17. $79.12

17._____

18. $28.39

18._____

19. $225.98

20. $4797.50

21. $11,839.73

22. $27,869.57

23. $276.49

19._____

20._____

21._____

22._____

23._____

Round to the nearest cent.

24. $1.2499

25. $1.0924

26. $112.0089

27. $134.20506

28. $1028.6666

29. $2096.0149

30. $62.179

24._____

25._____

26._____

27._____

28._____

29._____

30._____

Name: Date:

Instructor: Section:

Chapter 3 DECIMALS

3.3   Adding and Subtracting Decimals

Find each sum.

1. 43.96 + 48.53

                                                             1._____

2. 47.94 + 102.38 + 27.631

    2._____

3. 87.6 + 90.4

    3._____

4. 45.83 + 20.923 + 5.7

    4._____

5. 10.82 + 5.9 + 4.7 + 6.3 + 20.63

    5._____

Find the perimeter of (distance around) each geometric figure by adding the lengths of the sides.

6.

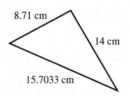

8.71 cm

14 cm

15.7033 cm

    6._____

7.

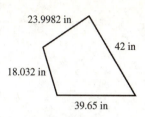

8.

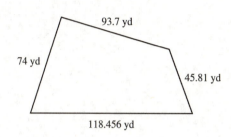

9.

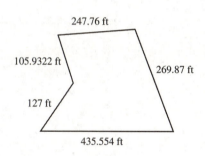

Find each difference.

10. $84.6 - 18.1$

11. $223.3 - 107.5$

12. 69.524 – 26.958

12._____

13. 23.104 – 6.98

13._____

14. 689 – 79.832

14._____

Find the value of x in each diagram.

15.

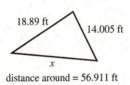

18.89 ft   14.005 ft

x

distance around = 56.911 ft

15._____

16.

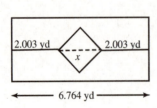

2.003 yd   2.003 yd

x

6.764 yd

16._____

17.

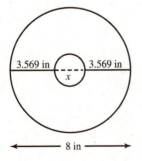

3.569 in   3.569 in

x

8 in

17._____

18.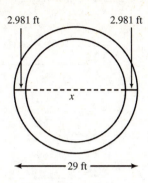

2.981 ft   2.981 ft

x

← 29 ft →

18. _____

First, use front end rounding and estimate each answer. Then add or subtract to find the exact answer.

19. 32.99
    41.72
+ 8.2

19. _____

20. 20.85
− 7.69

20. _____

21. 9.7
− 4.862

21. _____

22. 593.8
  27.93
+ 54.87

22. _____

23. 9
− 3.47

23. _____

First, use front end rounding and estimate each answer. Then add or subtract to find the exact answer for each application problem.

24. Kim spent $28.25 for books, $29.47 for a blouse, and $17.85 for a compact disk. How much did she spend?

24._____

25. Tom Rodriguiz made $365.29 at the regular rate of pay and $87.59 at the overtime rate. How much did he make?

25._____

26. Michael Lee worked 3.5 days one week, 5.1 days another week, and 4.8 days a third week. How many days did he work altogether?

26._____

27. A customer gives a clerk a $20 bill to pay for $11.29 in purchases. How much change should the customer get?

27._____

28. A man buys $37.57 worth of sporting goods and pays with a $50 bill. How much change should he get?

28._____

29. At a fruit stand, Lynn Knight bought $8.53 worth of apples, $11.10 worth of peaches, and $28.29 worth of pears. How much did she spend altogether?

29._____

30. A man receives a bill for $83.26 from Exxon. Of this amount, $53.29 is for a tune-up and the rest is for gas. How much did he pay for gas?

30._____

Chapter 3 DECIMALS

3.4   Multiplying Decimals

Multiply.

1.   0.053
     × 4.3

1._____

2.   0.682
     × 3.9

2._____

3.   19.3
     × 4.7

3._____

4.   96.5
     × 4.6

4._____

5.   67.6
     × 0.023

5._____

6.   906
     × 0.081

6._____

7. $0.074 \cdot 0.05$

7._____

8. $0.0009 \cdot 0.014$

8._____

9. $0.00321 \cdot 0.003$

9._____

10.   62.92
        $\underline{\times \, 0.032}$

10._____

In each of the following, find the amount of money earned on a job by multiplying the number of hours worked and the pay per hour. Round your answer to the nearest cent, if necessary.

11. 27 hours at $6.04 per hour

11._____

12. 31.6 hours at $9.83 per hour

12._____

13. 35 hours at $5.72 per hour

13._____

14. 27.6 hours at $7.21 per hour

14._____

Find the cost of each of the following.

15. 16 apples at $0.59 each

15._____

16. 7 quarts of oil at $1.05 each

16._____

17. 12 rolls of film at $1.72 each

17._____

18. 25 cans of soda at $0.29 each

18._____

First use front end rounding and estimate the answer. Then multiply to find the exact answer.

19.   49.7
    × 5.8

19._____

20.   29.8
    × 3.4

20._____

21.   58.73
    × 3.72

21._____

22.   32.53
         × 23.26

22. _____

23.   76.4
         × 0.57

23. _____

24.   2.99
         × 3.5

24. _____

25.   391.9
         × 7.74

25. _____

26.   27.5
         × 11.2

26. _____

Solve. If the problem involves money, round to the nearest cent, if necessary.

27. Mary Williams pays $32.96 per month for a television
payment. How much will she pay over 15 months?

27. _____

28. Steve's car payment is $309.56 per month for 48 months. How much will he pay altogether?

28._____

29. The Duncan family's state income tax is found by multiplying the family income of $32,906.15 by the decimal 0.064. Find their tax.

29._____

30. A recycling center pays $0.142 per pound of aluminum. How much would be paid for 176.3 pounds?

30._____

Chapter 3 DECIMALS

3.5   Dividing Decimals

Divide. Round answers to the nearest thousandth, if necessary.

1. $3\overline{)43.95}$ 1._____

2. $33\overline{)77.847}$ 2._____

3. $4\overline{)31.974}$ 3._____

4. $11\overline{)46.98}$ 4._____

5. $3\overline{)62.7}$ 5._____

6. $41\overline{)726.43}$ 6._____

Find the cost of each item. Round to the nearest cent.

7. 3 pairs of socks for $5.98   7._____

8. 5 pounds of applies for $3.99

8._____

9. 49 books for $154.84

9._____

10. 200 pencils for $10.20

10._____

Divide. Round answers to the nearest thousandth, if necessary.

11. $0.3\overline{)38.84}$

11._____

12. $0.6\overline{)78.59}$

12._____

13. $4.5\overline{)79.468}$

13._____

14. $3.1\overline{)726.43}$

14._____

15. $0.52\overline{)34.96}$

15._____

16. $0.07 \div 0.00043$

16._____

Solve each application problem. Round money answers to the nearest cent, if necessary.

17. Leon Williams drove 542.2 miles on the 16.3 gallons of gas in his Ford Taurus. How many miles per gallon did he get? Round to the nearest tenth.

17._____

18. Raymond Starr bought 7.4 yards of fabric, paying a total of $26.27. Find the cost per yard.

18._____

19. To build a barbecue, Diana Jenkins bought 589 bricks, paying $185.70. Find the cost per brick.

19._____

20. Cerise Montoya is a newspaper distributor. Last week she paid the newspaper $261.02 for 1684 copies. Find the cost per copy.

20._____

Decide if each answer is reasonable *or* unreasable *by rounding the numbers and estimating the answer.*

21. $49.8 \div 7.1 = 7.014$

21._____

22. $31.5 \div 8.4 = 37.5$

22._____

23. $624.7 \div 19.24 = 32.469$

23._____

24. $1092.8 \div 37.92 = 2.882$

24._____

25. $8695.15 \div 98.762 = 880.415$

25._____

Simplify by using the order of operations.

26. $3.7 + 5.1^2 - 9.4$

26._____

27. $42.92 \div 5.8 \cdot 7.3$

27._____

28. $18.5 + (37.1 - 29.8) \cdot 10.7$

    28. _____

29. $27.51 - 3.2 \cdot 9.8 \div 1.6$

    29. _____

30. $9.8 \cdot 4.76 + 17.94 \div 2.6$

    30. _____

Chapter 3 DECIMALS

3.6   Writing Fractions and Decimals

Write each fraction or mixed number as a decimal. Round to the nearest thousandth, if necessary.

1. $6\frac{1}{2}$

1._____

2. $\dfrac{1}{5}$

2._____

3. $2\frac{2}{3}$

3._____

4. $\dfrac{1}{8}$

4._____

5. $\dfrac{1}{11}$

5._____

6. $7\frac{1}{10}$

6._____

7. $\dfrac{3}{5}$

7._____

8. $\dfrac{7}{8}$

8._____

9. $4\frac{1}{9}$

9._____

10. $\dfrac{13}{25}$ 10._____

11. $\dfrac{3}{20}$ 11._____

12. $31\frac{3}{13}$ 12._____

13. $\dfrac{11}{18}$ 13._____

14. $19\frac{17}{24}$ 14._____

Find the smaller of the two given numbers. Write < or > to make a true statement.

15. $\dfrac{5}{8}$ ___ 0.634   15._____

16. $\dfrac{2}{5}$ ___ 0.401   16._____

17. $\dfrac{1}{5}$ ___ 0.19   17._____

18. $\dfrac{2}{3}$ ___ 0.67   **18.** _____

19. $\dfrac{3}{5}$ ___ 0.599   **19.** _____

20. $\dfrac{5}{6}$ ___ 0.83   **20.** _____

21. $\dfrac{1}{25}$ ___ 0.039   **21.** _____

22. $\dfrac{3}{8}$ ___ 0.38   **22.** _____

23. $\dfrac{5}{9}$ ___ 0.55   **23.** _____

24. $\dfrac{3}{16}$ ___ 0.188   **24.** _____

Arrange in order from smallest to largest.

25. $\dfrac{3}{11}, \dfrac{1}{3}, 0.29$ **25.** _____

26. $\dfrac{8}{9}, 0.88, 0.89$ **26.** _____

27. $\frac{1}{6}$, 0.166, 0.1666   **27.**_____

28. $\frac{7}{15}$, 0.466, $\frac{9}{19}$ **28.**_____

29. $\frac{1}{7}$, $\frac{3}{16}$, 0.187   **29.**_____

30. 0.8462, $\frac{11}{13}$, $\frac{6}{7}$ **30.**_____

Chapter 4 RATIO, PROPORTION, AND PERCENT

4.1   Ratios

Write each ratio as a fraction in lowest terms.

1. 7 to 8

1._____

2. 18 to 24

2._____

3. 76 to 101

3._____

4. 30 to 84

4._____

5. 95 cents to 125 cents

5._____

6. 80 miles to 30 miles

6._____

7. $85 to $135

7._____

Solve each application problem. Write each ratio as a fraction in lowest terms.

8. Mr. Williams is 42 years old, and his son is 18. Find the ratio of Mr. Williams' age to his son's age.

8._____

9. When using Roundup vegetation control, add 128 ounces of water for every 6 ounces of the herbicide. Find the ratio of herbicide to water.

9._____

10. The area of Gettysburg National Military Park is 3900
    acres, and the area of Chickamauga and Chattanooga
    National Military Park is 8100 acres. Find the ratio of
    the areas of the two parks.

10._____

Write each ratio as a fraction in lowest terms.

11. $6\frac{1}{2}$ to 2

11._____

12. $4\frac{1}{8}$ to 3

12._____

13. 3 to $2\frac{1}{2}$

13._____

14. 11 to $2\frac{4}{9}$

14._____

15. $1\frac{1}{4}$ to $1\frac{1}{2}$

15._____

16. $3\frac{1}{2}$ to $1\frac{3}{4}$

16._____

Solve each application problem. Write each ratio as a fraction in lowest terms.

17. One refrigerator holds $3\frac{3}{4}$ cubic feet of food, while
    another holds 5 cubic feet. Find the ratio of the amount
    of storage in the first refrigerator to the amount of
    storage in the second.

17._____

18. One car has a $15\frac{1}{2}$ gallon gas tank while another has a 22   **18.**_____
gallon gas tank. Find the ratio of the amount the first
tank holds to the amount the second tank holds.

19. A building is $42\frac{1}{2}$ feet tall. It casts a shadow $1\frac{3}{14}$ feet   **19.**_____
long. Find the ratio of the height of the building to the
length of its shadow.

For each triangle, find the ratio of the length of the longest side to the length of the shortest side. Write each ratio as a fraction in lowest terms.

20.

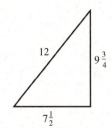

  20._____

21.

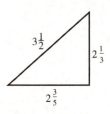

  21._____

Write each ratio as a fraction in lowest terms. Be sure to make all necessary conversions.

22. 4 days to 2 weeks   **22.**_____

23. 4 feet to 15 inches   **23.**_____

24. 6 yards to 10 feet   **24.**_____

25. 7 gallons to 8 quarts   **25.**_____

26. 40 ounces to 3 pounds

26._____

27. 80 cents to $3

27._____

Write each ratio as a fraction in lowest terms. Be sure to make all necessary conversions.

28. Find the ratio of $17\frac{1}{2}$ inches to $2\frac{1}{3}$ feet.

28._____

29. What is the ratio of $59\frac{1}{2}$ days to $4\frac{1}{4}$ weeks?

29._____

30. What is the ratio of $9\frac{1}{3}$ ounces to $3\frac{1}{2}$ pounds?

30._____

Chapter 4 RATIO, PROPORTION, AND PERCENT

4.2   Rates

Write each rate as a fraction in lowest terms.

1. 75 miles in 25 minutes

1._____

2. 85 feet in 17 seconds

2._____

3. 28 dresses for 4 people

3._____

4. 70 horses for 14 teams

4._____

5. 45 gallons in 3 hours

5._____

6. 225 miles on 15 gallons

6._____

7. 119 pills for 17 patients

7._____

8. 144 kilometers on 16 liters

8._____

9. 256 pages for 8 chapters

9._____

10. 990 miles in 18 hours

10._____

Find each unit rate.

11. $75 in 5 hours

11._____

12. $3500 in 20 days

12._____

13. $1540 in 14 days

13._____

14. $7875 for 35 pounds

14._____

15. $122.76 in 9 hours

15._____

16. 189.88 miles on 9.4 gallons

16._____

17. 262.08 miles on 9.6 gallons

17._____

18. $2\frac{3}{4}$ pounds for 11 people

18._____

19. $96.25 for 11 hours

19._____

Solve each application problem.

20. Eric can pack 12 crates of berries in 24 minutes. Give his rate in rate per minute and in minutes per crate.

20._____

21. Michelle can plow 7 acres in 14 hours. Give her rate in acres per hour and in hours per acre.

21._____

22. Candy makes $220.32 in 24 hours. What is her rate per hour?

22._____

23. The 4.6 yards of fabric needed for a dress costs $27.14. Find the cost of 1 yard.

23._____

24. The cost of 24.3 square yards of carpet is $872.37. Find the cost of 1 square yard.

24._____

25. Ms. Jordan bought 145 shares of stock for $1667.50. Find the cost of 1 share.

25._____

Find the best buy (based on cost per unit) for each item.

26. Beans: 12 ounces for $1.49, 16 ounces of $1.89

26._____

27. Orange juice: 16 ounces for $0.89, 32 ounces for $1.90

27._____

28. Cola: 6 cans for $1.98, 12 cans for $3.59, 24 cans for $8

28._____

29. Soup: 3 cans for $1.75, 5 cans for $2.75, 8 cans for $4.55

29._____

30. Cereal: 10 ounces for $1.34, 15 ounces for $1.76, 20 ounces for $2.29

30._____

Chapter 4 RATIO, PROPORTION, AND PERCENT

4.3   Proportions

Write each proportion.

1. 11 is to 15 as 22 is to 30.

   1._____

2. 50 is to 8 as 75 is to 12.

   2._____

3. 24 is to 30 as 8 is to 10.

   3._____

4. 36 is to 45 as 8 is to 10.

   4._____

5. 14 is to 21 as 10 is to 15.

   5._____

6. 3 is to 33 as 12 is to 132.

   6._____

7. 26 is to 4 as 39 is to 6.

   7._____

8. 9 is to 3 as 42 is to 14.

   8._____

9. $1\frac{1}{2}$ is to 4 as 21 is to 56.

   9._____

10. $3\frac{2}{3}$ is to 11 as 10 is to 30.

   10._____

Write each ratio in lowest terms in order to decide if the proportion is true *or* false.

11. $\dfrac{6}{100} = \dfrac{3}{50}$ **11.**_____

12. $\dfrac{48}{36} = \dfrac{3}{4}$ **12.**_____

13. $\dfrac{3}{8} = \dfrac{21}{28}$ **13.**_____

14. $\dfrac{30}{25} = \dfrac{6}{5}$ **14.**_____

15. $\dfrac{390}{100} = 27$ **15.**_____

16. $\dfrac{35}{21} = \dfrac{3}{4}$ **16.**_____

17. $\dfrac{28}{6} = \dfrac{42}{9}$ **17.**_____

18. $\dfrac{54}{30} = \dfrac{108}{60}$ **18.**_____

19. $\dfrac{15}{24} = \dfrac{25}{35}$ **19.**_____

20. $\dfrac{63}{18} = \dfrac{56}{14}$ 20._____

Cross multiply to see whether the proportion is true *or* false.

21. $\dfrac{12}{18} = \dfrac{20}{30}$ 21._____

22. $\dfrac{10}{45} = \dfrac{6}{27}$ 22._____

23. $\dfrac{28}{50} = \dfrac{49}{75}$ 23._____

24. $\dfrac{132}{24} = \dfrac{11}{3}$ 24._____

25. $\dfrac{210}{300} = \dfrac{14}{20}$ 25._____

26. $\dfrac{3\frac{1}{2}}{4} = \dfrac{14}{16}$ 26._____

27. $\dfrac{4\frac{3}{5}}{9} = \dfrac{23}{36}$ 27._____

28. $\dfrac{21}{28} = \dfrac{5\frac{3}{4}}{7}$ 28._____

29. $\dfrac{4}{4\frac{2}{3}} = \dfrac{30}{35}$ **29.** _____

30. $\dfrac{6\frac{1}{9}}{3\frac{2}{3}} = \dfrac{40}{24}$ **30.** _____

Chapter 4 RATIO, PROPORTION, AND PERCENT

4.4   Solving Proportions

Find the unknown number in each proportion.

1. $\dfrac{3}{2} = \dfrac{x}{6}$ 1._____

2. $\dfrac{9}{4} = \dfrac{36}{x}$ 2._____

3. $\dfrac{9}{7} = \dfrac{x}{28}$ 3._____

4. $\dfrac{x}{11} = \dfrac{44}{121}$ 4._____

5. $\dfrac{35}{x} = \dfrac{5}{3}$ 5._____

6. $\dfrac{x}{52} = \dfrac{5}{13}$ 6._____

7. $\dfrac{96}{60} = \dfrac{8}{x}$ 7._____

8. $\dfrac{7}{5} = \dfrac{98}{x}$ 8._____

9. $\dfrac{9}{14} = \dfrac{x}{70}$

9._____

10. $\dfrac{90}{x} = \dfrac{15}{8}$

10._____

11. $\dfrac{x}{110} = \dfrac{7}{10}$

11._____

12. $\dfrac{14}{x} = \dfrac{21}{18}$

12._____

13. $\dfrac{18}{81} = \dfrac{4}{x}$

13._____

14. $\dfrac{100}{x} = \dfrac{75}{30}$

14._____

15. $\dfrac{125}{75} = \dfrac{x}{33}$

15._____

Find the unknown number in each proportion. Write the answer as a whole number or a mixed number when possible.

16. $\dfrac{7}{8} = \dfrac{x}{2}$

16._____

17. $\dfrac{5}{x} = \dfrac{3}{7}$

17._____

18. $\dfrac{x}{6} = \dfrac{4}{9}$ **18.** _____

19. $\dfrac{2}{3\frac{1}{4}} = \dfrac{8}{x}$ **19.** _____

20. $\dfrac{3}{x} = \dfrac{5}{1\frac{2}{3}}$ **20.** _____

21. $\dfrac{x}{6} = \dfrac{5\frac{1}{4}}{7}$ **21.** _____

22. $\dfrac{1\frac{1}{5}}{\frac{1}{2}} = \dfrac{6}{x}$ **22.** _____

23. $\dfrac{0}{5\frac{1}{3}} = \dfrac{x}{5}$ **23.** _____

24. $\dfrac{x}{7\frac{1}{2}} = \dfrac{0}{9\frac{2}{3}}$ **24.** _____

25. $\dfrac{3}{x} = \dfrac{0.8}{5.6}$ **25.** _____

26. $\dfrac{16}{12} = \dfrac{2}{x}$ **26.** _____

27. $\dfrac{4.2}{x} = \dfrac{0.6}{2}$

27._____

28. $\dfrac{2\frac{1}{2}}{1\frac{2}{3}} = \dfrac{x}{2}$

28._____

29. $\dfrac{2\frac{5}{9}}{x} = \dfrac{23}{\frac{3}{5}}$

29._____

30. $\dfrac{x}{7.9} = \dfrac{0}{47.4}$

30._____

Chapter 4 RATIO, PROPORTION, AND PERCENT

4.5   Solving Application Problems with Proportions

Set up and solve a proportion for each problem.

1. A gardening service charges \$45 to install 50 square feet of sod. Find the charge to install 125 feet.

     1._____

2. On a road map, a length of 3 inches represents a distance of 8 miles. How many inches represent a distance of 32 miles?

     2._____

3. If 6 melons cost \$9, find the cost of 10 melons.

     3._____

4. If 22 hats cost \$198, find the cost of 12 hats.

     4._____

5. 6 pounds of grass seed cover 4200 square feet of ground. How many pounds are needed for 5600 square feet.

     5._____

6. Margie earns $168.48 in 26 hours. How much does she earn in 40 hours?

6._____

7. Juan makes $477.40 in 35 hours. How much does he make in 60 hours.

7._____

8. If 5 ounces of a medicine must be mixed with 12 ounces of water, how many ounces of medicine would be mixed with 132 ounces of water.

8._____

9. The distance between two cities on a road map is 5 inches. The two cities are really 600 miles apart. The distance between two other cities on the map is 8 inches. How many miles apart are these cities?

9._____

10. The distance between two cities is 600 miles. On a map the cities are 10 inches apart. Two other cities are 720 miles apart. How many inches apart are they on the map?

10._____

11. If 2 visits to a salon cost $80, find the cost of 11 visits.   **11.**_____

12. If a 4-minute phone call cots $0.96, find the cost of a 10-   **12.**_____
minute call.

13. If 150 square yards of carpet cost $3142.50, find the cost   **13.**_____
of 210 square yards of the carpet.

14. Scott paid $240,000 for a 5-unit apartment house. Find   **14.**_____
the cost of a 16-unit apartment house.

15. Brian plants his seeds early in the year. To keep them   **15.**_____
from freezing, he covers the ground with black plastic.
A piece with an area of 80 square feet costs $14. Find
the cost of a piece with an area of 700 square feet.

16. A taxi ride of 7 miles cots $9.45. Find the cost of a ride   **16.**_____
of 12 miles.

17. Dog food for 8 dogs cots $15. Find the cost of dog food for 12 dogs.

17._____

18. To make battery acid, Jeff mixes $9\frac{1}{2}$ gallons of pure acid with 25 gallons of water. How much acid would be needed for 75 gallons of water?

18._____

19. Tax on an $18,000 car is $1620. Find the tax on a $24,000 car.

19._____

20. If $18\frac{3}{4}$ yards of material are needed for 5 dresses, how much material is needed for 9 dresses?

20._____

21. If it takes 6 minutes to read 4 pages of a book, how long will it take to read 320 pages?

21._____

22. If a gallon of paint will cover 400 square feet, how many gallons are needed to cover 2200 square feet?

22._____

23. The height of the water in a fish tank rises at a constant rate of 2 inches every 5 minutes. How many minutes will it take to fill the tank if the height must reach 25 inches?

23._____

24. It costs $15 dollars to park for 4 hours. How long will you have parked a car if your cost is $25 dollars?

24._____

25. A ball that is dropped from a height of 60 inches will rebound to a height of 48 inches. How high will a ball rebound that is dropped from a height of 96 inches?

25._____

26. A person weighing 150 pounds on Earth weighs approximately 25 pounds on the moon. How much will a person weigh on Earth if their moon weight is 32 pounds?

26._____

27. Five apples cost $1.60. How much will 8 apples cost?

27._____

28. A biologist tags 50 deer and releases them in a wildlife preserve area. Over the course of a two-week period, she observes 80 deer, of which 12 are tagged. What is the estimate for the population of deer in this particular area?

28._____

29. A paving crew completes 10,000 feet of a road every 3 days. Approximately how many days will it take to pave a 7-mile stretch of road? (1 mile = 5280 feet)

29._____

30. A model airplane has a wingspan of 8 inches. The actual wingspan of the plane it represents is 38 feet. If the model's fuselage is 12 inches long, how long is the fuselage of the actual plane?

30._____

Chapter 4 RATIO, PROPORTION, AND PERCENT

4.6   Basics of Percent

Write as a percent.

1. 43 people out of 100 drive small cars.

1._____

2. The cost for labor was $45 for every $100 spent to manufacture an item.

2._____

3. 29 out of 100 gallons of gas were unleaded.

3._____

4. 32 out of 100 students majored in engineering.

4._____

5. 38 out of 100 planes departed on time.

5._____

Write each percent as a decimal.

6. 37%

6._____

7. 42%

7._____

8. 310%

8._____

9. 9%

9._____

10. 4%

10._____

11. 32.5%

11._____

12. 0.025%

12._____

Write each decimal as a percent.

13. 0.30

13._____

14. 0.40

14._____

15. 0.2   15._____

16. 0.86   16._____

17. 0.42   17._____

18. 0.09   18._____

19. 0.986   19._____

20. 3.47   20._____

Fill in the blanks.

21. 100% of $19 is _____.   21._____

22. 200% of 170 miles is _____.   22._____

23. 300% of $76 is _____.   23._____

24. 100% of 12 dogs is _____.   24._____

25. 300% of 29 days is _____.   25._____

Fill in the blanks.

26. 50% of 48 copies is _____.   26._____

27. 10% of 4920 televisions is _____.   27._____

28. 1% of 400 homes is _____.   28._____

29. 50% of 250 signs is _____.   29._____

30. 10% of 100 years is _____.   30._____

Chapter 4 RATIO, PROPORTION, AND PERCENT

4.7   Percents and Fractions

Write each percent as a fraction or mixed number in lowest terms.

1. 35%

                    1._____

2. 12%

                    2._____

3. 56%

                    3._____

4. 75%

                    4._____

5. 62.5%

                    5._____

6. 43.6%

                    6._____

7. $16\frac{1}{3}\%$

                    7._____

8. $22\frac{2}{9}\%$

                    8._____

9. $6\frac{2}{3}\%$

                    9._____

10. $46\frac{1}{3}\%$

                    10._____

Write each fraction or mixed number as a percent. Round percents to the nearest tenth if necessary.

11. $\dfrac{7}{10}$

11._____

12. $\dfrac{53}{100}$

12._____

13. $\dfrac{81}{100}$

13._____

14. $\dfrac{12}{25}$

14._____

15. $\dfrac{64}{75}$

15._____

16. $\dfrac{33}{50}$

16._____

17. $\dfrac{47}{50}$

17._____

18. $\dfrac{5}{9}$

18._____

19. $3\frac{4}{5}$

19._____

20. $2\frac{3}{4}$

20._____

Complete this chart. Round decimals to the nearest thousandth and percents to the nearest tenth if necessary.

	Fractions	Decimal	Percent	
21.	$\frac{1}{2}$	_____	_____	**21.**_____
22.	_____	0.125	_____	**22.**_____
23.	$\frac{1}{4}$	_____	_____	**23.**_____
24.	$\frac{5}{8}$	_____	_____	**24.**_____
25.	_____	_____	87.5%	**25.**_____
26.	$\frac{3}{8}$	_____	_____	**26.**_____
27.	_____	_____	$33\frac{1}{3}\%$	**27.**_____
28.	$\frac{2}{5}$	_____	_____	**28.**_____
29.	_____	0.325	_____	**29.**_____
30.	$\frac{2}{3}$	_____	_____	**30.**_____

Chapter 4 RATIO, PROPORTION, AND PERCENT

4.8   Using the Percent Proportion and identifying the Components in a Percent Problem

Use the percent proportion $\left(\dfrac{part}{whole} = \dfrac{percent}{100} \right)$ *and solve for the unknown value.*

1. part = 30, percent = 25

1._____

2. part = 160, percent = 20

2._____

3. part = 7, percent = 40

3._____

4. part = 18, percent = 150

4._____

5. whole = 48, percent = 25

5._____

6. whole = 36, percent = 15

6._____

7. whole = 25, percent = 14

7._____

8. whole = 50, percent = 17

8._____

9. part = 44, whole = 200

9._____

Identify the percent. Do not try to solve for any unknowns.

10. 35% of 1000 is 350

10._____

11. 71% of what number is 438?

11._____

12. 83% of what number is 21.5?

12._____

13. 63 is what percent of 218?

13._____

Identify the percent. Do not try to solve for any unknowns.

14. A team won 12 of the 18 games it played. What percent of its games did it win?

14._____

15. A chemical is 42% pure. Of 800 grams of the chemical, how much is pure?

15._____

16. 17% of Tom's check of $340 is withheld. How much is withheld?

16._____

Identify the whole. Do not try to solve for any unknowns.

17. 40% of 48 is 19.2.

17._____

18. 16 is 400% of 4.

18._____

19. What is 14% of 78?

19._____

20. 52 is 12% of what number?

20._____

21. What percent of 60 is 25?

21._____

Identify the whole in each application problem. Do not try to solve for any unknowns.

22. In one storm, Springbrook got 15% of the season's snowfall. Springbrook's total snowfall for that season was 30 inches. How many inches of snow fell in that one storm?

22._____

23. In one state, the sales tax is 8%. On a purchase, the amount of tax was $26. Find the cost of the item purchase.

23._____

Identify the part in each application problem. Do not try to solve for any unknowns.

24. 16% of 3500 is 560.

24._____

25. 29 is 25% of 116.

25._____

26. 29.81 is what percent of 508?

26._____

27. What number is 12.4% of 1408?

27._____

Identify the part in each application problem. Do not try to solve for any unknowns.

28. In a one-day storm, Odentown received 0.3% of the season's total rainfall. Odentown received 4 inches of rain on that day. How many inches of rain fell during the season?

28._____

29. A hatchery is notified that 7% of its shipment of baby salmon did not arrive healthy. Of 1500 salmon shipped, how many did not arrive healthy?

29._____

30. A teacher of English literature found that 15% of the students' papers are handed in late. If there are 40 students in a class, how many papers will be handed in late?

30._____

Chapter 4 RATIO, PROPORTION, AND PERCENT

4.9   Using Proportions to Solve Percent Problems

Use the percent proportion to find the part. Round to the nearest tenth if necessary.

1. 25% of 584

1._____

2. 20% of 1400

2._____

3. 9% of 42

3._____

Use multiplication to find the part. Round to the nearest tenth if necessary.

4. 135% of 35

4._____

5. 39.4% of 300

5._____

6. 22.5% of 1300

6._____

Solve each application problem. Round to the nearest tenth if necessary.

7. A library has 330 visitors of Saturday, 20% of whom are children. How many are children?

7._____

8. A survey at an intersection found that of 2200 drivers, 43% were wearing seat belts. How many drivers in the survey were wearing seat belts?

8._____

9. A family of four with a monthly income of $2100 spends 90% of its earnings and saves the balance. How much does the family save in one month?

9._____

Use the percent proportion to find the whole. Round to the nearest tenth if necessary.

10. 50 is 10% of what number?

10._____

11. 36% of what number is 74?

11._____

12. 50% of what number is 76?

12._____

13. 100 is 25% of what number?

13._____

14. 548 is 110% of what number?

14._____

15. 91 is 130% of what number?

15._____

Solve each application problem. Round to the nearest tenth if necessary.

16. On campus this semester there are 2028 married students, which is 26% of the total enrollment. Find the total enrollment.

16._____

17. There are 18 violin players in an orchestra. If this is 24% of the orchestra membership, find the number of members in the orchestra.

17._____

18. This year, there are 960 scholarship applications, which is 120% of the number of applications last year. Find the number of applications last year.

18._____

19. At Dee's Sandwich Shop, 20% of the customers order a dill pickle. If 465 dill pickles are sold, find the total number of customers.

19._____

Use the percent proportion to find the percent. Round to the nearest tenth if necessary.

20. 35 is what percent of 105?

20._____

21. 13 is what percent of 50?

21._____

22. 15 is what percent of 90?

22._____

23. 12 is what percent of 400?

23._____

24. 7 is what percent of 280?

24._____

25. What percent of 6000 is 12?

25._____

26. What percent of 100 is 150?

26._____

Solve each application problem. Round to the nearest tenth if necessary.

27. In one shipment, 695 out of 27,800 crates were damaged. What percent of the crates were damaged?

27._____

28. The number of ballots cast in a parish election is 12,969. If the number of registered voters in the parish is 19,800, what percent has voted?

28._____

29. G&G Pharmacy has a total payroll of $89,350, of which
   $19,657 goes towards employee fringe benefits. What
   percent of the total payroll goes to fringe benefits?

29._____

30. Vera's Antiquery says that of its 5100 items in stock,
   4233 are just plain junk, while the rest are antiques.
   What percent of the number of items in stock is
   antiques?

30._____

Chapter 4 RATIO, PROPORTION, AND PERCENT

4.10   Using the Percent Equation

Find the missing part using the percent equation. Round to the nearest tenth if necessary.

1. 40% of 480

1._____

2. 70% of 920

2._____

3. 65% of 1300

3._____

4. 99% of 300

4._____

5. 16% of 520

5._____

6. 22% of 960

6._____

7. 9% of 240

7._____

8. 12.4% of 8100

8._____

Solve each application problem. Round to the nearest tenth or cent if necessary.

9. A gardener has 56 clients, 25% of whom are residential.   9._____
   Find the number that are residential.

10. The total in sales at Hill's Market last month was   10._____
    $87,428. If the profit was $1\frac{1}{2}$ % of the sales, how much
    was the profit?

Find the whole using the percent equation. Round to the nearest tenth if necessary.

11. 80 is 20% of what number?

11._____

12. 50% of what number is 47?

12._____

13. 75% of what number is 1125?

13._____

14. 540 is 30% of what number?

14._____

15. $12\frac{1}{2}$% of what number is 270?

15._____

16. $1\frac{1}{4}$% of what number is 11.25?

16._____

17. $2\frac{1}{2}$% of what number is 15?

17._____

18. 200% of what number is 30?

18._____

Solve each application problem.

19. A tank of an industrial chemical is 25% full. The tank now contains 160 gallons. How many gallons will it contain when it is full?

19._____

20. Greg has completed 37.5% of the units needed for a degree. If he has completed 45 units, how many are needed for a degree?

20._____

Find the percent using the percent equation. Round to the nearest tenth if necessary.

21. 20 is what percent of 40?

21._____

22. 15 is what percent of 75?

22._____

23. 13 is what percent of 25?

23._____

24. 72 is what percent of 400?

24._____

25. What percent of 140 is 49?

25._____

26. What percent is 1.35 of 90?

26._____

27. 307.2 is what percent of 960?

27._____

28. 450 is what percent of 200?

28._____

Solve each application problem.

29. The Robinson family earns $2800 per month and saves $700 per month. What percent of the income is saved?

29._____

30. The Hogan family drove 145 miles of their 500-mile vacation. What percent of the total number of miles did they drive

30._____

Chapter 4 RATIO, PROPORTION, AND PERCENT

4.11   Solving Application Problems with Percent

Find the amount of sales tax and the total cost.

	Amount of Sale	Tax Rate	
1.	$100	3%	1._____
2.	$200	6%	2._____
3.	$50	7%	3._____
4.	$215	5%	4._____
5.	$15	2%	5._____
6.	$30	8%	6._____
7.	$67	9%	7._____

Solve the following application problems.

8. If the sales tax is 6.5% and the sales are $350, find the   8._____
amount of sales tax.

9. Find the sales tax on a car costing $14,900 if the sales   9._____
tax rate is 6.25%.

Find the commission earned. Round to the nearest cent if necessary.

	Sales	Rate of Commission
10.	$100	15%
11.	$1000	11%
12.	$5783	4%
13.	$3200	32%
14.	$6225	2.5%
15.	$75,000	4%
16.	$156,000	3%
17.	$25,000	15%

10._____

11._____

12._____

13._____

14._____

15._____

16._____

17._____

Find the amount of discount and the amount paid after the discount. Round to the nearest cent if necessary.

	Original Price	Rate of Discount
18.	$100	25%
19.	$200	15%
20.	$780	10%

18._____

19._____

20._____

Name: Date:
Instructor: Section:

	Original Price	Rate of Discount		
21.	$38	40%	**21.**	_____
22.	$17.50	50%	**22.**	_____
23.	$22.50	30%	**23.**	_____
24.	$125	35%	**24.**	_____
25.	$24.95	60%	**25.**	_____

Solve each application problem. Round to the nearest tenth of a percent if necessary.

26. Enrollment in secondary education courses increased   **26.** _____
from 1900 students last semester to 2280 students this
semester. Find the percent of increase.

27. The number of days employees of Prodex Manufacturing   **27.** _____
Company were absent from their jobs decreased from 96
days last month to 72 days this month. Find the percent
of decrease.

28. The earnings per share of Amy's Cosmetic Company   **28.** _____
decreased from $1.20 to $0.86 in the last year. Find the
percent of decrease.

29. The price of a certain model of calculator was $33.50
    five years ago. This calculator now costs $18.75. Find
    the percent of decrease in the price in the last five years.

29._____

30. Last year John Rivera planted 25 acres of corn. This
    year he planted 32 acres of corn. Find the percent of
    increase in corn acreage.

30._____

Chapter 4 RATIO, PROPORTION, AND PERCENT

4.12   Simple Interest

Find the interest.

	Principle	Rate	Time in Years	
1.	$200	10%	1	1._____
2.	$400	2%	3	2._____
3.	$300	12%	4	3._____
4.	$1000	12%	2	4._____
5.	$80	5%	1	5._____
6.	$175	13%	2	6._____
7.	$1500	3%	6	7._____
8.	$5280	8%	5	8._____

	Principle	Rate	Time in Months	
9.	$200	16%	3	9._____
10.	$400	9%	6	10._____

	Principle	Rate	Time in Months	
11.	$500	11%	9	11._____
12.	$1000	12%	12	12._____
13.	$820	3%	18	13._____
14.	$92	9%	10	14._____
15.	$780	6%	5	15._____
16.	$522	8%	9	16._____

Solve the following application problems.

17. Debbie Ondrika deposits $680 at 14% for 1 year. How much interest will she earn?

17._____

18. Bugby Pest Control invests $1500 at 16% for 6 months. What amount of interest will the company earn?

18._____

19. Diane DeGroot lends $6500 for 18 months at 12%. How much interest will she earn?

19._____

Find the total amount due on the following loans. Round to the nearest cent if necessary.

	Principle	Rate	Time	
20.	$200	11%	1 year	20._____
21.	$3000	5%	6 months	21._____
22.	$540	12%	3 months	22._____
23.	$1020	10%	2 years	23._____
24.	$1500	8%	18 months	24._____
25.	$5000	7%	5 months	25._____
26.	$2210	9%	6 months	26._____
27.	$5820	6%	1 year	27._____

Solve the following application problems. Round to the nearest cent if necessary.

28. Mary Ann borrows $1200 at 10% for 3 months. Find the total amount due.

28._____

29. An investor deposits $7000 at 16% for 2 years. If there are no withdrawals or further deposits, find the total amount in the account after 2 years.

29._____

30. An employee credit union pays 7% interest. If Mario
deposits $2100 in his account for $\frac{1}{3}$ year and makes no
withdrawals or further deposits, find the total amount in
Mario's account after that time.

30._____

Chapter 4 RATIO, PROPORTION, AND PERCENT

4.13   Compound Interest

Find the compound amount given the following deposits. Interest is compounded annually. Round to the nearest cent if necessary.

1. $7000 at 5% for 3 years

   1._____

2. $2500 at 8% for 4 years

   2._____

3. $1200 at 2% for 3 years

   3._____

4. $3200 at 7% for 2 years

   4._____

5. $4500 at 4% for 2 years

   5._____

6. $7000 at 6% for 3 years

   6._____

7. $8000 at 8% for 12 years

   7._____

8. $3200 at 6% for 5 years

   8._____

Name: Date:
Instructor: Section:

Use the table for compound interest to find the compound amount. Interest is compounded annually. Round to the nearest cent if necessary.

Time Periods	3.00%	3.50%	4.00%	4.50%	5.00%	5.50%	6.00%	8.00%	Time Periods
1	1.0300	1.0350	1.0400	1.0450	1.0500	1.0550	1.0600	1.0800	1
2	1.0609	1.0712	1.0816	1.0920	1.1025	1.1130	1.1236	1.1664	2
3	1.0927	1.1087	1.1249	1.1412	1.1576	1.1742	1.1910	1.2597	3
4	1.1255	1.1475	1.1699	1.1925	1.2155	1.2388	1.2625	1.3605	4
5	1.1593	1.1877	1.2167	1.2462	1.2763	1.3070	1.3382	1.4693	5
6	1.1941	1.2293	1.2653	1.3023	1.3401	1.3788	1.4185	1.5869	6
7	1.2299	1.2723	1.3159	1.3609	1.4071	1.4547	1.5036	1.7138	7
8	1.2668	1.3168	1.3686	1.4221	1.4775	1.5347	1.5938	1.8509	8
9	1.3048	1.3629	1.4233	1.4861	1.5513	1.6191	1.6895	1.9990	9
10	1.3439	1.4106	1.4802	1.5530	1.6289	1.7081	1.7908	2.1589	10
11	1.3842	1.4600	1.5395	1.6229	1.7103	1.8021	1.8983	2.3316	11
12	1.4258	1.5111	1.6010	1.6959	1.7959	1.9012	2.0122	2.5182	12

9. $1000 at 6% for 4 years   9._____

10. $1000 at 8% for 8 years   10._____

11. $4000 at 5% for 9 years   11._____

12. $7500 at 6% for 7 years   12._____

13. $60 at 5.5% for 2 years   13._____

14. $48 at 8% for 3 years

14._____

15. $37.50 at 8% for 3 years

15._____

16. $71.95 at 5% for 5 years

16._____

17. $8428.17 at $4\frac{1}{2}$% for 6 years

17._____

18. $10,422.75 at $5\frac{1}{2}$% for 12 years

18._____

19. $1000 at 4.5% for 6 years

19._____

20. $24,600 at 5% for 4 years

20._____

Find the compound amount and compound interest.. Round to the nearest cent if necessary. Use the table for compound interest in your textbook to find the compound amount.

	Principle	Rate	Time in Years	
21.	$1000	5%	10	21._____
22.	$1000	$3\frac{1}{2}$%	7	22._____

23. $8500   6%   12   **23.**_____

24. $12,800 $5\frac{1}{2}\%$ 9   **24.**_____

25. $9150   8%   8   **25.**_____

26. $45,000   4%   4   **26.**_____

27. $21,400 $4\frac{1}{2}\%$ 11   **27.**_____

28. $78,000   3%   12   **28.**_____

Solve each application problem. Use the table for compound interest in your textbook to find the compound amount.

29. Scott Williams lends $9000 to the owner of a new   **29.**_____
restaurant. He will be repaid at the end of 6 years at 8%
interest compounded annually. Find how much he will
be repaid and how much interest he will earn.

30. Michelle Roberts invests $2500 in a health spa. She will   **30.** _____
    be repaid at the end of 5 years at 6% interest
    compounded annually. Find how much she will be
    repaid and how much interest she will earn.

Chapter 5 GEOMETRY

5.1   Lines and Angles

Identify each line, line segment, or ray and name it using the appropriate symbol.

1.

1._____

2.

2._____

Label each pair of lines as appearing to be parallel or intersecting.

3.

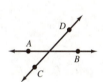

3._____

4.

4._____

5.

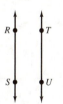

5._____

6.

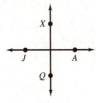

6._____

Name each highlighted angle by using the three-letter form of identification.

7.

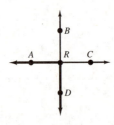

7._____

8.

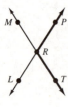

8._____

Label each angle as acute, right, obtuse, or straight. For right angles and straight angles, indicate the number of degrees in the angle.

9.

9._____

10.

10._____

In exercises 11 and 12, identify the pair of lines that is perpendicular. How can you describe the other pair of lines?

11.

11._____

12.

12._____

Identify each pair of complementary angles.

13.

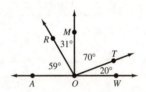

13._____

14.

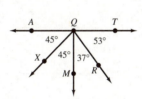

14._____

Identify each pair of supplementary angles.

15.

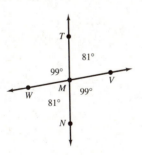

15._____

16.

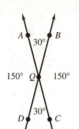

16._____

Find the complement of each angle.

17. 38°

17._____

18. 79°

18._____

19. 11°

19._____

20. 50°

20._____

Find the supplement of each angle.

21. 12°

21._____

22. 91°

22._____

23. 160°

23._____

24. 1°

24._____

In each figure, identify the angles that are congruent.

25.

25._____

26.

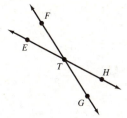

26. _____

In each figure, line m is parallel to line n. Find the measure of each angle.

27. ∠4 measures 100°.

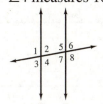

27. _____

28. ∠7 measures 37°.

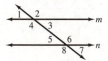

28. _____

29. ∠6 measures 125°.

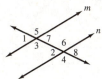

29. _____

30. ∠3 measures 44°.

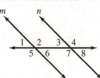

30. _____

Chapter 5 GEOMETRY

5.2   Rectangles and Squares

Find the perimeter and area of each rectangle.

1. 4 centimeters by 8 centimeters

1._____

2. 17 inches by 12 inches

2._____

3. 1 centimeter by 17 centimeter

3._____

4. 14.5 meters by 3.2 meters

4._____

5. $4\frac{1}{2}$ yards by $6\frac{1}{2}$ yards

5._____

6. 87.2 feet by 33 feet

6._____

7. 37.4 centimeters by 103.2 centimeters

7._____

Solve each application problem.

8. A picture frame measures 20 inches by 30 inches. Find the perimeter and area of the frame.

8._____

9. A lot is 114 feet by 212 feet. County rules require that nothing be built on land within 12 feet of any edge of the lot. Find the area on which you cannot build.

9._____

10. A room is 14 yards by 18 yards. Find the cost to carpet this room if carpet costs $23 per square yard.

10._____

Find the perimeter and area of each square.

11. 9 meters by 9 meters

11._____

12. A square 9.2 yards wide

12._____

13. A square 7.8 feet wide

13._____

14. 13 feet by 13 feet

14._____

15. $1\frac{2}{5}$ inches by $1\frac{2}{5}$ inches

15._____

16. 8.2 kilometers by 8.2 kilometers

16._____

17. 3.1 centimeters by 3.1 centimeters

17._____

18. 7.4 inches on each side

18._____

19. $4\frac{2}{3}$ miles by $4\frac{2}{3}$ miles

19._____

20. 21 meters by 21 meters

20._____

Name: Date:

Instructor: Section:

Find the perimeter and area of each figure. All angles that appear to be right angles are indeed right angles.

21.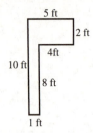

21._____

22.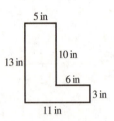

22._____

23.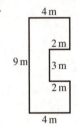

23._____

24.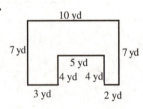

24._____

25.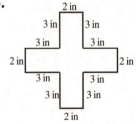

25._____

26.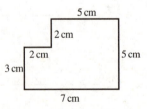

26._____

27.

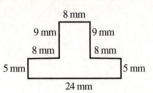

28.

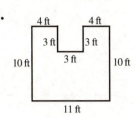

28._____

29.

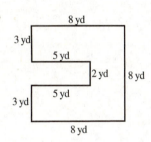

29._____

30.

30._____

Name: Date:

Instructor: Section:

Chapter 5 GEOMETRY

5.3   Parallelograms and Trapezoids

Find the perimeter of each parallelogram.

1.

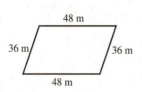

48 m
36 m   36 m
48 m

1._____

2.

10.1 in
6.3 in   6.3 in
10.1 in

2._____

3.

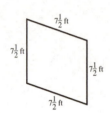

$7\frac{1}{2}$ ft

$7\frac{1}{2}$ ft

$7\frac{1}{2}$ ft

$7\frac{1}{2}$ ft

3._____

4.

10 m
4 m   4 m
10 m

4._____

5.

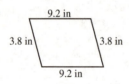

9.2 in
3.8 in   3.8 in
9.2 in

5._____

6.

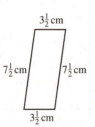

$3\frac{1}{2}$ cm

$7\frac{1}{2}$ cm $7\frac{1}{2}$ cm

$3\frac{1}{2}$ cm

6._____

Find the area of each parallelogram.

7.

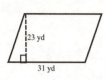

7._____

8.

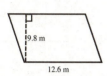

8._____

9.

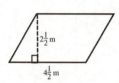

9._____

10.

10._____

11.

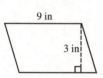

11._____

12.

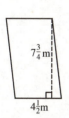

12._____

Solve each application problem.

13. A parallelogram has a height of 3.2 meters and a base of 4.6 meters. Find the area.

13._____

14. A parallelogram has a height of $15\frac{1}{2}$ feet and a base of 20 feet. Find the area.

14._____

15. A swimming pool is in the shape of a parallelogram with a height of 9.6 meters and base of 12 meters. Find the cost of a solar pool cover that sells for $5.10 per square meter.

15._____

16. An auditorium stage has a hardwood floor that is shaped like a parallelogram, having a height of 30 feet and a base of 40 feet. If a company charges $0.65 per square foot to refinish floors, find the cost of refinishing the stage floor.

16._____

Find the perimeter of each trapezoid.

17.

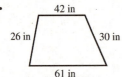

17._____

18.

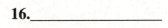

18._____

19.

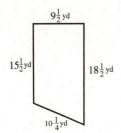

19._____

Find the area of each figure.

20.

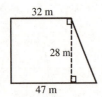

20._____

WS-182

21.

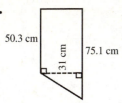

21._____

22.

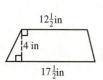

22._____

23.

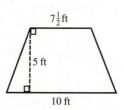

23._____

24.

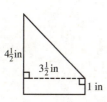

24._____

25.

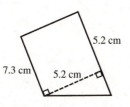

25._____

26.

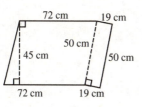

26._____

27.

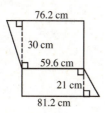

27._____

Solve each application problem.

28. The lobby in a resort hotel is in the shape of a trapezoid. The height of the trapezoid is 52 feet and the bases are 47 feet and 59 feet. Carpet that costs $2.75 per square foot is to be laid in the lobby. Find the cost of the carpet

28._____

.

29. The backyard of a new home is shaped like a trapezoid, having a height of 35 feet and bases of 90 feet and 110 feet. Find the cost of planting a lawn in the yard if the landscaper charges $0.20 per square foot.

29._____

30. A hot tub is in the shape below. Find the cost of a cover for the hot tub at a cost of $9.70 per square foot. Angles that appear to be right angles are indeed right angles.

30._____

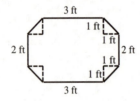

Name: Date:

Instructor: Section:

Chapter 5 GEOMETRY

5.4   Triangles

Find the perimeter of each triangle.

1.

1._____

2.

2._____

3.

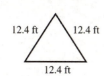

3._____

4.

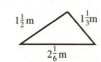

4._____

5.

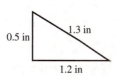

5._____

6.

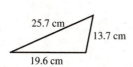

6._____

7. A triangle has sides $2\frac{1}{2}$ feet, 3 feet, and $5\frac{1}{4}$ feet

7._____

8. A triangle with two equal sides of 3.6 centimeters and the third side 4.1 centimeters

8._____

9. A triangle with three equal sides each 5.9 meters

9._____

10. A triangle with sides $13\frac{1}{8}$ inches, $11\frac{3}{4}$ inches, and $14\frac{1}{2}$ inches.

10._____

Find the area of each triangle.

11.

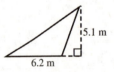

36 m

70 m

11._____

12.

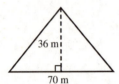

15.3 cm

30.4 cm

12._____

13.

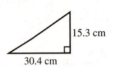

$7\frac{1}{4}$ ft

6 ft

13._____

14.

8 yd

7 yd

14._____

15.

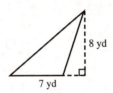

5.1 m

6.2 m

15._____

16.

$1\frac{3}{8}$ in

$\frac{7}{8}$ in

$1\frac{1}{4}$ in

16._____

Find the shaded area in each figure.

17.

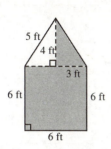

17._____

18.

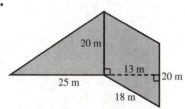

18._____

19.

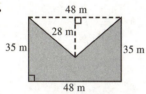

19._____

20.

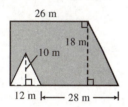

20._____

The measures of two angles of a triangle are given. Find the measure of the third angle.

21. 60°, 70°

21._____

22. 100°, 63°

22._____

23. 30°, 100°

23._____

24. 60°, 60°

24._____

25. 80°, 80°

25._____

26. $37^{o}, 62^{o}$

26._____

27. $49^{o}, 72^{o}$

27._____

28. $51^{o}, 72^{o}$

28._____

29. $77^{o}, 13^{o}$

29._____

30. $90^{o}, 45^{o}$

30._____

Chapter 5 GEOMETRY

5.5   Circles

Find the missing value in each circle.

1.

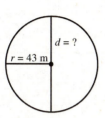

1._____

2.

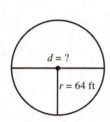

2._____

3.

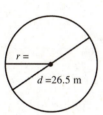

3._____

4.

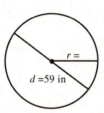

4._____

5.

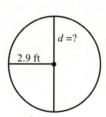

5._____

6.

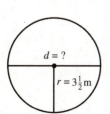

6._____

7.

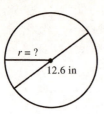

$r = ?$
12.6 in

7._____

8.

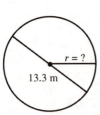

$r = ?$
13.3 m

8._____

9.

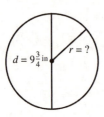

$d = 9\frac{3}{4}$ in $r = ?$

9._____

10. The radius of a circle is $\frac{1}{8}$ inch. Find the diameter.

10._____

Find the circumference of each circle. Use 3.14 as an approximation value for π. Round the answer to the nearest tenth.

11.

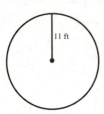

11 ft

11._____

12.

23 cm

12._____

13.

30 m

13. _____

14. A circle with a diameter of 13 inches

14. _____

15. A circle with a radius of 17 feet

15. _____

16. A circle with a radius of 4.5 yards

16. _____

17. A circle with a diameter of 20 centimeters

17. _____

18. A circle with a diameter of $4\frac{3}{4}$ inches

18. _____

19. A circle with a radius of $\frac{3}{4}$ mile

19. _____

20. A circle with a diameter of 12.78 centimeters

20. _____

Find the area of each circle. Use 3.14 as an approximation value for π. Round the answer to the nearest tenth.

21.

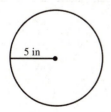

5 in

21. _____

22.

3.7 m

22._____ \`_____

23.

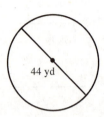

44 yd

23._____

24. A circle with diameter of 51 feet

24._____

25. A circle with diameter of $5\frac{1}{3}$ yards

25._____

26. A circle with diameter of 9.8 centimeters

26._____

Find each shaded area. Use 3.14 as an approximation value for π. Round the answer to the nearest tenth if necessary.

27.

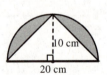

10 cm

20 cm

27._____

28.

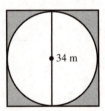

34 m

28._____

29.

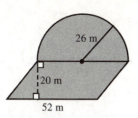

29. _____

30.

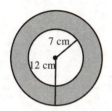

30. _____

Chapter 5 GEOMETRY

5.6   Volume and Surface Area

Find the volume of each rectangular solid. Round the answer to the nearest tenth, if necessary.

1.

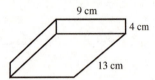

1._____

2.

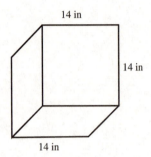

2._____

3.

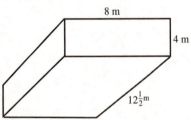

3._____

4.

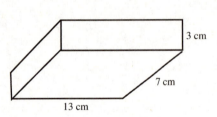

4._____

5.

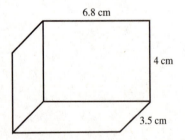

5._____

6.

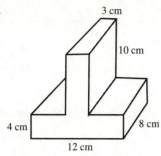

6._____

Find the volume of each sphere or hemisphere. Use 3.14 as an approximation for π. Round the answer to the nearest tenth if necessary.

7.

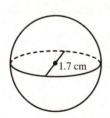

7._____

8.

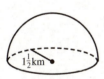

8._____

9.

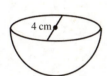

9._____

10.

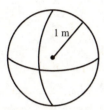

10._____

11.

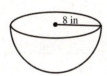

11._____

12. Find the volume of a sphere that has a diameter of $3\frac{1}{4}$ inches.

12._____

Find the volume of each cylinder. Use 3.14 as an approximation value for π. Round the answer to the nearest tenth.

13.

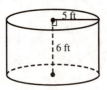

5 ft
6 ft

13._____

14.

10 ft
3 ft

14._____

15.

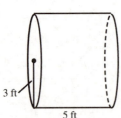

3 ft
5 ft

15._____

16.

9 m
2 m

16._____

17. Find the volume of the figure. Use 3.14 as an approximation value for π. Round the answer to the nearest tenth.

17._____

23.2 in
4 in

18. Find the volume of the shaded part. Use 3.14 as an approximation value for π. Round the answer to the nearest tenth.

18._____

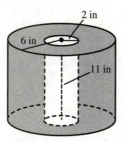
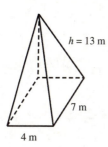

Find the volume of each figure. Use 3.14 as an approximation value for π. Round the answer to the nearest tenth.

19.

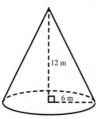

19._____

20.

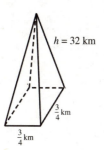

20._____

21.

h = 32 km

$\frac{3}{4}$ km

$\frac{3}{4}$ km

21._____

22.

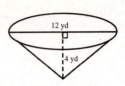

12 yd

4 yd

22._____

23. Find the volume of a pyramid with square base 42 meters on a side and height 38 meters.

23._____

24. Find the volume of a cone with base diameter 3.2 centimeters and height 5.8 centimeters.

24._____

Find the surface area of each rectangular solid. Round your answers to the nearest tenth.

25.

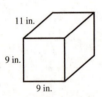

11 in.

9 in.

9 in.

25._____

26.

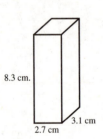

8.3 cm.

2.7 cm

3.1 cm

26._____

Find the surface area of each cylinder. Use 3.14 as the approximate value for π. Round your answers to the nearest tenth.

27.

28.

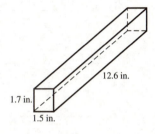

29.

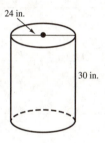

30.

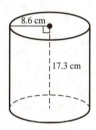

Chapter 5 GEOMETRY

5.7   Pythagorean Theorem

Find each square root. Use a calculator with a square root key. Round the answer to the nearest thousandth, if necessary.

1. $\sqrt{17}$ 1._____

2. $\sqrt{27}$ 2._____

3. $\sqrt{62}$ 3._____

4. $\sqrt{55}$ 4._____

5. $\sqrt{24}$ 5._____

6. $\sqrt{10}$ 6._____

7. $\sqrt{2}$ 7._____

8. $\sqrt{37}$ 8._____

9. $\sqrt{13}$ 9._____

10. $\sqrt{28}$ 10._____

Find the unknown length in each right triangle. Use a calculator with a square root key. Round the answer to the nearest thousandth, if necessary.

11. 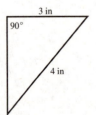   11._____

12.

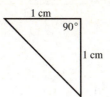

13.

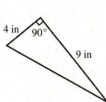

14.

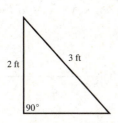

15.

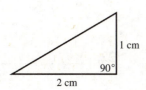

16.

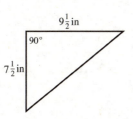

17.

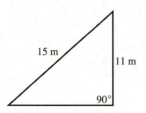

18.

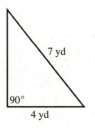

19.

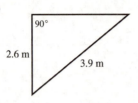

19._____

20.

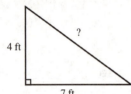

20._____

Solve each application problem. Use a calculator with a square root key. Round the answer to the nearest thousandth, if necessary.

21. Find the length of this loading dock.

21._____

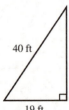

22. Find the unknown length in this roof plan.

22._____

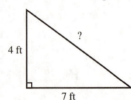

23. A boat goes 10 miles south and then 15

23._____

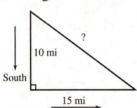

24. Find the height of this telephone pole

24._____

Name: Date:
Instructor: Section:

*Find the unknown lengths if ladders are leaning against a building as shown. Round the
answer to the nearest thousandth, if necessary.*

25.

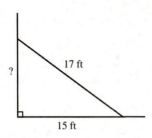

25._____

26.

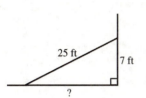

26._____

27.

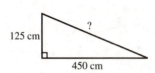

27._____

28.

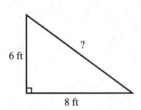

28._____

*Find the distance between the centers of each pair of holes in each metal plate. Round the
answer to the nearest thousandth, if necessary.*

29.

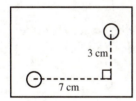

29._____

30.

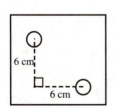

30._____

Name: Date:
Instructor: Section:

Chapter 5 GEOMETRY

5.8   Congruent and Similar Triangles

Each pair of triangles is congruent. List the corresponding angles and the corresponding sides.

1.   1._____

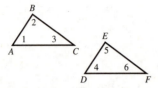

2.   2._____

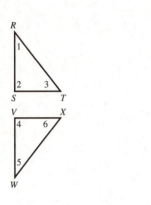

3.   3._____

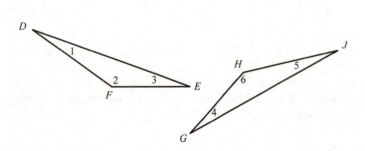

Determine which of these methods can be used to prove that each pair of triangles is congruent:
Angle-Side-Angle (ASA), Side-Side-Side (SSS), or Side-Angle-Side (SAS).

4.   4._____

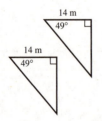

5.

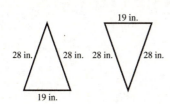

5._____

6.

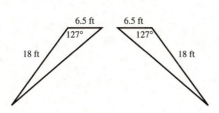

6._____

7.

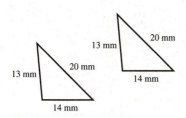

7._____

8.

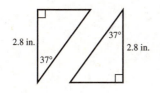

8._____

9.

9._____

10.

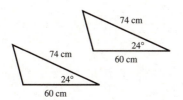

10._____

11.

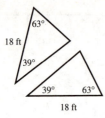

12.

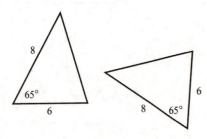

Write the ratio for each pair of corresponding sides in the similar triangles shown below. Write the ratios as fractions in lowest terms.

13. $\dfrac{ML}{AB}; \dfrac{MS}{AC}; \dfrac{LS}{BC}$

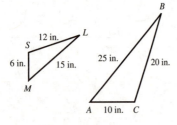

14. $\dfrac{BM}{FT}; \dfrac{BQ}{FA}; \dfrac{MQ}{TA}$

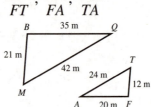

15. $\dfrac{LM}{PS}$; $\dfrac{MK}{ST}$; $\dfrac{LK}{PT}$

15._____

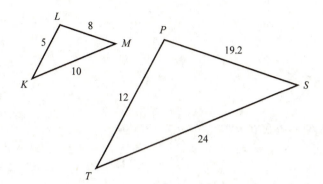

Find the unknown lengths in each pair of similar triangles.

16.

16._____

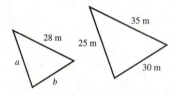

17.

17._____

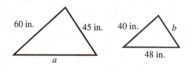

18.

18._____

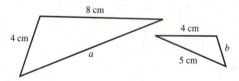

19.

19._____

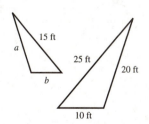

20.

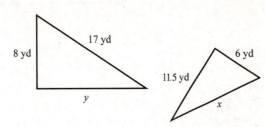

Find the perimeter of each triangle. Assume the triangles are similar.

21.

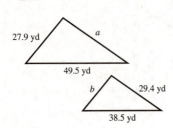

22.

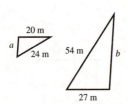

23.

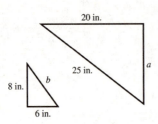

24.

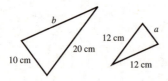

25.

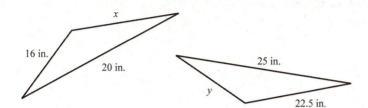

25._____

Solve each application problem.

26. A flagpole casts a shadow 52 m long at the same time that a pole 9 m tall casts a shadow 12 m long. Find the height of the flagpole.

26._____

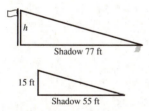

27. A flagpole casts a shadow 77 feet long at the same time that a pole 15 feet tall casts a shadow 55 ft long. Find the height of the flagpole.

27._____

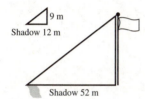

28. The height of the house shown here can be found by comparing its shadow to the shadow cast by a 5 ft stick. Find the height of the house by writing a proportion and solving it.

28._____

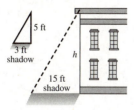

29. A fire lookout tower provides an excellent view of the surrounding countryside. The height of the tower can be found by lining up the top of the tower with the top of a 3-meter stick. Use similar triangles to find the height of the tower.

29._____

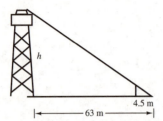

63 m   4.5 m

30. The ratio of the rise of a roof to the run of a roof is 5 to 12. Use this information to find the height of the roof indicated by *h* in the diagram.

30._____

h

18 ft

Name: Date:

Instructor: Section:

Chapter 6 STATISTICS

6.1   Circle Graphs

The circle graph shows the enrollment by major at a small college.

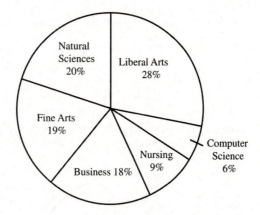

The total enrollment at the college is 3200 students. Use the circle graph to find the number of students with each of the following majors.

1. Liberal Arts

1._____

2. Natural Sciences

2._____

3. Fine Arts

3._____

4. Business

4._____

5. Nursing

5._____

6. Computer Science

6._____

7. What is the most popular major at the college?

7._____

8. What major has the fewest students?

8._____

9. Find the ratio of business majors to computer science majors.

9._____

10. Find the ratio of natural science majors to liberal arts majors.

10._____

The circle graph shows the costs of remodeling a kitchen. Use the graph to solve Problems 11–15.

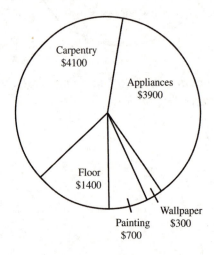

Carpentry $4100

Appliances $3900

Floor $1400

Painting $700

Wallpaper $300

11. Find the total cost of remodeling the kitchen.

11._____

12. What is the largest single expense in remodeling kitchen?

12._____

13. Find the ratio of the cost of appliances to the total remodeling cost.

13._____

14. Find the ratio of the cost of painting to the total remodeling cost.

14._____

15. Find the ratio of the cost of wallpaper to the cost of the floor.

15._____

The following circle graph shows the number of students at a college enrolled in certain majors. Use the graph to solve Problems 16–19.

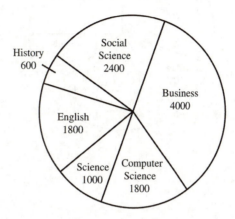

16. Which major has the most number of students enrolled?

16._____

17. Find the ratio of the number of business majors to the total number of students.

17._____

18. Find the ratio of the number of English majors to the total number of students.

18._____

19. Find the ratio of the number of science majors to the number of English majors.

19._____

The circle graph shows the expenses involved in keeping a sales force on the road. Each expense item is expressed as a percent of the total sales force cost of $950,000. Find the number of dollars of expense for each category in Problems 20–23.

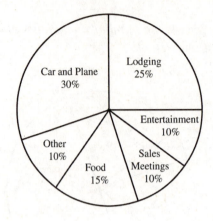

20. Car and plane

20._____

21. Lodging

21._____

22. Entertainment

22._____

23. Sales meetings

23._____

During one semester, Evie Allsot, a student spent $1600 for expenses as shown in the following chart. Complete the chart.

Item	Dollar Amount	Percent of Total	Degrees of a Circle	
Food	$400	25%	90°	
24. Rent	$320	_____	72°	24._____
25. Clothing	$240	_____	_____	25._____

26. Books $160   10%   _____   26._____

27. Entertainment $240   _____ 54° 27._____

28. Savings $80   _____   _____   28._____

29. Other   _____   _____ 36° 29._____

30. Draw a circle graph using the above information.   30._____

6.2   Bar Graphs and Line Graphs

The bar graph sows the enrolment for the fall semester at a small college for the past five year.

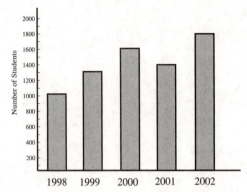

Use the bar graph above for Problems 1–7. Find the enrollment for the fall semester for the following years.

1. 1998

1._____

2. 2000

2._____

3. 2002

3._____

4. How many more students were enrolled in 2000 than in 1999?

4._____

5. What year had the greatest enrollment?

5._____

6. Which year showed a decrease in enrollment?

6._____

7. By how many students did the enrollment increase from 1998 to 1999?

7._____

Name: Date:
Instructor: Section:

The double-bar graph shows the enrollment by gender in each class at a small college. Use the double-bar graph for Problems 8–14.

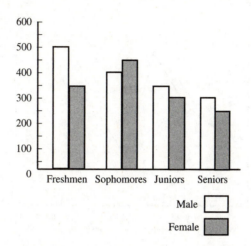

8. Which class has a greater female enrollment than male enrollment?

8._____

9. How many female freshmen are enrolled?

9._____

10. Find the total number of juniors enrolled.

10._____

11. Find the ratio of freshmen males to freshmen females.

11._____

12. Find the total number of students enrolled.

12._____

13. Find the ratio of freshmen students to senior students.

13._____

14. Which class has the greatest difference between male students and female students?

14._____

The line graph gives the value of one share of stock of Microchip Computer Corporation on the first trading day of the month for six consecutive months. Use the line graph for Problems 15–21.

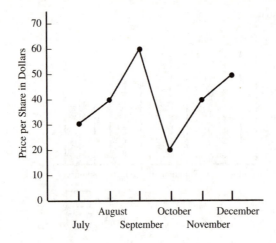

15. In which month was the value of the stock highest?

15._____

16. Find the value of one share on the first trading day October.

16._____

17. Find the increase in the value of one share from October to November.

17._____

18. What is the largest monthly decrease in the value of one share?

18._____

19. Find the ratio of the value of one share on the first trading day in September to the value of one share on the first trading day of October.

19._____

20. Comparing the value of one share on the first trading day in July to the first trading day in November, has the value increased, decreased, or remained unchanged?

20._____

21. By how much did the value of one share increase from July to September?

21._____

Name: Date:

Instructor: Section:

The comparison line graph shows annual sales for two different stores for each of the past few years. Use the graph to solve Problems 22–30.

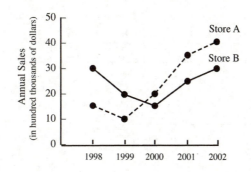

Find the annual sales for store A in each of the following years.

22. 2001

22._____

23. 1999

23._____

24. 1998

24._____

Find the annual sales for store B in each of the following years.

25. 2001

25._____

26. 1999

26._____

27. 1998

27._____

28. In which years did the sales of store A exceed the sales of store B?

28._____

29. Which year showed the least difference between the sales of store A and the sales of store B?

29._____

30. Find the ratio of the sales of store A to the sales of store B in 1998.

30._____

6.3 Frequency Distributions and Histograms

The following scores were earned by students on an algebra exam. Use the data to complete the table.

84	90	83	72	84	93	83	90	83
90	72	64	90	83	72	83	83	64

Score	Tally	Frequency

1. 64   1._____

2. 72   2._____

3. 83   3._____

4. 84   4._____

5. 90   5._____

6. 93   6._____

The following list of numbers represents IQ scores of 18 students. Use these numbers to complete the following table.

98	121	112	99	105	112
110	100	92	109	104	106
105	88	92	103	98	118

IQ Scores	*Tally*	*Frequency*

7. 80–89

7._____

8. 90–99

8._____

9. 100–109

9._____

10. 110–119

10._____

11. 120–129

11._____

12. What was the most common range of IQ scores?

12._____

13. What was the least common range of IQ scores

13._____

The following list of numbers represents systolic blood pressure of 21 patients. Use these numbers to complete the table.

120	98	180	128	143	98	105
136	115	190	118	105	180	112
160	110	138	122	98	175	118

Systolic Blood Pressure	*Tally*	*Frequency*

14. 90–109   14._____

15. 110–129   15._____

16. 130–149   16._____

17. 150–169   17._____

18. 179–189   18._____

19. 190–209   19._____

20. What was the most common range of systolic blood   20._____
pressure?

21. What was the least common range of systolic blood   21._____
pressure?

22. The following list of numbers represents systolic blood pressures of 21 patients. Construct a histogram from this data. Use intervals 90–109, 110–129, 130–149, 150–169, 170–189, and 190–209.

22._____

120	98	180	128
136	115	190	118
160	110	138	122

143	98	105
105	180	102
98	175	118

23. The following list of numbers represents IQ scores of 18 students. Construct a histogram from this data. Use intervals 80–89, 90–99, 100–109, 110–119, and 120–129.

23._____

98	121	112
110	100	92
105	88	92

99	105	112
109	104	106
103	98	118

A local chess club recorded the ages of their members and constructed a histogram. Use the histogram to solve Problems 24–30.

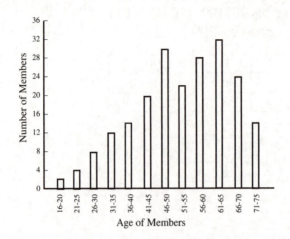

24. The greatest number of members is in which age group?   **24.**_____

25. The fewest number of members are in which age group?   **25.**_____

26. Find the number of members 30 years of age or younger.   **26.**_____

27. Find the number of members 51 years and older.   **27.**_____

28. How many members are 51–65 years of age?   **28.**_____

29. How many members are 46–50 years of age?   **29.**_____

30. Which age range contains the least number of members?   **30.**_____

6.4   Mean, Median, and Mode

Find the mean for each list of numbers. Round the answer to the nearest tenth, if necessary.

1. 7, 12, 3, 5, 9

1._____

2. 51, 47, 33, 43, 79, 58

2._____

3. 39, 50, 59, 61, 69, 73, 51, 80

3._____

4. 31, 37, 44, 51, 52, 74, 69, 83

4._____

5. 3.8, 9.2, 6.7, 3.5, 4.9, 8.8

5._____

6. 62.7, 59.6, 71.2, 65.8, 63.1

6._____

7. 19900, 23850, 25930, 27710, 29340, 41000

7._____

Find the weighted mean. Round the answer to the nearest tenth, if necessary.

8.

Value	Frequency
2	4
6	2
9	1
13	3

8._____

9.

Value	Frequency
17	4
12	5
15	3
19	1

9._____

10.

Value	Frequency
13	4
12	2
19	5
15	3
21	1
27	5

10._____

11.

Value	Frequency
35	1
36	2
39	5
40	4
42	3
43	5

11._____

Find the grade point average for each of the following students. Assume A = 4, B = 3, C = 2, D = 1, F = 0. Round to the nearest tenth, if necessary.

12.

Units	Grade
4	C
2	B
5	C
1	D
3	F

12._____

13.

Units	Grade
3	C
3	A
4	B
5	B
2	A

13._____

14.

Units	Grade
5	B
4	C
3	B
2	C
2	C

14._____

15.

Units	Grade
3	A
4	B
2	C
5	C
2	D

15._____

Find the median for each list of numbers.

16. 13, 19, 33, 52, 93, 107

16._____

17. 199, 472, 312, 298, 254

17._____

18. 200, 215, 226, 238, 250, 283

18._____

19. 30.0, 28.2, 28.8, 32.6

19._____

20. .002, .004, .012, .008

20._____

21. 1.6, 1.8, 1.7, 1.1, 1.7

21._____

22. 389, 464, 521, 610, 654, 672, 682, 712

22._____

23. 43, 69, 108, 32, 51, 49, 83, 57, 64

23._____

Find the mode.

24. 3, 8, 7, 5, 8, 1, 6, 2

24._____

25. 5, 4, 5, 1, 3, 6, 9, 7, 5, 2

25._____

26. 85, 79, 79, 79, 86, 86, 85, 85, 81

26._____

27. 4, 8, 16, 2, 1, 7, 18, 9, 3, 19

27._____

28. 13, 16, 18, 19, 22, 30, 33, 85, 90

28._____

29. 37, 24, 35, 35, 24, 38, 39, 28, 27, 39

29._____

30. 172.6, 199.7, 182.4, 167.1, 172.6, 183.4, 187.6

30._____

Chapter 7 THE REAL NUMBER SYSTEM

7.1   Exponents, Order of Operations, and Inequality

Find the value of each exponential expression.

1. 3^3

1._____

2. 2^5

2._____

3. $\left(\dfrac{3}{4}\right)^2$

3._____

4. $\left(\dfrac{2}{5}\right)^3$

4._____

5. $(.03)^3$

5._____

Find the value of each expression.

6. $3 \cdot 4 + 5$

6._____

7. $5 + 16 \div 4$

7._____

8. $3 \cdot 15 - 10^2$

8._____

9. $\dfrac{1}{2} \cdot \dfrac{2}{3} + \dfrac{3}{4} \cdot \dfrac{1}{3}$

9._____

10. $(1.2)(2.3) - (.4)(.8)$

10._____

11. $\dfrac{3 \cdot 15 + 10^2}{12^2 - 8^2}$

Find the value of each expression.

12. $6\left[5 + 3(4)\right]$

12._____

13. $8 + 4\left[2(4) + 3\right]$

13._____

14. $4 + 3\left[7 - 3(3)\right]$

14._____

15. $4\left[3 + 2(9 - 2)\right]$

15._____

16. $19 - 3\left[8(5 - 2) + 6\right]$

16._____

17. $4^2\left[(8 - 3) + 6\right]$

17._____

Tell whether each statement is true *or* false.

18. $95 > 97$

                                              18._____

19. $3 \cdot 4 \div 2^2 \neq 3$

                                              19._____

20. $2\big[7(4) - 3(5)\big] \geq 45$

                                              20._____

21. $4\frac{1}{2} + 2\frac{3}{4} < 7$

                                              21._____

22. $4 \geq \dfrac{2(3+1) - 3(2+1)}{3 \cdot 2 - 1}$

                                              22._____

Write each word statement in symbols.

23. Seven equals thirteen minus five.

                                              23._____

24. Twenty-two is greater than seventeen.

                                              24._____

25. The difference between thirty and seven is greater than twenty.

                                              25._____

26. Seven is greater than the quotient of fifteen and five.

                                              26._____

Write each statement with the inequality symbol reversed.

27. $9 < 12$

                                              27._____

28. $\dfrac{1}{2} < \dfrac{2}{3}$

                                              28._____

29. .21 > .19

29._____

30. .1 > .01

30._____

7.2   Variables, Expressions, and Equations

Find the value of each expression if $x = 2$ and $y = -4$.

1. $x + 3$

1._____

2. $8x^2 - 6x$

2._____

3. $2x^3 - y^2$

3._____

4. $5x + 2 - 4y$

4._____

5. $\dfrac{x^2 + y}{x + 1}$

5._____

6. $\dfrac{2x - 4}{2}$

6._____

7. $\dfrac{3y^2 + 2x^2}{5x + y^2}$

7._____

Change the word phrases to algebraic expressions. Use x as the variable.

8. Four added to a number

8._____

9. A number subtracted from three

9._____

10. The product of four less than a number and two

10._____

11. Ten times a number, added to 21

11._____

12. The sum of a number and 4 is divided by twice the number

12._____

13. Half a number is subtracted from two-thirds of the number

13._____

Determine whether the given number is a solution of the equation.

14. $x + 2 = 11$; 9

14._____

15. $6b + 2(b + 3) = 14$; 2

15._____

16. $\dfrac{p + 4}{p - 2} = 2$; 6

16._____

17. $\dfrac{x^2 - 7}{x} = 6$; 2

17._____

18. $(a - 2)^3 = 27$; 6

18._____

19. $7r + 5r - 8 = 16$; 2

19._____

Change each word sentence to an equation. Use x as the variable.

20. The sum of a number and six is ten.

20._____

21. A number minus five is equal to nine.

21._____

22. The sum of five times a number and two is 23.

                                              22._____

23. The quotient of a number and nine is 17.

                                              23._____

24. Four times a number is equal to two more than three times the number.

                                              24._____

25. The product of six and a number is 18.

                                              25._____

Decide whether each of the following is an equation or an expression.

26. $3x + 2y$

                                              26._____

27. $9x + 2y = 2$

                                              27._____

28. $y^2 - 4y - 3$

                                              28._____

29. $\dfrac{x + 4}{5}$

                                              29._____

30. $y = 2x^2 + 4$

                                              30._____

7.3   Real Numbers and the Number Line

Use a real number to express each number in the following applications.

1. Carl withdrew $273 from his checking account. Then he
deposited a check for $103.

1._____

2. Last year Nina lost 75 pounds.

2._____

3. Last year Lauren's savings account decreased by $72.

3._____

4. Between 1970 and 1982, the population of Norway
increased by 279,867.

4._____

5. When Stan graduated, he owed $42,500 on his college
loans.

5._____

Graph each group of rational numbers on a number line.

6. −2, −1, 0, 2, 4

6._____

7. 6, 3, −3, −6, 0

7._____

8. $\frac{1}{2}$, 0, −3, $-\frac{5}{2}$

8._____

9. $-\frac{3}{2}$, 0, $\frac{3}{2}$, $\frac{7}{2}$

9._____

10. $2\frac{1}{3}$, 4, $6\frac{2}{3}$, 8

10._____

Select the smaller number in each pair.

11. −15, −12

11._____

12. −.802, −.820

12._____

13. −2, 1

13._____

14. $\frac{2}{3}, -\frac{1}{2}$

14._____

Decide whether each statement is true *or* false.

15. $-76 < 45$

15._____

16. $-5 > -5$

16._____

17. $-12 > -10$

17._____

Find the opposite of each number.

18. 23

18._____

19. -25

19._____

20. $-\frac{5}{7}$

20._____

21. $-(-4)$

21._____

Simplify by removing absolute value symbols.

22. $|-4|$

22._____

23. $|0|$

23._____

24. $-|-25|$

24._____

25. $-|49 - 39|$

25._____

26. $-|-.9|$

26._____

27. $\left|1\frac{1}{2} - 2\frac{1}{4}\right|$

27._____

Select the smaller number in each pair.

28. $|-8|, |2|$

28._____

29. $-|-2|, -|5|$

29._____

30. $-|-7|, -|10|$

30._____

7.4   Adding Real Numbers

Use a number line to find the sums.

1. $8+7$

1._____

2. $7+(-12)$

2._____

3. $-4+(-6)$

3._____

4. $-9+(-9)$

4._____

Find the following sums.

5. $7+(-5)$

5._____

6. $-9+4$

6._____

7. $\dfrac{7}{12}+\left(-\dfrac{3}{4}\right)$

7._____

8. $3\frac{5}{8}+\left(-2\frac{1}{4}\right)$

8._____

9. $14.1+(-14.1)$

9._____

Name: Date:
Instructor: Section:

Perform each operation and then determine whether the statement is true or false. Try to do all work in your head.

10. $19 + (-13) = 6$ 10._____

11. $4.9 + (-2.6) = -2.3$ 11._____

12. $(-14) + 15 + (-2) = -3$ 12._____

13. $\dfrac{3}{5} + \left(-\dfrac{3}{10}\right) = -\dfrac{3}{10}$ 13._____

14. $5\frac{3}{8} + \left(-4\frac{1}{2}\right) = 2\frac{1}{8}$ 14._____

15. $\left|14 + (-22)\right| = -14 + 22$ 15._____

Find the following sums.

16. $6 + \left[2 + (-7)\right]$ 16._____

17. $-9 + \left[5 + (-19)\right]$ 17._____

18. $-14 + 3 + \left[8 + (-13)\right]$ 18._____

19. $\left[(-7)+14\right]+\left[(-16)+3\right]$

19._____

20. $\dfrac{3}{8}+\left[-\dfrac{2}{3}+\left(-\dfrac{7}{12}\right)\right]$

20._____

21. $\left[\dfrac{7}{10}+\left(-\dfrac{3}{5}\right)\right]+\dfrac{1}{2}$

21._____

Write a numerical expression for each phrase, and then simplify the expression.

22. The sum of –9 and 14

22._____

23. –2 increased by 16

23._____

24. 10 added to the sum of –4 and –3

24._____

25. 10 increased by the sum of –20 and 9

25._____

26. The sum of –14 and –29, increased by 27

26._____

Solve each problem by writing a sum of real numbers and adding. No variables are needed.

27. A football team gained 4 yards from scrimmage on the first play, lost 21 yards on the second play, and gained 9 yards on the third play. How many yards did the team gain or lose altogether?

27._____

28. Pablo has $723 in his checking account. He write two checks, one for $358 and the other for $75. Finally, he deposits $205 in the account. How much does he now have in his account?

28._____

29. Charlie starts to climb a mountain at an altitude of 2324 feet. He climbs so that he gains 247 feet in altitude. Then he finds that, because of an obstruction, he must descend 15 feet. Then he climbs 98 feet up. What is his final altitude?

29._____

30. Penny owes $48 to a credit card company. She makes a purchase of $45 with her card, and then pays $77 to the company. How much does she still owe?

30._____

7.5   Subtracting Real Numbers

Use a number line to find the difference.

1. $10 - 5$

1._____

2. $7 - 10$

2._____

3. $-3 - 9$

3._____

4. $-7 - 2$

4._____

Find each difference.

5. $14 - 20$

5._____

6. $-9 - 4$

6._____

7. $13 - (-9)$

7._____

8. $-4 - (-20)$

8._____

9. $-3.2 - (-7.6)$

9._____

10. $4.5 - (-2.8)$

10._____

11. $-\frac{3}{10} - \left(-\frac{4}{15}\right)$

11._____

12. $3\frac{3}{4} - \left(-2\frac{1}{8}\right)$ **12.**_____

Work each problem.

13. $4 - \left[6 + (-9)\right]$ **13.**_____

14. $-.2 - \left[.6 + (-.9)\right]$ **14.**_____

15. $\left[6 - (-14)\right] - 26$ **15.**_____

16. $\left[3 + (-9)\right] - (-6)$ **16.**_____

17. $\left[\frac{1}{3} - \left(-\frac{1}{5}\right)\right] - \left(-\frac{4}{15}\right)$ **17.**_____

18. $-4 + \left[(-5 - 3) - (-2 + 4)\right]$ **18.**_____

19. $-2 + \left[(-12 + 10) - (-4 + 2)\right]$ **19.**_____

Write a numerical expression for each phrase, and then simplify the expression.

20. The difference between –9 and 3

20._____

21. The difference between –6 and –2

21._____

22. 4 less than -4

22._____

23. –8 decreased by 2 less than –1

23._____

24. –4 decreased by 1 less than –4

24._____

25. –6 subtracted from the sum of 4 and –7

25._____

26. –12 subtracted from the sum of –4 and –2

26._____

Solve each of the following problems by writing a difference between real numbers and subtracting. No variables are needed.

27. Dr. Somers runs an experiment at –43.3°C. He then lowers the temperature by 7.9°C. What is the new temperature for the experiment?

27._____

28. David has a checking account balance of $439.42. He overdraws his account by writing a check for $702.58. Write his new balance as a negative number.

28._____

29. At 1:00 A.M., the temperature on the top of Mt. Washington in New Hampshire was –12°F. At 11:00 A.M., the temperature was 25°F. What was the rise in temperature?

29._____

30. The highest point in a country has an elevation of 1408 meters. The lowest point is 396 meters below sea level. Using zero as sea level, find the difference between the two elevations.

30._____

7.6   Multiplying and Dividing Real Numbers

Find the products.

1. $7(-4)$

1._____

2. $(-80)(4)$

2._____

3. $\left(-\frac{2}{7}\right)\left(\frac{21}{26}\right)$

3._____

Find the products.

4. $(-3)(-4)$

4._____

5. $(-4)(-10)$

5._____

6. $(-1.3)(-2.1)$

6._____

Find the reciprocal, if one exists, for each number.

7. -2

7._____

8. $\frac{3}{5}$

8._____

9. $-.125$

9._____

Find the quotients.

10. $\frac{40}{-5}$

10._____

11. $\frac{-120}{-20}$

12. $-\frac{3}{16} \div \frac{9}{8}$

Perform the indicated operations.

13. $24 - 5 \cdot 7$

14. $4(-8) - (-2)(-7)$

15. $-7\left[-4 - (-2)(-3)\right]$

Simplify the numerators and denominators separately. Then find the quotients.

16. $\dfrac{9(-4)}{-6 - (-2)}$

17. $\dfrac{5(-8 + 3)}{13(-2) + (-6 - 1)(-4 + 1)}$

18. $\dfrac{-4\left[8 - (-3 + 7)\right]}{-6\left[3 - (-2)\right] - 3(-3)}$

19. $\dfrac{4^3 - 3^3}{-5(-4 + 2)}$

Evaluate the following expressions if $x = -3$, $y = 2$, and $a = 4$.

20. $-3x + 4y - (a - x)$

20.＿＿＿＿＿＿＿＿＿

21. $-x^2 + 3y$

21.＿＿＿＿＿＿＿＿＿

22. $2(x-3)^2 + 2y^2$

22.＿＿＿＿＿＿＿＿＿

23. $\dfrac{2x^2 - 3y}{4a}$

23.＿＿＿＿＿＿＿＿＿

Write a numerical expression for each phrase and simplify.

24. The product of –7 and 3, added to –7

24.＿＿＿＿＿＿＿＿＿

25. Twice the sum of 14 and –4, added to –2

25.＿＿＿＿＿＿＿＿＿

26. –34 subtracted from two-thirds of the sum of 16 and –10

26.＿＿＿＿＿＿＿＿＿

27. The product of 40 and –3, divided by the difference between 5 and –10

27.＿＿＿＿＿＿＿＿＿

Write each statement in symbols, using x as the variable.

28. The quotient of a number and –2 is –9.

28._____

29. The difference between a number and –7 is 12.

29._____

30. When a number is divided by –4, the result is 1.

30._____

7.7   Properties of Real Numbers

Complete each statement. Use a commutative property.

  1. $y+4=\underline{\hspace{1cm}}+y$

  1._____

  2. $(ab)(2)=(2)\underline{\hspace{1cm}}$

  2._____

  3. $7m=\underline{\hspace{1cm}}(7)$

  3._____

  4. $-4(p+9)=\underline{\hspace{1cm}}(-4)$

  4._____

  5. $2+\left[10+(-9)\right]=\underline{\hspace{1cm}}+2$

  5._____

Complete each statement. Use an associative property.

  6. $x(9y)=\underline{\hspace{1cm}}(y)$

  6._____

  7. $\left[-4+(-2)\right]+y=\underline{\hspace{1cm}}+(-2+y)$

  7._____

  8. $4(ab)=\underline{\hspace{1cm}}\cdot b$

  8._____

  9. $(-12x)(-y)=(-12)\underline{\hspace{1cm}}$

  9._____

  10. $(-r)\left[(-p)(-1)\right]=\underline{\hspace{1cm}}(-q)$

  10._____

11. $\left[x+(-4)\right]+3y=x+$ _____   **11.** _____

Simplify.

12. $4+0=$ **12.** _____

13. $-7+0=$ **13.** _____

14. $1(-4)=$ **14.** _____

15. $7(1)=$ **15.** _____

Complete the statements so that they are examples of either an identity property or an inverse property. Identify which property is used.

16. $-4+$ _____ $=0$ **16.** _____

17. _____ $+\frac{1}{7}=0$ **17.** _____

18. $1\cdot$ _____ $=1$ **18.** _____

19. $-\frac{3}{5}\cdot$ _____ $=1$ **19.** _____

20. $-14+$ _____ $=0$ **20.** _____

21. _____ $+0=0$ **21.** _____

22. _____$\cdot -2\frac{5}{6} = 1$

22._____

Use the distributive property to rewrite each expression. In Exercise 23, simplify the result.

23. $6y + 7y$

23._____

24. $a(z + 2)$

24._____

25. $4 \cdot r + 4 \cdot p$

25._____

26. $3(a + b)$

26._____

27. $n(2a - 4b + 6c)$

27._____

28. $-2(5y - 9z)$

28._____

29. $-(-2k + 7)$

29._____

30. $2(7x) + 2(8z)$

30._____

7.8   Simplifying Expressions

Simplify each expression.

1. $14+3y-8$ 1._____

2. $4(2x+5)+7$ 2._____

3. $-(9-4b)-8$ 3._____

4. $7-6y+(5-2)$ 4._____

5. $-2(-5x+2)+7$ 5._____

Give the numerical coefficient of each term.

6. $4x$ 6._____

7. $-2y^2$ 7._____

8. $.3a^2b$ 8._____

9. 125 9._____

10. z^5 10._____

11. $\dfrac{7x}{9}$ 11._____

Identify each group of terms as like *or* unlike.

12. $2x, 7x$ 12._____

13. $-7q^2, 2q^2$ 13._____

14. $2w,\ 4w,\ -w$

14._____

15. $-5y,\ -4y,\ 2$

15._____

16. $4x,\ -10x^2,\ -9x^2$

16._____

17. $2,\ -4,\ 16$

17._____

Simplify each expression by combining like terms.

18. $23a - 16a$

18._____

19. $2.3r + 6.9 + 2.8 + 3.6r$

19._____

20. $\frac{1}{3} + \frac{3}{4}y - \frac{5}{6} - \frac{2}{3}y$

20._____

Use the distributive property and combine like terms to simplify the following expressions.

21. $2(3x + 5)$

21._____

22. $7r - (2r + 4)$

22._____

23. $4(2q + 7) - (3q - 2)$

23._____

24. $-5(s+4)+4(2s+2)$

24._____

25. $2.5(3y+1)-4.5(2y-3)$

25._____

Write each phrase as a mathematical expression and simplify by combining like terms. Use x as the variable.

26. Seven times a number, added to twice the number

26._____

27. The sum of six times a number and 12, added to four times the number

27._____

28. The sum of seven times a number and 2, subtracted from three times the number

28._____

29. The sum of ten times a number and 7, subtracted from the difference between 2 and nine times the number

29._____

30. Four times the difference between twice a number and six times the number, added to six times the sum of the number and 9

30._____

Chapter 8 EQUATIONS, INEQUALITIES, AND APPLICATIONS

8.1   The Addition Property of Equality

Tell whether each of the following is a linear equation.

1. $9x + 2 = 0$

    1._____

2. $3x^2 + 4x + 3 = 0$

    2._____

3. $7x^2 = 10$

    3._____

4. $3x^3 = 2x^2 + 5x$

    4._____

5. $\frac{5}{x} - \frac{3}{2} = 0$

    5._____

6. $4x - 2 = 12x + 9$

    6._____

Solve each equation by using the addition property of equality. Check each solution.

7. $y - 4 = 16$

    7._____

8. $r + 9 = 8$

    8._____

9. $3x + 2 = 5x + 12$

    9._____

10. $3y = 7y - 4$

10.＿＿＿＿＿＿＿＿＿＿＿

11. $p - \frac{2}{3} = \frac{5}{6}$

11.＿＿＿＿＿＿＿＿＿＿＿

12. $y + 4\frac{1}{2} = 3\frac{3}{4}$

12.＿＿＿＿＿＿＿＿＿＿＿

13. $\frac{2}{3}t - 5 = \frac{5}{3}t$

13.＿＿＿＿＿＿＿＿＿＿＿

14. $\frac{9}{8}p - \frac{1}{2} = \frac{1}{8}p$

14.＿＿＿＿＿＿＿＿＿＿＿

15. $5.7x + 12.8 = 4.7x$

15.＿＿＿＿＿＿＿＿＿＿＿

16. $9.5y - 2.4 = 10.5y$

16.＿＿＿＿＿＿＿＿＿＿＿

Solve each equation. First simplify each side of the equation as much as possible. Check each solution.

17. $6x - 3x + 10 = -2$

17.＿＿＿＿＿＿＿＿＿＿＿

18. $6x + 3x - 7x + 4 = 10$

18.＿＿＿＿＿＿＿＿＿＿＿

19. $3(t+3)-(2t+7)=9$

19._____

20. $5x+4(2x+1)-(5x-1-2)=9$

20._____

21. $-4(5g-7)+3(8g-3)=15-4+3g$

21._____

22. $10x+4x-11x+4-7=2-4x-3+8x$

22._____

23. $4(3a-2)-6(2+a)=5(2a-5)$

23._____

24. $2(4t+6)-3(2t-3)=-3(3t-4)+5-t$

24._____

25. $-7(1+2b)-6(3-5b)=5(4+3b)-45$

25._____

26. $8(2-4b)+3(5-b)=4(1-9b)+22$

26._____

27. $\frac{8}{5}t+\frac{1}{3}=\frac{5}{6}+\frac{3}{5}t-\frac{1}{6}$

27._____

28. $\frac{5}{12}+\frac{7}{6}s-\frac{1}{6}=\frac{5}{6}s+\frac{1}{4}-\frac{2}{3}s$

28._____

29. $3.6p+4.8+4.0p=8.6p-3.1+.7$

29._____

30. $.03x+0.6+.09x-.9=2.1$

30._____

8.2   The Multiplication Property of Equality

Solve each equation and check your solution.

1. $8x = 24$

1._____

2. $-3w = 42$

2._____

3. $-16a = -48$

3._____

4. $\frac{b}{5} = 4$

4._____

5. $\frac{3p}{7} = -6$

5._____

6. $\frac{b}{-2} = 21$

6._____

7. $-9k = 81$

7._____

8. $\frac{3}{4}r = -27$

8._____

9. $\frac{y}{4} = \frac{1}{3}$

9._____

10. $\frac{6}{7}y = \frac{2}{3}$

10._____

11. $.9x = 5.4$

11._____

12. $2.1a = 9.03$

12._____

13. $1.9k = 11.02$

13._____

14. $7.5p = -61.5$

14._____

15. $-2.7v = -17.28$

15._____

16. $4.3r = -11.61$

16._____

Solve each equation and check your solution.

17. $4r + 3r = 63$

17._____

18. $3x + 6x = 72$

18._____

19. $7y - 2y = 45$

19._____

20. $9z - 3z = 24$

20._____

21. $10a - 7a = -24$

21._____

22. $14m - 6m = -56$

22._____

23. $8f + 4f - 3f = 72$

23._____

24. $-y = 3.9$

24._____

25. $-t = -26$

25._____

26. $-h = \frac{7}{4}$

26._____

27. $3b - 4b = 8$

27._____

28. $3w - 7w = 20$

28._____

29. $4x - 8x + 2x = 16$

29._____

30. $-11h - 6h + 14h = -21$

30._____

8.3   More on Solving Linear Equations

Solve each equation and check your solution.

1. $7t + 6 = 11t - 4$ 1._____

2. $7x + 11 = 9x + 25$ 2._____

3. $7j + 1 = 10j - 29$ 3._____

4. $4(z - 2) - (3z - 1) = 2z - 6$ 4._____

5. $3 - (1 - y) = 3 + 5y$ 5._____

6. $4w - 5w + 3(w - 7) = -4(w + 4) + 7$ 6._____

7. $3a - 6a + 4(a - 4) = -2(a + 2)$ 7._____

8. $4r - 3(3r - 2) = 8 - 3(r - 4)$

8._____

9. $6f - 8f + 4(f - 3) = -2(f + 4)$

9._____

Solve each equation.

10. $\frac{1}{5}(z - 5) = \frac{1}{3}(z + 2)$

10._____

11. $\frac{2}{3}y - \frac{1}{4}y = -\frac{5}{12}y + \frac{1}{2}$

11._____

12. $\frac{1}{3}(2m - 1) - \frac{3}{4}m = \frac{5}{6}$

12._____

13. $\frac{5}{6}(r - 2) - \frac{2}{9}(r + 4) = \frac{7}{18}$

13._____

14. $\frac{1}{8}(t-3)+\frac{3}{8}(t+2)=t-2$

14._____

15. $.90x=.40(30)+.15(100)$

15._____

16. $.12x+.24(x-5)=.56x$

16._____

17. $.24x-.38(x+2)=-.34(x+4)$

17._____

18. $.07(10,000)+.02x=.03(10,000+x)$

18._____

Solve each equation.

19. $3(6x-7)=2(9x-6)$

19._____

20. $6y - 3(y + 2) = 3(y - 2)$

20._____

21. $-1 - (2 + y) = -(-4 + y)$

21._____

22. $8k + 14 = 2(k + 2) + 3(2k + 1)$

22._____

23. $6(6t + 1) = 9(4t - 3) + 11$

23._____

24. $4b - 2b - 16 - b = 4b + 16 - 3b$

24._____

25. $8(2d - 4) - 3(7d + 8) = -5(d + 2)$

25._____

26. $2(5w-3)+11=3(3w+1)+5(w-3)-4w$

26._____

27. $7(3b-4)-4(2b+2)=4-2(b-3)+5(3b-6)$

27._____

Write an expression for the two related unknown quantities.

28. Two numbers have a sum of 36. One is m. Find the other number.

28._____

29. The product of two numbers is 17. One number is p. What is the other number?

29._____

30. A cashier has q dimes. Find the value of the dimes in cents.

30._____

8.4   An Introduction to Applications of Linear Equations

Write an equation for each of the following and then solve the problem. Use x as the variable.

1. If 4 is added to 3 times a number, the result is 7. Find the number.

   1._____

2. If 2 is subtracted from four times a number, the result is 3 more than six times the number. What is the number?

   2._____

3. If –2 is multiplied by the difference between 4 and a number, the result is 24. Find the number.

   3._____

4. Six times the difference between a number and 4 equals the product of the number and –2. Find the number.

   4._____

5. When the difference between a number and 4 is multiplied by –3, the result is two more than –5 times the number. Find the number.

   5._____

6. If four times a number is added to 7, the result is five less than six times the number. Find the number.

   6._____

7. A rope 116 inches long is cut into three pieces. The middle-sized piece is 10 inches shorter than twice the shortest piece. The longest piece is $\frac{5}{3}$ as long as the shortest piece. What is the length of the shortest piece?

7._____

8. George and Al were opposing candidates in the school board election. George received 21 more votes than Al, with 439 votes cast. How many votes did Al receive?

8._____

9. On a psychology test, the highest grade was 38 points more than the lowest grade. The sum of the two grades was 142. Find the lowest grade.

9._____

10. Mount McKinley is Alaska is 5910 feet higher than Mount Rainier in Washington. Together, their heights total 34,730 feet. How high is each mountain?

10._____

11. Charles bought five general admission tickets and four student tickets for a movie. He paid $35.25. If each student ticket cost $3.50, how much did each general admission ticket cost?

11._____

12. Penny is making punch for a party. The recipe requires twice as much orange juice as cranberry juice and 8 times as much ginger ale as cranberry juice. If she plans to make 176 ounces of punch, how much of each ingredient should she use

12._____

13. Pablo, Faustino, and Mark swim at a public pool each day for exercise. One day Pablo swam five more than three times as many laps as Mark, and Faustino swam four times as many laps as Mark. If the men swam 29 laps altogether, how many laps did each one swim?

13._____

14. Linda wishes to build a rectangular dog pen using 52 feet of fence and the back of her house, which is 36 feet long to enclose the pen. How wide will the dog pen be if the pen is 36 feet long?

14._____

Solve each problem.

15. Find the measure of an angle if the measure of the angle is 8° less than three times the measure of its supplement.

15._____

16. Find the measure of an angle whose supplement measures 20° more than twice its complement.

16._____

17. Find the measure of an angle such that the sum of the measures of its complement and its supplement is 138°.

17._____

18. Find the measure of an angle such that the difference between the measure of its supplement and twice the measure of its complement is 49°.

18._____

19. Find the measure of an angle whose complement is 9° more than twice its measure.

19._____

20. Find the measure of an angle whose supplement measures 3 times its complement.

20._____

21. Find the measure of an angle whose supplement measures 6° more than 7 times its complement.

21._____

22. Find the measure of an angle such that the difference between the measures of an angle and its complement is 20°.

22._____

23. Find the measure of an angle if its supplement measures 15° less than four times its complement.

23._____

24. Find the measure of an angle if its supplement measures 4° less than three times its complement.

24._____

Solve each problem.

25. Find two consecutive even integers whose sum is 154.

25._____

26. Find two consecutive even integers such that the smaller, added to twice the larger, is 292.

26._____

27. Find two consecutive integers such that the larger, added to three times the smaller, is 109.

27._____

28. Find two consecutive odd integers such that if three times the smaller is added to twice the larger, the sum is 69.

28._____

29. Find two consecutive odd integers such that the larger, added to eight times the smaller, equals 119.

29._____

30. Find three consecutive odd integers whose sum is 363.   30._____

8.5   Formulas and Applications from Geometry

In the following exercises, a formula is given, along with the values of all but one of the variables in the formula. Find the value of the variable that is not given.

1. $V = LWH$; $L = 2$, $W = 4$, $H = 3$

     1._____

2. $P = 2L + 2W$; $P = 42$; $W = 6$

     2._____

3. $A = \frac{1}{2}bh$; $b = 8$, $h = 2.5$

     3._____

4. $V = \frac{1}{3}Bh$; $B = 27$, $V = 63$

     4._____

5. $C = 2\pi r$; $C = 43.96$, $\pi = 3.14$

     5._____

6. $I = prt$; $I = 288$, $r = .04$, $t = 3$

     6._____

7. $F = \frac{9}{5}C + 32$; $C = 35$

     7._____

8. $V = \frac{4}{3}\pi r^3$; $r = 3$, $\pi = 3.14$

     8._____

Use a formula to write an equation for each of the following applications; then solve the application.

9. Find the length of a rectangular garden if its perimeter is 96 feet and its width is 12 feet.

9._____

10. Find the height of a triangular banner whose area is 48 square inches and base is 12 inches.

10._____

11. Ruth has 42 feet of binding for a rectangular rug that she is weaving. If the rug is 9 feet wide, how long can she make the rug if she wishes to use all the binding on the perimeter of the rug?

11._____

12. The radius of a pizza is 8 inches. Find the area of the pizza. (Use 3.14 as an approximation for π.)

12._____

13. A water tank is a right circular cylinder. The tank has a radius of 6 meters and a volume of 1356.48 cubic meters. Find the height of the tank. (Use 3.14 as an approximation for π.)

13._____

14. A spherical balloon has a radius of 9 centimeters. Find the amount of air required to fill the balloon. (Use 3.14 as an approximation for π.)

14._____

15. Find the height of an ice cream cone if the diameter is 6 centimeters and the volume is 37.68 cubic centimeters. (Use 3.14 as an approximation for π. Round answer to the nearest hundredth.)

15._____

Find the measure of each marked angle.

16.

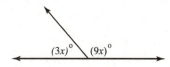

16._____

17.

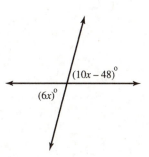

17._____

18.

18._____

19.

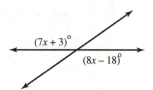

19._____

20.

20._____

21.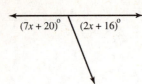

21._____

22.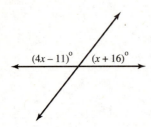

22._____

Solve each formula for the specified variable.

23. $V = LWH$ for H

23._____

24. $S = \dfrac{a}{1-r}$ for r

24._____

25. $A = \frac{1}{2}bh$ for h

25._____

26. $S = 2\pi rh + 2\pi r^2$ for h

26._____

27. $P = A - Art$ for A

27._____

28. $A = \frac{1}{2}(b + B)h$ for b

28._____

29. $C = \frac{5}{9}(F - 32)$ for F

29._____

30. $V = \frac{1}{3}\pi r^2 h$ for h

30._____

8.6   Solving Linear Inequalities

Graph each inequality on a number line.

1. $x \geq 3$

                                                  1._____

2. $7 < a$

                                                  2._____

3. $y \geq -2$

                                                  3._____

4. $-5 \leq r$

                                                  4._____

5. $-4 \leq x < 4$

                                                  5._____

6. $-6 < m < -2$

                                                  6._____

Solve each inequality and graph the solutions.

7. $j + 6 \leq 11$

                                                  7._____

8. $-2 + 8b \geq 7b - 1$

                                                  8._____

9. $y - 7 > -12$

                                                  9._____

10. $14 + 23a \geq 24a + 18$

                                                  10._____

11. $3 + 5p \leq 4p + 3$

11._____

Solve each inequality and graph the solutions.

12. $-7q \geq 5$

12._____

13. $5r > -20$

13._____

14. $-6k \leq 0$

14._____

15. $-\frac{2}{3}z > 4$

15._____

16. $-.04t \leq .2$

16._____

17. $-5t \leq -35$

17._____

Solve each inequality.

18. $8p + 8 \geq 9p$

18._____

19. $4(y - 3) + 2 > 3(y - 2)$

19._____

20. $-3(m + 4) + 1 \leq -4(m - 2)$

20._____

21. $5(2z+2)-2(z-3)>3(2z+5)+z$

21._____

Solve each inequality and graph the solutions.

22. $7m-8\geq 5m$

22._____

23. $5p-5-p>7p-2$

23._____

24. $5(y+3)-5y>3(y+1)+4$

24._____

25. $4-\frac{1}{3}y\leq 6+\frac{2}{3}y$

25._____

Use an inequality to solve each problem.

26. Faustino sold two antique desks for \$280 and \$305. How much should he charge for the third in order to average at least \$300 per desk?

26._____

27. Lauren has grades of 98 and 86 on her first two chemistry quizzes. What must she score on her third quiz to have an average of at least 91 on the three quizzes?

27._____

28. Nina has a budget of $230 for gifts for this year. So far she has bought gifts costing $47.52, $38.98, and $26.98. If she has three more gifts to buy, find the average amount she can spend on each gift and still stay within her budget.

28._____

29. Ruth tutors mathematics in the evenings in an office for which she pays $600 per month rent. If rent is her only expense and she charges each student $40 per month, how many students must she teach to make a profit of at least $1600 per month?

29._____

30. Two sides of a triangle are equal in length, with the third side 8 feet longer than one of the equal sides. The perimeter of the triangle cannot be more than 38 feet. Find the largest possible value for the length of the equal sides.

30._____

Chapter 9 GRAPHS OF LINEAR EQUATIONS AND INEQUALITIES IN TWO VARIABLES

9.1   Reading Graphs: Linear Equations in Two Variables

The graphs below show the total number of degrees awarded by Jefferson University for the years 1990 – 1995 and the distribution of degrees awarded over this period. Use these graphs to answer the questions in Exercises 1 – 5.

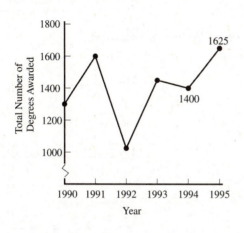
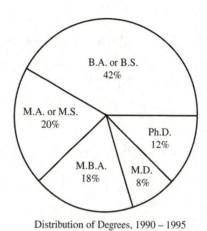

Distribution of Degrees, 1990 – 1995

1. Between which two years did the total number of degrees awarded show the greatest decline?

   1._____

2. About how many students received master's degrees (M.A. or M.S.) in 1994?

   2._____

3. About how many more students received M.B.A. degrees in 1995 than 1994?

   3._____

4. Between which two years did the total number of degrees awarded show the smallest change?

   4._____

5. About how many students received doctoral degrees (M.D., or Ph.D.) in 1994?

   5._____

Name: Date:

Instructor: Section:

Write each solution as an ordered pair.

6. $x = 4$ and $y = 7$

6._____

7. $x = -4$ and $y = 0$

7._____

8. $x = \frac{1}{3}$ and $y = -9$

8._____

9. $y = \frac{1}{3}$ and $x = 0$

9._____

Decide whether the given ordered pair is a solution of the given equation.

10. $3x + 2y = 4;\ (0, 2)$

10._____

11. $4x - 3y = 10;\ (1, 2)$

11._____

12. $2x = -4y;\ (0, 0)$

12._____

13. $x = 1 - 2y;\ \left(0, -\frac{1}{2}\right)$

13._____

For each of the given equations, complete the ordered pairs beneath it.

14. $y = 2x - 5$

14._____

      (a) $(2,\ \ )$

      (b) $(0,\ \ )$

      (c) $(\ \ , 3)$

      (d) $(\ \ , -7)$

      (e) $(\ \ , 9)$

15. $y = 3 + 2x$

       (a) $(-4, \quad)$

       (b) $(2, \quad)$

       (c) $(\quad, 0)$

       (d) $(-2, \quad)$

       (e) $(\quad, -7)$

15._____

16. $x = -2$

       (a) $(\quad, -2)$

       (b) $(\quad, 0)$

       (c) $(\quad, 19)$

       (d) $(\quad, 3)$

       (e) $\left(\quad, -\frac{2}{3}\right)$

16._____

Complete the table of ordered pairs for each equation.

17. $4x + y = 6$

x	2		1
y		4	

17._____

18. $3x + 2y = 4$

x	0		4
y		0	

18._____

19. $3x + y = 9$

x	0		-1
y		0	

19._____

20. $x = -2$

x			
y	0	4	-5

20._____

21. $3x - 4y = -6$

x	y
0	
	0
2	

21. _____

Plot the following ordered pairs on a coordinate system.

22. $(6, 1)$

22. _____

23. $(-2, 4)$

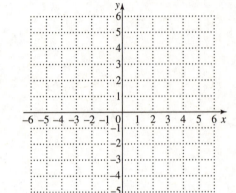

23. _____

24. $(-2, -6)$

24. _____

25. $(4, -2)$

25. _____

26. $(0, -4)$

26. _____

Without plotting the given point, name the quadrant in which it lies.

27. $(1, -2)$

27. _____

28. $(5, -4)$

28. _____

29. $(-7, -2)$

29. _____

30. $(3, 0)$

30. _____

Chapter 9 GRAPHS OF LINEAR EQUATIONS AND INEQUALITIES IN TWO VARIABLES

9.2   Graphing Linear Equations in Two Variables

Complete the ordered pairs for each equation. Then graph the equation by plotting the points and drawing a line through them.

1. $x + y = 3$

(0,   )

(, 0)

(2,   )

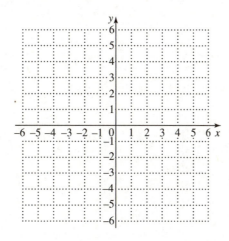

1._____

2. $y + 3 = 0$

(0,   )

(4,   )

(−3,   )

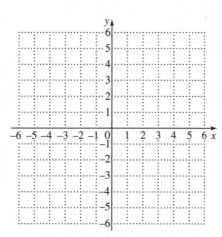

2._____

3. $2y - 4 = x$

(0,   )

(, 0)

(−2,   )

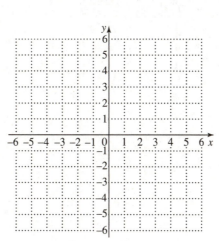

3._____

4. $x - y = 4$

   $(0, \quad)$

   $(\quad, 0)$

   $(-2, \quad)$

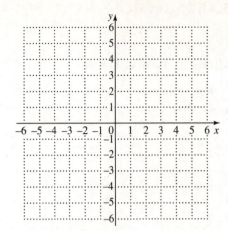

4. _____

5. $x = -2y + 2$

   $(0, \quad)$

   $(\quad, 0)$

   $(-2, \quad)$

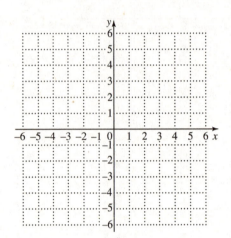

5. _____

6. $x - y = -1$

   $(0, \quad)$

   $(\quad, 0)$

   $(4, \quad)$

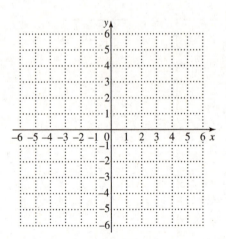

6. _____

7. $y = 2x - 3$

$(0, \quad)$

$(\quad, 0)$

$(1, \quad)$

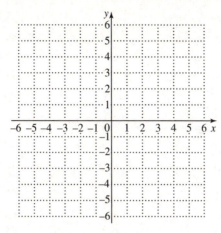

7.＿＿＿＿＿＿＿＿＿

Find the intercepts for the graph of each equation.

8. $-5x + 2y = 10$

8.＿＿＿＿＿＿＿＿＿

9. $3x + 2y = 12$

9.＿＿＿＿＿＿＿＿＿

10. $2x + 4y = 0$

10.＿＿＿＿＿＿＿＿＿

11. $4x + 5y = 8$

11.＿＿＿＿＿＿＿＿＿

Find the intercepts and graph the equation.

`12. $3x + y = 6$

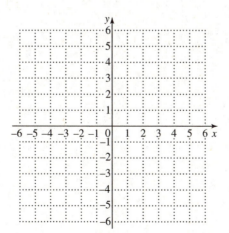

12.＿＿＿＿＿＿＿＿＿

13. $6x + 5y = 15$

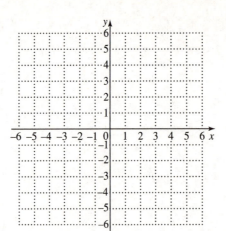

13._____

14. $x + 2y = -3$

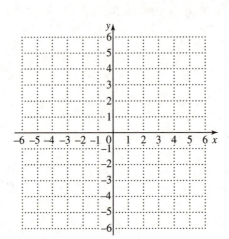

14._____

15. $5x + 6y = -30$

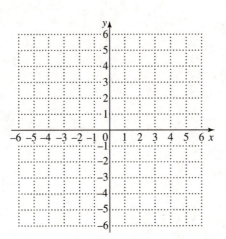

15._____

Graph each equation.

16. $3x - y = 0$

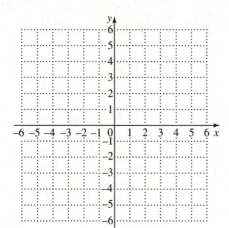

16. _____

17. $2x + y = 0$

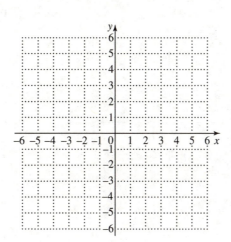

17. _____

18. $3x + 4y = 0$

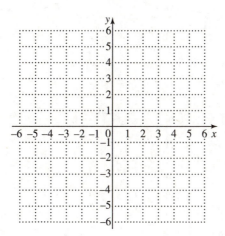

18. _____

19. $3x - 2y = 0$

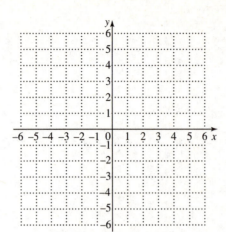

19. _____

20. $-4x + 5y = 0$

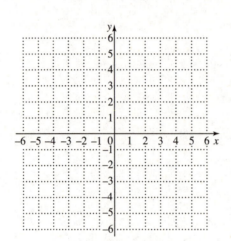

20. _____

21. $x + y = 0$

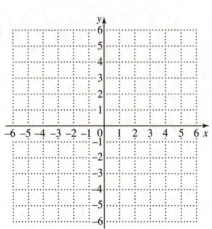

21. _____

Name: Date:
Instructor: Section:

Graph each equation.

22. $x = 3$

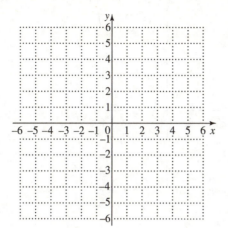

22._____

23. $x + 4 = 0$

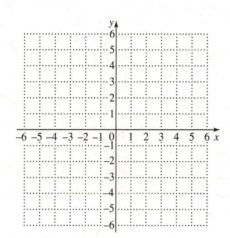

23._____

24. $y = 0$

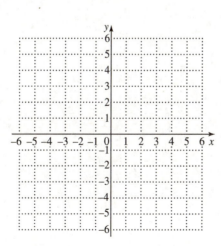

24._____

25. $y = 2$

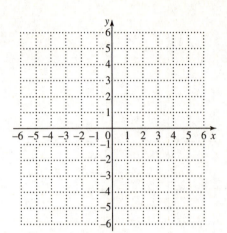

26. $y = -2$

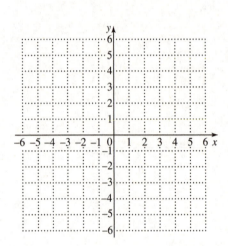

`**27.** $x - 1 = 0$

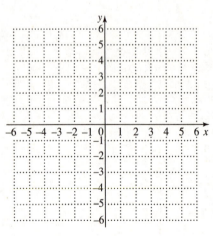

Solve each problem.

28. The enrollment at Lincolnwood High School decreased
    during the years 1990 to 1995. If $x = 0$ represents 1990,
    $x = 1$ represents 1991, and so on, the number of students
    enrolled in the school can be approximated by the
    equation

    $$y = -85x + 2435.$$

    Use this equation to approximate the number of students
    in each year from 1990 through 1995.

28._____

29. The profit y in millions of dollars earned by a small
    computer company can be approximated by the linear
    equation

    $$y = .63x + 4.9,$$

    where $x = 0$ corresponds to 1994, $x = 1$ corresponds to
    1995, and so on. Use this equation to approximate the
    profit in each year from 1994 through 1997.

29._____

30. The number of band instruments sold by Elmer's Music
    Shop can be approximated by the equation

    $$y = 325 + 42x,$$

    where y is the number of instruments sold and x is the
    time in years, with $x = 0$ representing 1993. Use this
    equation to approximate the number of instruments sold
    in each year from 1993 through 1996.

30._____

Chapter 9 GRAPHS OF LINEAR EQUATIONS AND INEQUALITIES IN TWO VARIABLES

9.3   Slope of a Line

Find the slope of each line.

1. Through $(4,3)$ and $(3,5)$

1._____

2. Through $(2,3)$ and $(6,7)$

2._____

3. Through $(-3,2)$ and $(7,4)$

3._____

4. Through $(5,-2)$ and $(2,7)$

4._____

5. Through $(2,-4)$ and $(-3,-1)$

5._____

6. Through $(7,2)$ and $(-7,3)$

6._____

7. Through $(0,-5)$ and $(7,2)$

7._____

8. Through $(-7,-7)$ and $(2,-7)$

8._____

9. Through $(-4,-4)$ and $(-2,-2)$

9._____

10. Through $(0,0)$ and $(6,-7)$

10._____

Find slope of each line.

11. $y = -5x$

11._____

12. $y = \frac{1}{2}x + 5$

12._____

13. $y = -\frac{2}{5}x - 4$

13._____

14. $y = -\frac{4}{7}x + 9$

14._____

15. $4y = 3x + 7$

15._____

16. $3y = 2x - 1$

16._____

17. $2x + 7y = 7$

17._____

18. $7y - 4x = 11$

18._____

19. $4x - 3y = 0$

19._____

20. $y = -4$

20._____

In each pair of equations, give the slope of each line, and then determine whether the two lines are parallel, perpendicular, *or* neither.

21. $y = -5x - 2$

$\ y = 5x + 11$

21. _____

22. $y = 4x + 4$

$\ y = 3 - \frac{1}{4}x$

22. _____

23. $-x + y = -7$

$\ x - y = -3$

23. _____

24. $2x + 2y = 7$

$\ 2x - 2y = 5$

24. _____

25. $4x + 2y = 8$

$\ x + 4y = -3$

25. _____

26. $9x + 3y = 2$

$\ x - 3y = 5$

26. _____

27. $4x + 2y = 7$
$5x + 3y = 11$

27._____

28. $8x + 2y = 7$
$x = 3 - y$

28._____

29. $y + 4 = 0$
$y - 7 = 0$

29._____

30. $y = 9$
$x = 0$

30._____

Chapter 9 GRAPHS OF LINEAR EQUATIONS AND INEQUALITIES IN TWO VARIABLES

9.4   Equations of Lines

Write an equation in slope-intercept form for each of the following lines.

1. $m = \frac{2}{3}; \ b = -4$

    1._____

2. $m = -2; \ b = 0$

    2._____

3. $m = \frac{1}{3}; \ b = -\frac{1}{2}$

    3._____

4. $m = 1; \ b = -2$

    4._____

5. Slope $\frac{-1}{2}$; y-intercept $(0, -3)$

    5._____

6. Slope -1; y-intercept $(0, -2)$

    6._____

7. Slope 0; y-intercept $(0, -4)$

    7._____

Graph the line passing through the given point and having the given slope.

8. $(4, -2); \ m = -1$

    8._____

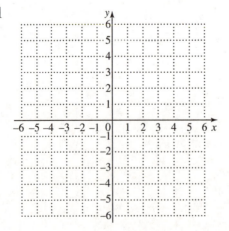

9. $(-3,-2)$; $m = \frac{2}{3}$

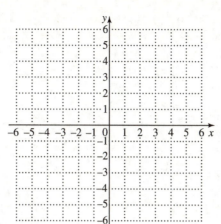

9._____

10. $(2,4)$; undefined slope

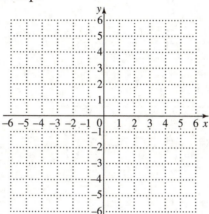

10._____

11. $(1,-3)$; $m = -\frac{5}{2}$

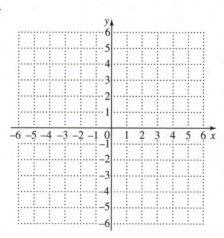

11._____

12. $(-2, 1)$; $m = 1$

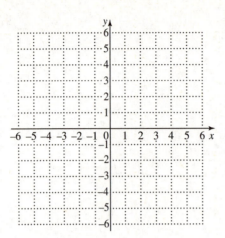

12. _____

13. $(3, -1)$; $m = 2$

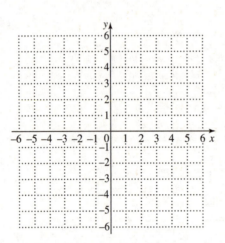

13. _____

14. $(-3, 3)$; $m = -\frac{1}{5}$

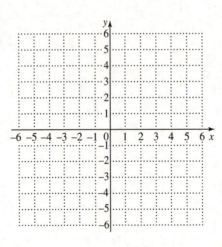

14. _____

Write an equation for the line passing through the given point and having the given slope.
Write the equations in the standard form $Ax + By = C$.

15. $(5,4)$; $m = \frac{1}{3}$ **15.**_____

16. $(-2,4)$; $m = 2$ **16.**_____

17. $(-3,-1)$; $m = -\frac{2}{3}$ **17.**_____

18. $(-4,-3)$; $m = -2$ **18.**_____

19. $(-1,2)$; $m = \frac{2}{3}$ **19.**_____

20. $(-3,4)$; $m = -\frac{3}{5}$ **20.**_____

21. $(3,-7)$; $m = \frac{5}{2}$ **21.**_____

22. $(1,-5);\ m=-\frac{1}{3}$ 22._____

Write an equation for the line passing through each pair of points. Write the equations in standard form $Ax+By=C$.

23. $(2,3)$ and $(7,5)$ 23._____

24. $(3,-4)$ and $(2,7)$ 24._____

25. $(-7,2)$ and $(4,0)$ 25._____

26. $(1,-2)$ and $(-2,8)$ 26._____

27. $(-2,1)$ and $(3,11)$ 27._____

28. $(7,2)$ and $(-2,-4)$ 28._____

29. $(2,3)$ and $(-2,-3)$ **29.**_____

30. $\left(\frac{1}{2},\frac{2}{3}\right)$ and $\left(-\frac{3}{2},2\right)$ **30.**_____

Name: Date:

Instructor: Section:

Chapter 9 GRAPHS OF LINEAR EQUATIONS AND INEQUALITIES IN TWO VARIABLES

9.5   Graphing Inequalities in Two Variables

Graph each linear inequality.

1. $y \geq x - 1$

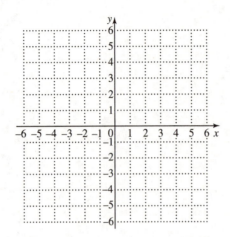

1._____

2. $x + y \geq 2$

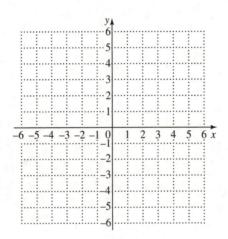

2._____

3. $3x - 2y \leq 6$

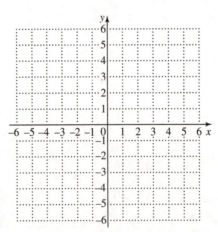

3._____

4. $y \geq -1$

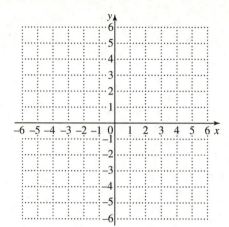

5. $x - 4 \leq -1$

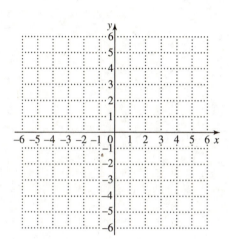

5._____

6. $y \leq -\frac{2}{5}x + 2$

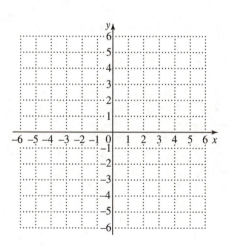

6._____

7. $y \geq 3x$

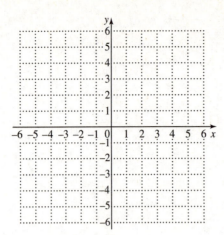

7. _____

8. $x - y \leq -3$

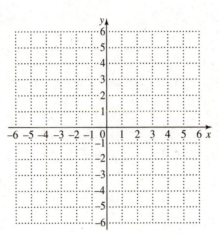

8. _____

9. $2x + 5y \leq -8$

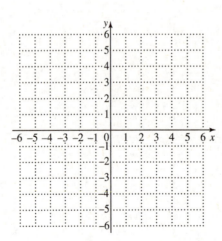

9. _____

10. $y \le x + 4$

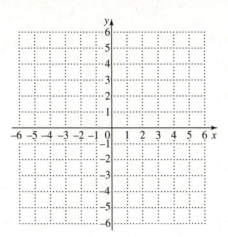

10. _____

Graph each linear inequality.

11. $x + 3y < 3$

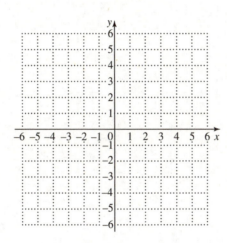

11. _____

12. $3x - 5y > -15$

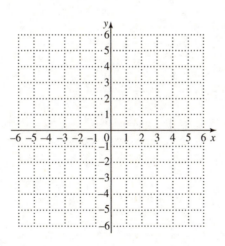

12. _____

13. $2x + 5y > -10$

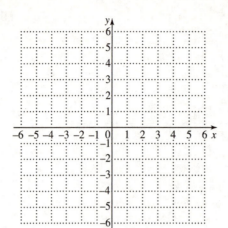

13._____

14. $y < x - 3$

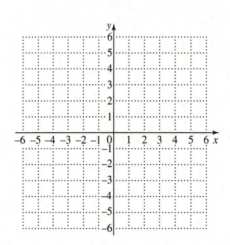

14._____

15. $y > -x + 2$

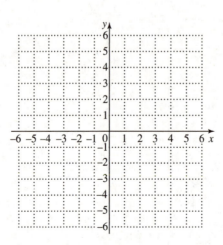

15._____

16. $x < 2y + 4$

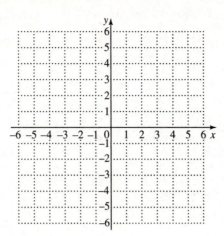

16. _____

17. $5x + 4y > 20$

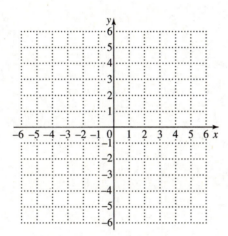

17. _____

18. $2x - 3y < 6$

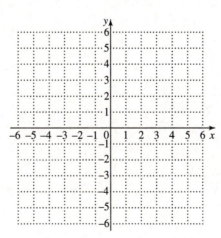

18. _____

19. $5x - 2y + 10 < 0$

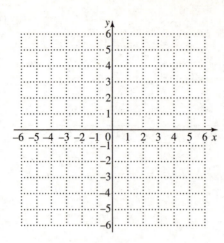

19. _____

20. $2 - 3y > x$

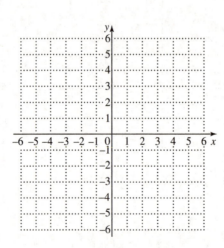

20. _____

Graph each linear inequality.

21. $y \geq 3x$

21. _____

22. $y \leq \frac{2}{5}x$

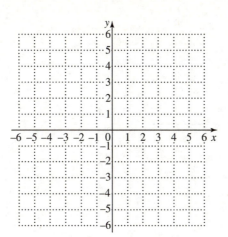

_____

23. $y \geq \frac{1}{3}x$

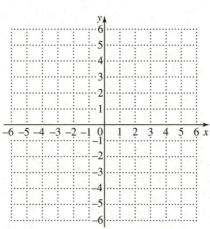

23._____

24. $y \geq x$

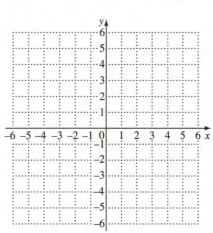

24._____

25. $3x - 4y \geq 0$

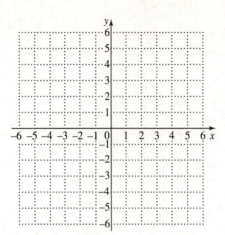

25. _____

26. $x \geq -4y$

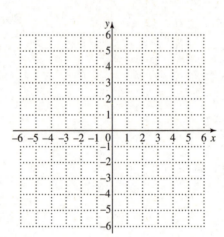

26. _____

27. $x < 2y$

27. _____

28. $x > -2y$

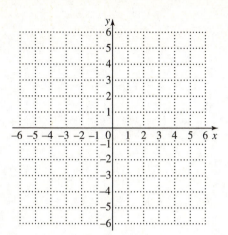

28. _____

29. $x > 4y$

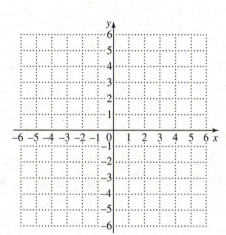

29. _____

30. $3x - 2y < 0$

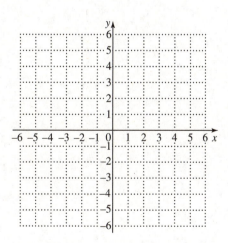

30. _____

Chapter 10 EXPONENTS AND POLYNOMIALS

10.1 Adding and Subtracting Polynomials

For each of the following, determine the number of terms in the polynomial and name the coefficients of the terms.

1. $-7y^4$

1._____

2. $9x^3 + 3x^3 - 4x + 2$

2._____

3. $\frac{1}{2} - \frac{3}{4}m$

3._____

In each polynomial, add like terms whenever possible. Write the result in descending powers of the variable.

4. $-6s^3 + 12s^3$

4._____

5. $7z^3 - 4z^3 + 5z^3 - 11z^3$

5._____

6. $6c^3 - 9c^2 - 2c^2 + 14 + 3c^2 - 6c - 8 + 2c^3$

6._____

7. $-\frac{1}{2}r^3 + \frac{1}{3}r + \frac{1}{4}r^3 - \frac{1}{3}r$

7._____

Choose one or more of the following descriptions for each expression: (a) polynomial, (b) polynomial written in descending order, (c) not a polynomial.

8. $-2w^3 + 9w^2 + 4w - 10$

8._____

9. $7y^4 - 3y^2 + \frac{2}{y}$

9._____

10. $j^{-2} - 4j^{-1}$

10._____

For each polynomial, first simplify, if possible, and write the resulting polynomial in descending powers of the variable. Then give the degree of this polynomial, and tell whether it is a monomial, *a* binomial, *a* trinomial, *or* none of these.

11. $3z^4 + 5z^3 - 2z^3$

11._____

12. $3n^8 - n^2 - 2n^8$

12._____

13. $\frac{7}{8}x^2 - \frac{3}{4}x - \frac{3}{8}x^2 + \frac{1}{4}x$

13._____

Add each pair of polynomials.

14. $5m^4 + 2m^3 - 4$
         $\underline{-3m^4 + 5m^3 - 3}$

14._____

15. $9m^3 + 4m^2 - 2m + 3$
         $\underline{-4m^3 - 6m^2 - 2m + 1}$

15._____

16. $7p^4 \qquad + 5p^2 - 2p + \ 8$
       $\underline{2p^4 - 3p^3 - 8p^2 - 3p + 14}$

16._____

17. $\left(3x^2 + 2x^4 - 3\right) + \left(8x^3 - 5x^4 - 6x^2\right)$

17._____

18. $\left(m^4 - 3m^3 + 6m^2 + m + 3\right) + \left(m^3 + 3m^2 - m - 7\right)$

18._____

Subtract.

19. $9x^2 + 2x$
$\underline{3x^2 + 4x}$

19._____

20. $2m^2 - 5m + 1$
$\underline{-2m^2 - 5m + 3}$

20._____

21. $8y^4 - 3y^3 + 2y^2 - 2y - 1$
$\underline{-11y^4 + 5y^3 - 2y^2 - 1}$

21._____

22. $\left(9x^3 + 7x^2 - 6x + 3\right) - \left(6x^3 - 6x + 1\right)$

22._____

23. $\left(8p^2 + 7p - 2\right) - \left(3p^2 - 3p + 7\right) - \left(2p^2 - 3\right)$

23._____

Add or subtract as indicated. Give the degree of the answer.

24. $\left(4x^2y + 5xy + 7y^2\right) + \left(-2x^2y - 5xy + 3y^2\right)$

24._____

25. $\left(-4m^2n + 3n - 6m\right) - \left(2m + 7n + 4nm^2\right)$

25._____

26. $\left(2x^2y + 2xy - 4xy^2\right) + \left(6xy + 9xy^2\right) - \left(9x^2y + 5xy\right)$ **26.**_____

Add.

27. $5x^3y - 3x^2y^2 + 5xy^3$ **27.**_____
          $\underline{-3x^3y + 3x^2y^2 - 7xy^3}$

28. $6c^3d + 4c^2d - 7d^2$ **28.**_____
         $\underline{-8c^3d + 3c^2d - 2d^2}$

Subtract.

29. $13a^3b \qquad\quad + 2ab^3 + 7$ **29.**_____
        $\underline{16a^3b - 6a^2b^2 - 7ab^3 - 4}$

30. $-6rs + 2rt - 5st$ **30.**_____
         $\underline{5rs - 7rt - \ st}$

Chapter 10 EXPONENTS AND POLYNOMIALS

10.2   The Product Rule and Power Rules for Exponents

Write each expression in exponential form and evaluate.

1. $2 \cdot 2 \cdot 2 \cdot 2 \cdot 2 \cdot 2$

1._____

2. $(-1)(-1)(-1)(-1)(-1)(-1)(-1)(-1)$

2._____

3. $\left(\frac{1}{3}\right)\left(\frac{1}{3}\right)\left(\frac{1}{3}\right)\left(\frac{1}{3}\right)\left(\frac{1}{3}\right)$

3._____

Write each expression in exponential form.

4. $r \cdot r \cdot r \cdot r \cdot r$

4._____

5. $(-2y)(-2y)(-2y)(-2y)(-2y)$

5._____

Evaluate each exponential expression. Name the base and the exponent.

6. $(-4)^4$

6._____

7. -3^4

7._____

Use the product rule to simplify each expression, if possible. Write each answer in exponential form.

8. $4^7 \cdot 4^4$

8._____

9. $(-3)^4 \cdot (-3)^9$

9._____

10. $10^2 + 10^3$

10._____

Multiply.

11. $10n^4 \cdot 10n \cdot n^7$

11._____

12. $12b \cdot \left(-12b^{11}\right)$

12._____

In each of the following exercises, first add the given terms. Then start over and multiply them.

13. $7y^2,\ 9y^2$

13._____

14. $-5a^3,\ 4a^3,\ -a^3$

14._____

Simplify each expression. Write all answers in exponential form.

15. $\left(3^4\right)^3$

15._____

16. $\left(6^4\right)^6$

16._____

17. $\left[(-3)^3\right]^7$

17._____

18. $-\left(13^3\right)^8$

18._____

Simplify each expression.

19. $\left(yz^4\right)^3$

19._____

20. $\left(p^2q^3\right)^4$

20._____

21. $\left(2w^3z^7\right)^4$

21._____

22. $\left(4c^3d^4\right)^3$

22._____

Simplify each expression. Assume all variables represent nonzero real numbers.

23. $\left(\dfrac{3}{5}\right)^3$

23._____

24. $\left(\dfrac{-2a}{b^2}\right)^7$

24._____

25. $\left(\dfrac{xy}{z^2}\right)^4$

25._____

26. $\left(-\dfrac{2x}{5}\right)^3$

26._____

Name: Date:
Instructor: Section:

Simplify. Write all answers in exponential form.

27. $(7a)^4 (7a)^5$

27._____

28. $\left(\dfrac{4^2}{7^3}\right)^3 \cdot 4^7$

28._____

29. $\left(-x^3\right)^2 \left(-x^5\right)^4$

29._____

30. $\left(\dfrac{7a^2 b^3}{2}\right)^7$

30._____

Chapter 10 EXPONENTS AND POLYNOMIALS

10.3   Multiplying Polynomials

Find each product.

1. $\left(3y^3\right)\left(4y^2\right)$ 1._____

2. $7z\left(5z^3+2\right)$ 2._____

3. $-2x^4\left(3+6x+2x^2\right)$ 3._____

4. $-6z\left(z^5+3z^3+4z+2\right)$ 4._____

5. $4k^2\left(3+2k^3+6k^4\right)$ 5._____

6. $7b^2\left(-5b^2+1-4b\right)$ 6._____

7. $8mn\left(4m^2+2mn+7n^2\right)$ 7._____

Find the product.

8. $\left(x+3\right)\left(x+9\right)$ 8._____

9. $\left(3p+4\right)\left(p+2\right)$ 9._____

10. $(5n+1)(2n+5)$

10._____

11. $(2x+5)(3x+4)$

11._____

12. $(3m-5)(2m+4)$

12._____

13. $(y+4)(y^2-4y+16)$

13._____

14. $(3y-4)(3y^3-2y^2-y+4)$

14._____

15. $(2z^2-3)(z^4+2z^3+z^2+3z+2)$

15._____

Find each product, using the vertical method of multiplication.

16. $(2x+3)(2x^2-3x+2)$

16._____

17. $(2y-3)(3y^3+2y^2-y+2)$

17._____

18. $\left(2a^2 + 1\right)\left(3a^3 - 2a^2 + a\right)$

18._____

19. $\left(2x^2 + 3x + 2\right)\left(4x^3 + 2x + 3\right)$

19._____

20. $\left(a + 5\right)\left(a + 5\right)$

20._____

21. $\left(2m^3 + 3m - 3\right)\left(-2m^3 - 4m + 1\right)$

21._____

22. $\left(2x^2 - x + 2\right)\left(2x^2 - x + 2\right)$

22._____

Use the FOIL method to find each product.

23. $\left(4m + 3\right)\left(m - 7\right)$

23._____

24. $\left(3x + 2y\right)\left(2x - 3y\right)$

24._____

25. $\left(5a - b\right)\left(4a + 3b\right)$

25._____

26. $(11k - 4)(11k + 4)$ **26.** _____

27. $(4x - 3y)(x + 2y)$ **27.** _____

28. $(2v^2 + w^2)(v^2 - 3w^2)$ **28.** _____

29. $(2y + .1)(2y - .5)$ **29.** _____

30. $\left(z + \frac{4}{5}\right)\left(z - \frac{2}{5}\right)$ **30.** _____

Chapter 10 EXPONENTS AND POLYNOMIALS

10.4   Special Products

Find each square by using the pattern for the square of a binomial.

1. $(z+3)^2$ 1._____

2. $(t+4)^2$ 2._____

3. $(a+2b)^2$ 3._____

4. $(5y-3)^2$ 4._____

5. $(2m+5)^2$ 5._____

6. $(5m+3n)^2$ 6._____

7. $(7+x)^2$ 7._____

8. $(5+2y)^2$ 8._____

9. $(4y-.7)^2$ 9._____

10. $\left(4x-\frac{1}{4}y\right)^2$ 10._____

Find each product by using the pattern for the sum and difference of two terms.

11. $(z-6)(z+6)$ **11.**_____

12. $(12+x)(12-x)$ **12.**_____

13. $(7x-3y)(7x+3y)$ **13.**_____

14. $(8k+5p)(8k-5p)$ **14.**_____

15. $(4p+7q)(4p-7q)$ **15.**_____

16. $(2+3x)(2-3x)$ **16.**_____

17. $(9-4y)(9+4y)$ **17.**_____

18. $\left(y+\frac{4}{3}\right)\left(y-\frac{4}{3}\right)$ **18.**_____

19. $\left(2a+\frac{4}{3}b\right)\left(2a-\frac{4}{3}b\right)$ **19.**_____

20. $\left(y^2+2\right)\left(y^2-2\right)$ **20.**_____

Find each product.

21. $(x+2)^3$

21._____

22. $(a-3)^3$

22._____

23. $(y+4)^3$

23._____

24. $(2x-3)^3$

24._____

25. $(2x+1)^3$

25._____

26. $(k+2)^4$

26._____

27. $(t-3)^4$

27._____

28. $(x+2y)^4$

28._____

29. $(3b-2)^4$

29._____

30. $(4s+3t)^4$

30._____

Chapter 10 EXPONENTS AND POLYNOMIALS

10.5   Integer Exponents and the Quotient Rule

Evaluate each expression.

1. 7^0

1._____

2. $-(-8)^0$

2._____

3. -12^0

3._____

4. $2^0 + 6^0$

4._____

5. $\left(\frac{2}{3}\right)^0 + \left(\frac{1}{3}\right)^0 - 2^0$

5._____

6. $-12^0 + (-12)^0$

6._____

7. $-r^0 \ \ (r \neq 0)$

7._____

Evaluate each expression.

8. 4^{-2}

8._____

9. $(-3)^{-3}$

9._____

10. $\left(\frac{3}{5}\right)^{-2}$

10._____

11. $8^{-1} + 4^{-1}$

11._____

Simplify by using the definition of negative exponents. Write each expression with only positive exponents. Assume all variables represent nonzero real numbers.

12. r^{-7}

12._____

13. $\dfrac{2}{r^{-7}}$

13._____

14. $\dfrac{2x^{-4}}{3y^{-7}}$

14._____

Use the quotient rule to simplify each expression. Write answers with only positive exponents. Assume that all variables represent nonzero real numbers.

15. $\dfrac{2^9}{2^5}$

15._____

16. $\dfrac{9^{12}}{9^7}$

16._____

17. $\dfrac{(-2)^8}{(-2)^3}$

17._____

18. $\dfrac{2^4 \cdot x^2}{2^5 \cdot x^8}$

18._____

19. $\dfrac{12x^9 y^5}{12^4 x^3 y^7}$

19._____

20. $\dfrac{12^{-7}}{12^{-6}}$

20._____

21. $\dfrac{3^{-1} m^{-4} p^{6}}{3^{4} m^{-1} p^{-2}}$

21._____

Simplify each expression. Write answers with only positive exponents. Assume that all variables represent nonzero real numbers.

22. $\dfrac{\left(7^{2}\right)^{6}}{7^{6}}$

22._____

23. $\dfrac{\left(9^{3}\right)^{2}}{9^{5}}$

23._____

24. $\dfrac{6^{-2} \cdot 6^{3}}{6^{-3}}$

24._____

25. $\left(2^{-4}\right)^{4}$

25._____

26. $\left(5w^{2} y^{2}\right)^{-2} \left(4wy^{-3}\right)^{2}$

26._____

27. $\dfrac{\left(2y\right)^{-4}}{\left(3y\right)^{-2}}$

27._____

28. $\dfrac{c^{10}\left(c^2\right)^3}{\left(c^3\right)^3\left(c^2\right)^{-9}}$

28._____

29. $\left(\dfrac{x^8 y^2}{x^6 y^{-4}}\right)^{-2}$

29._____

30. $\dfrac{\left(3^{-2} x^{-5} y\right)^{-4}\left(2 x^2 y^{-4}\right)^2}{\left(2 x^{-2} y^2\right)^{-2}}$

30._____

Chapter 10 EXPONENTS AND POLYNOMIALS

10.6   Dividing a Polynomial by a Monomial

Divide each polynomial by $4m^2$.

1. $16m^3 + 8m^2$ 1._____

2. $32m^4 - 8m^3 + 12m^2$ 2._____

3. $28m^5 - 4m^2$ 3._____

4. $36m^6 + 24m^4 - 44m^2 - 8$ 4._____

5. $28m^3 + 48m^2 - 16m$ 5._____

6. $32m^3 - 2m^2$ 6._____

7. $-84m^3 + 36m^2 + 8m$ 7._____

8. $4m^2 - 4$

8._____

9. $32m^3 - 48m - 12$

9._____

10. $16m^5 - 20m^4 + 4m^2 - 2m + 3$

10._____

Perform each division.

11. $\dfrac{6p^4 + 18p^7}{6p^4}$

11._____

12. $\dfrac{12x^6 + 28x^5 + 20x^3}{4x^2}$

12._____

13. $\left(20a^3 - 9a\right) \div \left(4a\right)$

13._____

14. $\left(6z^5 + 27z^3 - 12z + 10\right) \div \left(3z\right)$

14._____

15. $\left(20x^4 - 10x^2\right) \div \left(2x\right)$

15._____

16. $\left(9x^4 + 24x^3 - 48x + 12\right) \div (3x)$

16._____

17. $\left(m^2 + 7m - 42\right) \div (2m)$

17._____

18. $\dfrac{40p^4 - 35p^3 - 15p}{5p^2}$

18._____

19. $\dfrac{70q^4 - 40q^2 + 10q}{10q^2}$

19._____

20. $\dfrac{2y^9 + 8y^6 - 41y^3 - 12}{y^3}$

20._____

21. $\dfrac{12z^5 + 28z^4 - 8z^3 + 3z}{4z^3}$

21._____

22. $\dfrac{48x + 64x^4 + 2x^8}{4x}$

22._____

23. $\dfrac{-25u^3v + 20u^2v^2 - 45uv^3}{5uv}$

23._____

24. $\dfrac{21y^2 - 14y + 42}{-7y^2}$

24._____

25. $\dfrac{39m^4 - 12m^3 + 15}{-3m^2}$

25._____

26. $\dfrac{-20d^4 - 8d^3 + 14d^2 + 8}{-2d^2}$

26._____

27. $\left(15x^5 - 10x^4 - 10x^2 + 4\right) \div \left(-5x\right)$

27._____

28. $\dfrac{16a^5 - 24a^3}{8a^2}$

28._____

29. $\dfrac{-12y^4 - 15y^3 + 2y}{-3y^2}$

29._____

30. $\dfrac{18r^4 - 12r^3 + 36r^2 - 12}{6r}$

30._____

Chapter 10 EXPONENTS AND POLYNOMIALS

10.7   Dividing a Polynomial by a Polynomial

Perform each division.

1. $\dfrac{x^2 - x - 6}{x - 3}$

1._____

2. $\dfrac{y^2 - 2y - 24}{y + 4}$

2._____

3. $\dfrac{18a^2 - 9a - 5}{3a + 1}$

3._____

4. $\dfrac{p^2 + 5p - 24}{p - 3}$

4._____

5. $\left(x^2 + 16x + 64\right) \div \left(x + 8\right)$

5._____

6. $\left(r^2 - 2r - 20\right) \div \left(r - 5\right)$

6._____

7. $\left(2a^2 - 11a + 16\right) \div \left(2a + 3\right)$

7._____

8. $\left(9w^2 + 12w + 4\right) \div \left(3w + 2\right)$ 8._____

9. $\dfrac{5w^2 - 22w + 4}{w - 4}$ 9._____

10. $\dfrac{5b^2 + 32b + 3}{b + 7}$ 10._____

11. $\dfrac{9m^2 - 18m + 16}{3m - 4}$ 11._____

12. $\dfrac{81a^2 - 1}{9a + 1}$ 12._____

13. $\dfrac{4x^2 - 25}{2x - 5}$ 13._____

14. $\dfrac{12y^3 - 11y^2 + 9y + 18}{4y + 3}$ 14._____

15. $\dfrac{2z^3 - 7z^2 + 3z + 2}{2z + 3}$

15. _____

16. $\dfrac{6m^3 + 7m^2 - 13m + 16}{3m + 2}$

16. _____

17. $\left(27p^4 - 36p^3 - 6p^2 + 26p - 24\right) \div \left(3p - 4\right)$

17. _____

18. $\left(3x^3 - 11x^2 + 25x - 25\right) \div \left(x^2 - 3x - 5\right)$

18. _____

19. $\dfrac{6x^4 - 12x^3 + 13x^2 - 5x - 1}{2x^2 + 3}$

19. _____

20. $\dfrac{12y^5 - 8y^4 - y^3 + 2y^2 - 5}{4y^2 - 3}$

20. _____

21. $\dfrac{2a^4 + 5a^2 + 3}{2a^2 + 3}$

21._____

22. $\dfrac{3x^4 + 2x^3 - 2x^2 - 2x - 2}{x^2 - 1}$

22._____

23. $\dfrac{y^3 + 1}{y + 1}$

23._____

24. $\dfrac{b^4 - 1}{b^2 - 1}$

24._____

25. $\dfrac{6x^5 + 7x^4 - 7x^3 + 7x + 4}{3x + 2}$

25._____

26. $\dfrac{32x^5 - 243}{2x - 3}$

26._____

27. $\dfrac{-6x^2 + 23x - 20}{2x - 5}$

27. _____

28. $\dfrac{12p^3 - 28p^2 + 21p - 5}{6p - 5}$

28. _____

29. $\dfrac{6y^6 + 8y^5 - 8y^4 - 3y^3 - 4y^2 + 4y + 3}{2y^3 - 1}$

29. _____

30. $\dfrac{125r^3 - 8}{5r - 2}$

30. _____

Chapter 10 EXPONENTS AND POLYNOMIALS

10.8   An Application of Exponents: Scientific Notation

Write each number in scientific notation.

1. 325

1.＿＿＿＿＿＿＿＿＿

2. 4579

2.＿＿＿＿＿＿＿＿＿

3. 23,651

3.＿＿＿＿＿＿＿＿＿

4. 209,907

4.＿＿＿＿＿＿＿＿＿

5. 7.42

5.＿＿＿＿＿＿＿＿＿

6. 429,600,000,000

6.＿＿＿＿＿＿＿＿＿

7. .0257

7.＿＿＿＿＿＿＿＿＿

8. .246

8.＿＿＿＿＿＿＿＿＿

9. .00000413

9.＿＿＿＿＿＿＿＿＿

10. .00426

10.＿＿＿＿＿＿＿＿＿

Write each number without exponents.

11. 2.5×10^{4}

11.＿＿＿＿＿＿＿＿＿

12. 7.2×10^7 **12.**_____

13. -2.45×10^6 **13.**_____

14. 4.045×10^0 **14.**_____

15. 2.3×10^4 **15.**_____

16. 7.24×10^{-4} **16.**_____

17. 4.007×10^{-2} **17.**_____

18. 4.752×10^{-1} **18.**_____

19. -4.02×10^0 **19.**_____

20. -9.11×10^{-4} **20.**_____

Perform the indicated operations with the numbers in scientific notation, and then write your answers without exponents.

21. $\left(7 \times 10^7\right) \times \left(3 \times 10^0\right)$ **21.**_____

22. $\left(3 \times 10^6\right) \times \left(4 \times 10^{-2}\right) \times \left(2 \times 10^{-1}\right)$ **22.**_____

23. $\left(2.3\times10^4\right)\times\left(1.1\times10^{-2}\right)$

23._____

24. $\left(2.3\times10^{-4}\right)\times\left(3.1\times10^{-2}\right)$

24._____

25. $\dfrac{4.6\times10^{-3}}{2.3\times10^{-1}}$

25._____

26. $\dfrac{9.39\times10^1}{3\times10^3}$

26._____

27. $\left(6\times10^4\right)\times\left(3\times10^5\right)\div\left(9\times10^7\right)$

27._____

28. $\left(3\times10^4\right)\times\left(4\times10^2\right)\div\left(2\times10^3\right)$

28._____

29. $\dfrac{\left(7.5\times10^6\right)\times\left(4.2\times10^{-5}\right)}{\left(6\times10^4\right)\times\left(2.5\times10^{-3}\right)}$

29._____

30. $\dfrac{\left(2.1\times10^{-3}\right)\times\left(4.8\times10^4\right)}{\left(1.6\times10^6\right)\times\left(7\times10^{-6}\right)}$

30._____

Chapter 11 FACTORING AND APPLICATIONS

11.1   The Greatest Common Factor

Find the greatest common factor for each group of numbers.

1. 9, 12, 18

1._____

2. 40, 25, 70

2._____

3. 60, 75, 120

3._____

4. 9, 18, 24, 48

4._____

5. 96, 480, 128

5._____

Find the greatest common factor for each list of terms.

6. $30x^3$, $40x^2$, $25x^2$

6._____

7. $18b^3$, $36b^6$, $45b^4$

7._____

8. $12ab^3$, $18a^2b^4$, $26ab^2$, $32a^2b^2$

8._____

9. $24a^6$, $18a^7$, $42a^9$

9._____

10. $6k^2m^4n^5$, $8k^3m^7n^4$, k^4m^8,n^7

10._____

11. $45a^7y^4$, $75a^3y^2$, $90a^2y$, $30a^4y^3$ 11._____

Complete the factoring.

12. $84 = 4(\quad)$ 12._____

13. $18x^3 = 3x^2(\quad)$ 13._____

14. $-18y^8 = -3y^5(\quad)$ 14._____

Factor out the greatest common factor.

15. $18r + 24t$ 15._____

16. $18q^2 - 27q$ 16._____

17. $20x^2 + 40x^2y - 70xy^2$ 17._____

18. $42tw + 21t - 63t^2$ 18._____

19. $15a^7 - 25a^3 - 40a^4$ 19._____

20. $26x^8 - 13x^{12} + 52x^{10}$ 20._____

21. $56x^2y^4 - 24xy^3 + 32xy^2$

21._____

22. $9(x+8) + a(x+8)$

22._____

Factor each polynomial by grouping.

23. $x^2 + 2x + 5x + 10$

23._____

24. $x^4 + 2x^2 + 5x^2 + 10$

24._____

25. $x^2 + 5x - 4x - 20$

25._____

26. $3x^2 - 9x + 12x - 36$

26._____

27. $xy - 2x - 2y + 4$

27._____

28. $5x + 15 - xy - 3y$

28._____

29. $2a^3 - 3a^2b + 2ab^2 - 3b^3$

29._____

30. $2x^4 + 4x^2y^2 + 3x^2y + 6y^3$

30._____

Chapter 11 FACTORING AND APPLICATIONS

11.2   Factoring Trinomials

List all pairs of integers with the given product. Then find the pair whose sum is given.

1. Product: 42; sum: 17

1._____

2. Product: 28; sum: -11

2._____

3. Product: –64; sum: 12

3._____

4. Product: –54; sum –3

4._____

Complete the factoring.

5. $x^2 + 7x + 12 = (x+3)(\quad)$

5._____

6. $x^2 + 3x - 28 = (x-4)(\quad)$

6._____

7. $x^2 + 4x + 4 = (x+2)(\quad)$

7._____

8. $x^2 - x - 30 = (x+5)(\quad)$

8._____

Factor completely. If a polynomial cannot be factored, write prime.

9. $x^2 + 11x + 18$

9._____

10. $x^2 - 11x + 28$

10._____

11. $x^2 - x - 2$

11._____

12. $x^2 + 14x + 49$

12._____

13. $x^2 - 2x - 35$

13._____

14. $x^2 - 8x - 33$

14._____

15. $x^2 + 6x + 5$

15._____

16. $x^2 - 15xy + 56y^2$

16._____

17. $x^2 - 4xy - 21y^2$

17._____

18. $m^2 - 2mn - 3n^2$

18._____

Factor completely.

19. $2x^2 + 10x - 28$

19._____

20. $3x^2 + 6x - 24$

20._____

21. $3h^3k - 21h^2k - 54hk$

21._____

22. $7b^2 - 42b + 56$

22._____

23. $4a^2 - 24b + 5$

23._____

24. $3p^6 + 18p^5 + 24p^4$

24._____

25. $2a^3b - 10a^2b^2 + 12ab^3$

25._____

26. $3y^3 + 9y^2 - 12y$

26._____

27. $5r^2 + 35r + 60$

27._____

28. $3xy^2 - 24xy + 36x$

28._____

29. $10k^6 + 70k^5 + 100k^4$

29._____

30. $x^5 - 3x^4 + 2x^3$

30._____

Chapter 11 FACTORING AND APPLICATIONS

11.3   Factoring Trinomials by Grouping

Complete the factoring.

1. $2x^2 + 5x - 3 = (2x - 1)(\quad)$

1._____

2. $6x^2 + 19x + 10 = (3x + 2)(\quad)$

2._____

3. $16x^2 + 4x - 6 = (4x + 3)(\quad)$

3._____

4. $24y^2 - 17y + 3 = (3y - 1)(\quad)$

4._____

Factor each trinomial by grouping.

5. $8b^2 + 18b + 9$

5._____

6. $3x^2 + 13x + 14$

6._____

7. $15a^2 + 16a + 4$

7._____

8. $6n^2 + 11n + 4$

8._____

9. $3b^2 + 8b + 4$

9._____

10. $3m^2 - 5m - 12$

10._____

11. $3p^3 + 8p^2 + 4p$

11._____

12. $8m^2 + 26mn + 6n^2$

12._____

13. $7a^2b + 18ab + 8b$

13._____

14. $2s^2 + 5st - 3st^2$

14._____

15. $9c^2 + 24cd + 12d^2$

15._____

16. $25a^2 + 30ab + 9b^2$

16._____

17. $10c^2 - 29ct + 21t^2$

17._____

18. $24s^2 - 14st - 5t^2$

18._____

19. $12x^2 + 32xy - 35y^2$

19._____

20. $24c^2 + 90cd - 81d^2$ 20._____

21. $6m^3 + 2m^2n - 8mn^2$ 21._____

22. $40x^2 + 18xy - 9y^2$ 22._____

23. $18f^2 + 27fg - 5g^2$ 23._____

24. $16p^2 + 8pq + q^2$ 24._____

25. $40a^3 - 82a^2 + 40a$ 25._____

26. $4x^2 + 32x + 55$ 26._____

27. $8x^2 - 4xy - 4y^2$ 27._____

28. $7m^2 + 3mn - 22n^2$ **28.**_____

29. $10c^2 + 39cd + 36d^2$ **29.**_____

30. $9x^3 - 30x^2 y + 24xy^2$ **30.**_____

Chapter 11 FACTORING AND APPLICATIONS

11.4   Factoring Trinomials using FOIL

Factor each trinomial using the FOIL method.

1. $10x^2 + 19x + 6$

1._____

2. $4y^2 + 3y - 10$

2._____

3. $2a^2 + 13a + 6$

3._____

4. $5w^2 - 9w - 2$

4._____

5. $8q^2 + 10q + 3$

5._____

6. $8m^2 - 10m - 3$

6._____

7. $14b^2 + 3b - 2$

7._____

8. $15q^2 - 2q - 24$

8._____

9. $3a^2 + 8ab + 4b^2$

9._____

10. $9w^2 + 12wz + 4z^2$

10._____

11. $10c^2 - cd - 2d^2$

11._____

12. $6x^2 + xy - 12y^2$

12._____

13. $18x^2 - 27xy + 4y^2$

13._____

14. $12y^2 + 11y - 15$

14._____

15. $3x^2 - 11x - 4$

15._____

16. $2p^2 + 11p + 5$

16._____

17. $6y^2 + y - 1$

17._____

18. $9y^2 - 16y - 4$

18._____

19. $3p^2 + 17p + 10$

19._____

20. $9r^2 + 12r - 5$

20._____

21. $7x^2 + 27x - 4$

21._____

22. $4c^2 + 14cd - 8d^2$

22._____

23. $2x^4 + 5x^3 - 12x^2$

23._____

24. $12a^3 + 26a^2b + 12ab^2$

24._____

25. $27r^2 + 6rt - 8t^2$

25._____

26. $2y^5z^2 - 5y^4z^3 - 3y^3z^4$

26._____

27. $6x^4y^2 + x^2y - 15$

27._____

28. $28c^2 + 23cd - 15d^2$

28._____

29. $8x^3 - 10x^2y + 3xy^2$

29._____

30. $-30a^4b + 3a^3b + 6a^2b$

30._____

Chapter 11 FACTORING AND APPLICATIONS

11.5   Special Factoring Techniques

Factor each binomial completely. If a binomial cannot be factored, write prime.

1. $x^2 + 16$

1._____

2. $x^2 - 49$

2._____

3. $100r^2 - 9s^2$

3._____

4. $y^2 - 64$

4._____

5. $25a^2 - 36$

5._____

6. $9j^2 - \frac{16}{49}$

6._____

7. $36 - 121d^2$

7._____

8. $121m^2 - 9n^2$

8._____

9. $x^4 - 81$

9._____

10. $9m^4 - 1$

10._____

11. $9x^4 - 16$

11._____

12. $81y^4 - 1$

12._____

13. $9x^2 + 16$

13._____

14. $m^4 n^2 - m^2$

14._____

15. $9p^2 - 121$

15._____

Factor each trinomial completely. It may be necessary to factor out the greatest common factor first.

16. $y^2 + 6y + 9$

16._____

17. $q^2 + 14q + 49$

17._____

18. $m^2 - 8m + 16$

18._____

19. $c^2 + 22c + 121$

19._____

20. $z^2 - \frac{4}{3}z + \frac{4}{9}$

20._____

21. $4w^2 + 12w + 9$

21._____

22. $16q^2 - 40q + 25$

22._____

23. $9j^2 + 12j + 4$

23._____

24. $64p^4 + 48p^2q^2 + 9q^4$

25. $100p^2 - \frac{25}{2}pr + \frac{25}{64}r^2$

26. $9m^2 + .6m + .01$

27. $-16x^2 - 48x - 36$

28. $-12a^2 + 60ab - 75b^2$

29. $18x^2 + 84xy + 98y^2$

30. $\frac{1}{9}x^2 - 4xy + 36y^2$

Chapter 11 FACTORING AND APPLICATIONS

11.6   Solving Quadratic Equations by Factoring

Solve each equation. Check your answers.

1. $(y+9)(2y-3)=0$

1._____

2. $(3k+4)(5k-7)=0$

2._____

3. $3x^2+7x+2=0$

3._____

4. $b^2-49=0$

4._____

5. $2x^2-3x-20=0$

5._____

6. $x^2-2x-63=0$

6._____

7. $8r^2=24r$

7._____

8. $3x^2-7x-6=0$

8._____

9. $3 - 5x = 8x^2$

9._____

10. $9x^2 + 12x + 4 = 0$

10._____

11. $25x^2 = 20x$

11._____

12. $9y^2 = 16$

12._____

13. $12x^2 + 7x - 12 = 0$

13._____

14. $14x^2 - 17x - 6 = 0$

14._____

15. $c(5c + 17) = 12$

15._____

Solve each equation.

16. $3x(x + 7)(x - 2) = 0$

16._____

17. $x\left(2x^2-7x-15\right)=0$ 17._____

18. $z\left(4z^2-9\right)=0$ 18._____

19. $z^3-49z=0$ 19._____

20. $25a=a^3$ 20._____

21. $x^3+2x^2-8x=0$ 21._____

22. $2m^3+m^2-6m=0$ 22._____

23. $\left(4x^2-9\right)\left(x-2\right)=0$ 23._____

24. $z^4+8z^3-9z^2=0$ 24._____

25. $3z^3+z^2-4z=0$ 25._____

26. $(x+4)(x^2+7x+10)=0$

26._____

27. $(y^2-5y+6)(y^2-36)=0$

27._____

28. $15x^2=x^3+56x$

28._____

29. $(y-7)(2y^2+7y-15)=0$

29._____

30. $(x-\frac{3}{2})(2x^2+11x+15)=0$

30._____

Chapter 11 FACTORING AND APPLICATIONS

11.7   Applications of Quadratic Equations

Solve each problem.

1. The length of a rectangle is 8 centimeters more than the
   width. The area is 153 square centimeters. Find the
   length and width of the rectangle.

   1._____

2. The length of a rectangle is three times its width. If the
   width were increased by 4 and the length remained the
   same, the resulting rectangle would have an area of 231
   square inches. Find the dimensions of the original
   rectangle.

   2._____

3. The area of a rectangular room is 252 square feet. Its
   width is 4 feet less than its length. Find the length and
   width of the room.

   3._____

4. Two rectangles with different dimensions have the same
   area. The length of the first rectangle is three times its
   width. The length of the second rectangle is 4 meters
   more than the width of the first rectangle, and its width is
   2 meters more than the width of the first rectangle. Find
   the lengths and widths of the two rectangles.

   4._____

5. Each side of one square is 1 meter less than twice the length of each side of a second square. If the difference between the areas of the two squares is 16 meters, find the lengths of the sides of the two rectangles.

5._____

6. The area of a triangle is 42 square centimeters. The base is 2 centimeters less than twice the height. Find the base and height of the triangle.

6._____

7. A rectangular bookmark is 6 centimeters longer than it is wide. Its area is numerically 3 more than its perimeter. Find the length and width of the bookmark.

7._____

8. A book is three times as long as it is wide. Find the length and width of the book in inches if its area is numerically 128 more than its perimeter.

8._____

9. The volume of a box is 192 cubic feet. If the length of the box is 8 feet and the width is 2 feet more than the height, find the width of the box.

9._____

10. Mr. Fixxall is building a box which will have a volume of 60 cubic meters. The height of the box will be 4 meters, and the length will be 2 meters more than the width. Find the width of the box.

10._____

Solve each problem.

11. The product of two consecutive integers is four less than four times their sum. Find the integers.

11._____

12. Find two consecutive integers such that the square of their sum is 169.

12._____

13. The product of two consecutive positive odd integers is 1 less than four times their sum. Find the integers.

13._____

14. Find two consecutive integers such that the sum of the squares of the two integers is 3 more than the opposite (additive inverse) of the smaller integer.

14._____

15. If the square of the sum of two consecutive integers is reduced by twice their product, the result is 25. Find the integers.

15._____

16. The product of two consecutive even integers is 24 more than three times the larger integer. Find the integers.

16._____

17. Find all possible pairs of consecutive odd integers whose sum is equal to their product decreased by 47.

17._____

18. Find two consecutive positive even integers whose product is 168.

18._____

19. Find two consecutive positive even integers whose product is 6 more than three times its sum.

19._____

20. Find two consecutive integers whose product is three   **20.**＿＿＿＿＿＿＿＿＿＿
more than three times its sum.

Solve each problem.

21. The hypotenuse of a right triangle is 4 inches longer than   **21.**＿＿＿＿＿＿＿＿＿＿
the shorter leg. The longer leg is 4 inches shorter than
twice the shorter leg. Find the length of the shorter leg.

22. A flag is shaped like a right triangle. The hypotenuse is   **22.**＿＿＿＿＿＿＿＿＿＿
6 meters longer than twice the length of the shortest side
of the flag. If the length of the other side is 2 meters less
than the hypotenuse, find the lengths of the sides of the
flag.

23. A field has a shape of a right triangle with one leg 10   **23.**＿＿＿＿＿＿＿＿＿＿
meters longer than twice the length of the other leg. The
hypotenuse is 4 meters longer than three times the length
of the shorter leg. Find the dimensions of the field.

24. A train and a car leave a station at the same time, the train traveling due north and the car traveling west. When they are 100 miles apart, the train has traveled 20 miles farther than the car. Find the distance each has traveled.

24._____

25. The hypotenuse of a right triangle is 1 foot larger than twice the shorter leg. The longer leg is 7 feet larger than the shorter leg. Find the length of the longer leg.

25._____

26. Mark is standing directly beneath a kite attached to a string which Nina is holding, with her hand touching the ground. The height of the kite at that instant is 12 feet less than twice the distance between Mark and Nina. The length of the kite string is 12 feet more than that distance. Find the length of the kite string.

26._____

27. A 30-foot ladder is leaning against a building. The distance from the bottom of the ladder to the building is 6 feet less than the distance from the top of the ladder to the ground. How far is the bottom of the ladder from the building?

27._____

28. A field is in the shape of a right triangle. The shorter leg measures 45 meters. The hypotenuse measures 45 meters less than twice the longer the leg. Find the dimensions of the lot.

28._____

29. Two ships left a dock at the same time. When they were 25 miles apart, the ship that sailed due south had gone 10 miles less than twice the distance traveled by the ship that sailed due west. Find the distance traveled by the ship that sailed due south.

29._____

30. A ladder is leaning against a building. The distance from the bottom of the ladder to the building is 8 feet less than the length of the ladder. How high up the side of the building is the top of the ladder if that distance is 4 feet less than the length of the ladder?

30._____

Chapter 12 RATIONAL EXPRESSIONS AND APPLICATIONS

12.1   The Fundamental Property of Rational Expressions

Find all values for which the following expressions are undefined.

1. $\dfrac{9}{4x}$ 1._____

2. $\dfrac{4x^2}{x+7}$ 2._____

3. $\dfrac{x-4}{4x^2-16x}$ 3._____

4. $\dfrac{4x+3}{x^2+x-12}$ 4._____

5. $\dfrac{2x^2}{x^2+4}$ 5._____

6. $\dfrac{5x}{x^2-25}$ 6._____

7. $\dfrac{2y-5}{2y^2+4y-16}$ 7._____

8. $\dfrac{x+5}{x^3+9x^2+18x}$ 8._____

Find the numerical value of each expression when (a) $x = 4$ and (b) $x = -3$.

9. $\dfrac{3x^2 - 2x}{2x}$ 9._____

10. $\dfrac{-3x + 1}{2x + 1}$ 10._____

11. $\dfrac{(-4 - x)^2}{x + 3}$ 11._____

12. $\dfrac{4 - 3x}{4 + x}$ 12._____

13. $\dfrac{3x^2}{3x + 2}$ 13._____

14. $\dfrac{3x}{(3x - 2)^2}$ 14._____

15. $\dfrac{3x^2 - 8}{x^2 - x - 4}$ 15._____

16. $\dfrac{2x - 5}{2 + x - x^2}$ 16._____

Write each rational expression in lowest terms.

17. $\dfrac{9k}{24k^2}$

17._____

18. $\dfrac{24a^6b^9}{8a^2b^3}$

18._____

19. $\dfrac{15ab^3c^9}{-24ab^2c^{10}}$

19._____

20. $\dfrac{16r^2-4s^2}{8r+4s}$

20._____

21. $\dfrac{3a^2-2a-1}{9a^2-1}$

21._____

22. $\dfrac{16-x^2}{2x-8}$

22._____

23. $\dfrac{12k^3+12k^2}{3k^2+3k}$

23._____

24. $\dfrac{6r^2-7rs-10s^2}{r^2+3rs-10s^2}$

24._____

25. $\dfrac{5x^2 - 17xy - 12y^2}{x^2 - 7xy + 12y^2}$

25._____

26. $\dfrac{vw - 5v + 3w - 15}{vw - 5v - 2w + 10}$

26._____

Write four equivalent forms of the following rational expressions.

27. $-\dfrac{2x - 3}{x + 2}$

27._____

28. $\dfrac{4x + 1}{5x - 3}$

28._____

29. $-\dfrac{2x - 1}{3x + 5}$

29._____

30. $-\dfrac{2x + 6}{x - 5}$

30._____

Chapter 12 RATIONAL EXPRESSIONS AND APPLICATIONS

12.2   Multiplying and Dividing Rational Expressions

Multiply. Write each answer in lowest terms.

1. $\dfrac{8m^4n^3}{3} \cdot \dfrac{5}{4mn^2}$

1._____

2. $\dfrac{6}{9y+36} \cdot \dfrac{4y+16}{9}$

2._____

3. $\dfrac{12-4z}{6} \cdot \dfrac{9}{4z-12}$

3._____

4. $\dfrac{x^2+x-12}{x^2+7x+10} \cdot \dfrac{x^2+3x-10}{x^2+2x-8}$

4._____

5. $\dfrac{a^2-3a+2}{a^2-1} \cdot \dfrac{a^2+2a-3}{a^2+a-6}$

5._____

6. $\dfrac{3m^2-m-10}{2m^2-7m-4} \cdot \dfrac{4m^2-1}{6m^2+7m-5}$

6._____

7. $\dfrac{2x^2+5x-12}{x^2-2x-24} \cdot \dfrac{x^2-9x+18}{9-4x^2}$

7._____

8. $\dfrac{3x^2-12}{x^2-x-6} \cdot \dfrac{x^2-6x+9}{2x-4}$

8._____

9. $\dfrac{2x+1}{16-x^2} \cdot \dfrac{x-4}{4x+2}$

9._____

10. $\dfrac{4r+4p}{8z^2} \cdot \dfrac{36z^6}{r^2+rp}$

10._____

11. $\dfrac{3x+12}{6x-30} \cdot \dfrac{x^2-x-20}{x^2-16}$

11._____

12. $\dfrac{m^2-16}{m-3} \cdot \dfrac{m^2-9}{m+4}$

12._____

Find the reciprocal of each rational expression

13. $\dfrac{7}{2y}$

13._____

14. $\dfrac{5a+7}{9b^2c^3}$

14._____

15. $4r^2+2r+3$

15._____

16. $\dfrac{3x+4y}{5x-2y}$ **16.**_____

17. $\dfrac{x^2-2x+3}{2x^2+9}$ **17.**_____

18. $\dfrac{4}{2x^3+5x^2+x-7}$ **18.**_____

Divide. Write each answer in lowest terms.

19. $\dfrac{9}{7m^3} \div \dfrac{15}{28m^6}$ **19.**_____

20. $\dfrac{6z^3}{9zw} \div \dfrac{z^7}{21zw^2}$ **20.**_____

21. $\dfrac{b-7}{16} \div \dfrac{7-b}{8}$ **21.**_____

22. $\dfrac{2x+2y}{8z} \div \dfrac{x^2\left(x^2-y^2\right)}{24}$ **22.**_____

23. $\dfrac{4m-12}{2m+10} \div \dfrac{m^2-9}{m^2-25}$ **23.**_____

24. $\dfrac{6a(a+3)}{3a+1} \div \dfrac{a^2(a+3)}{9a^2-1}$

24._____

25. $\dfrac{m^2+2mn+n^2}{m^2+m} \div \dfrac{m^2-n^2}{m^2-1}$

25._____

26. $\dfrac{9a^2-1}{9a^2-6a+1} \div \dfrac{3a^2-11a-4}{12a^2+5a-3}$

26._____

27. $\dfrac{12k^2-5k-3}{9k^2-1} \div \dfrac{16k^2-9}{12k^2+13k+3}$

27._____

28. $\dfrac{2z^2-11z-21}{z^2-5z-14} \div \dfrac{4z^2-9}{z^2-6z-16}$

28._____

29. $\dfrac{y^2 + yz - 12z^2}{y^2 + yz - 20z^2} \div \dfrac{y^2 + 9yz + 20z^2}{y^2 - yz - 30z^2}$

29. _____

30. $\dfrac{4(b-3)(b+2)}{b^2 + 3b + 2} \div \dfrac{b^2 - 6b + 9}{b^2 + 4b + 4}$

30. _____

Chapter 12 RATIONAL EXPRESSIONS AND APPLICATIONS

12.3   Least Common Denominator

Find the least common denominator for each list of rational expressions.

1. $\dfrac{5}{12}, \dfrac{9}{16}$

1._____

2. $\dfrac{5}{9}, \dfrac{7}{15}$

2._____

3. $\dfrac{8}{12}, \dfrac{7}{20}, \dfrac{11}{18}$

3._____

4. $\dfrac{9}{14}, \dfrac{17}{20}, \dfrac{7}{15}$

4._____

5. $\dfrac{5}{8ab^2}, \dfrac{7}{6a^2b}$

5._____

6. $\dfrac{13}{36b^4}, \dfrac{17}{27b^2}$

6._____

7. $\dfrac{4}{5r-25}, \dfrac{7}{15r^3}$

7._____

8. $\dfrac{15}{7t-28}, \dfrac{21}{6t-24}$

8._____

9. $\dfrac{7}{x-y}, \dfrac{3}{y-x}$

9._____

10. $\dfrac{4}{a^2-b^2}, \dfrac{8}{b^2-a^2}$

10. _____

11. $\dfrac{3m}{2m^2+9m-5}, \dfrac{4}{m^2+5m}$

11. _____

12. $\dfrac{v-4}{3v^4-6v^3}, \dfrac{v+2}{v^2+2v-8}$

12. _____

13. $\dfrac{3z+1}{z^4+2z^3-8z^2}, \dfrac{5z+2}{z^3+8z^2+16z}$

13. _____

14. $\dfrac{3p+2}{p^2-9}, \dfrac{2p+7}{p^2-p-12}$

14. _____

15. $\dfrac{11q-3}{2q^2-q-10}, \dfrac{21-q}{2q^2-9q+10}$

15. _____

16. $\dfrac{17r}{9r^2-6r-8}, \dfrac{-13r}{9r^2-9r-4}$

16. _____

Rewrite each rational expression with the indicated denominator. Give the numerator of the new fraction.

17. $\dfrac{5}{6} = \dfrac{?}{18}$

17._____

18. $\dfrac{4}{r} = \dfrac{?}{7r}$

18._____

19. $\dfrac{7m}{8n} = \dfrac{?}{24n^6}$

19._____

20. $\dfrac{11a+1}{2a-6} = \dfrac{?}{8a-24}$

20._____

21. $\dfrac{-3y}{4y+12} = \dfrac{?}{4(y+3)^2}$

21._____

22. $\dfrac{8z}{3z+3} = \dfrac{?}{12z^2+15z+3}$

22._____

23. $\dfrac{5}{2r+8} = \dfrac{?}{2(r+4)(r^2+2r-8)}$

23._____

24. $\dfrac{9}{y^2-4} = \dfrac{?}{(y+2)^2(y-2)}$

24._____

25. $\dfrac{2}{7p-35} = \dfrac{?}{14p^3-70p^2}$

25._____

26. $\dfrac{3}{5r-10} = \dfrac{?}{50r^2-100r}$ **26.** _____

27. $\dfrac{3}{k^2+3k} = \dfrac{?}{k^3+10k^2+21k}$ **27.** _____

28. $\dfrac{3x+1}{x^2-4} = \dfrac{?}{2x^2-8}$ **28.** _____

29. $\dfrac{5a}{7a-14} = \dfrac{?}{28a^2-56a}$ **29.** _____

30. $\dfrac{r+1}{r^2+2r} = \dfrac{?}{2r^3+3r^2-2r}$ **30.** _____

Chapter 12 RATIONAL EXPRESSIONS AND APPLICATIONS

12.4   Adding and Subtracting Rational Expressions

Find each sum. Write each answer in lowest terms.

1. $\dfrac{4x}{x-2}+\dfrac{2x}{x-2}$ 1._____

2. $\dfrac{4}{3w^2}+\dfrac{7}{3w^2}$ 2._____

3. $\dfrac{x}{x^2-4}+\dfrac{7x}{x^2-4}$ 3._____

4. $\dfrac{5t+4}{2t+1}+\dfrac{4t+2}{2t+1}$ 4._____

5. $\dfrac{b}{b^2-4}+\dfrac{2}{b^2-4}$ 5._____

6. $\dfrac{3m+4}{2m^2-7m-15}+\dfrac{m+2}{2m^2-7m-15}$ 6._____

7. $\dfrac{6x}{\left(2x+1\right)^2}+\dfrac{3}{\left(2x+1\right)^2}$ 7._____

8. $\dfrac{2x}{9x^2-25y^2}+\dfrac{x-5y}{9x^2-25y^2}$ 8._____

Name: Date:

Instructor: Section:

Find each sum. Write each answer in lowest terms.

9. $\dfrac{x}{3} + \dfrac{2}{5}$

9._____

10. $\dfrac{2}{2x-3} + \dfrac{9}{4x-6}$

10._____

11. $\dfrac{2}{a^2-4} + \dfrac{3}{a+2}$

11._____

12. $\dfrac{-4}{h+1} + \dfrac{2h}{1-h^2}$

12._____

13. $\dfrac{2y+9}{y^2+6y+8} + \dfrac{y+3}{y^2+2y-8}$

13._____

14. $\dfrac{2m+3}{m^2-3m-4} + \dfrac{3m-2}{m^2-16}$

14._____

15. $\dfrac{3z-2}{5z+20} + \dfrac{2z+1}{3z+12}$

15._____

16. $\dfrac{3p}{4p^2-9} + \dfrac{4}{6p+9}$

16._____

17. $\dfrac{5p-2}{2p^2+9p+9}+\dfrac{p+7}{6p^2+13p+6}$

17._____

18. $\dfrac{1-3x}{4x^2-1}+\dfrac{3x-5}{2x^2+5x+2}$

18._____

Find each difference. Write each answer in lowest terms.

19. $\dfrac{6p}{p-4}-\dfrac{p+20}{p-4}$

19._____

20. $\dfrac{2z^2}{z+y}-\dfrac{2y^2}{z+y}$

20._____

21. $\dfrac{2x}{x^2+3x-10}-\dfrac{x+2}{x^2+3x-10}$

21._____

22. $\dfrac{4x}{9x^2-16y^2}-\dfrac{x+4y}{9x^2-16y^2}$

22._____

23. $\dfrac{7}{x+4}-\dfrac{5}{3x+12}$

23._____

24. $\dfrac{6+2k}{9}-\dfrac{2+k}{18}$

24._____

25. $\dfrac{5-10s}{6}-\dfrac{5-5s}{9}$

25._____

26. $\dfrac{1}{m^2-16}-\dfrac{1}{m-4}$

26._____

27. $\dfrac{6}{x-y}-\dfrac{4+y}{y-x}$

27._____

28. $\dfrac{-4}{x^2-4}-\dfrac{3}{2x-4}$

28._____

29. $\dfrac{6}{2q^2+5q+2}-\dfrac{5}{2q^2-3q-2}$

29._____

30. $\dfrac{4y}{y^2+4y+3}-\dfrac{3y+1}{y^2-y-2}$

30._____

Name: Date:

Instructor: Section:

Chapter 12 RATIONAL EXPRESSIONS AND APPLICATIONS

12.5   Complex Fractions

Simplify each complex fraction by writing it as a division problem

1. $\dfrac{-\frac{3}{5}}{\frac{9}{10}}$

1._____

2. $\dfrac{\frac{3}{4}-\frac{1}{2}}{\frac{1}{4}+\frac{5}{8}}$

2._____

3. $\dfrac{\frac{49m^3}{18n^5}}{\frac{21m}{27n^2}}$

3._____

4. $\dfrac{\frac{r-s}{12}}{\frac{r^2-s^2}{6}}$

4._____

5. $\dfrac{2-\frac{1}{y-2}}{3-\frac{2}{y-2}}$

5._____

6. $\dfrac{3-\frac{5}{m}}{\frac{2}{m}+2}$

6._____

7. $\dfrac{\dfrac{p}{2}-\dfrac{1}{3}}{\dfrac{p}{3}+\dfrac{1}{6}}$

7. _____

8. $\dfrac{\dfrac{4}{z}+2}{\dfrac{1+z}{2}}$

8. _____

9. $\dfrac{3+\dfrac{4}{s}}{2s+\dfrac{2}{3}}$

9. _____

10. $\dfrac{\dfrac{4}{p}-2p}{\dfrac{3-p^2}{6}}$

10. _____

11. $\dfrac{\dfrac{a+2}{a-2}}{\dfrac{1}{a^2-4}}$

11. _____

12. $\dfrac{\dfrac{2}{a+2}-4}{\dfrac{1}{a+2}-3}$

12. _____

13. $\dfrac{\dfrac{3}{w-4}-\dfrac{3}{w+4}}{\dfrac{1}{w+4}+\dfrac{1}{w^2-16}}$

13._____

14. $\dfrac{\dfrac{5}{rs^2}-\dfrac{2}{rs}}{\dfrac{3}{rs}-\dfrac{4}{r^2s}}$

14._____

15. $\dfrac{\dfrac{12a}{2a-3}}{\dfrac{21a^2}{4a^2-9}}$

15._____

Simplify each complex fraction by multiplying by the least common denominator.

16. $\dfrac{\frac{9}{20}}{-\frac{11}{25}}$

16._____

17. $\dfrac{\frac{1}{2}+\frac{3}{8}}{\frac{3}{4}-\frac{9}{8}}$

17._____

18. $\dfrac{\dfrac{a^2}{b}}{\dfrac{a^2}{b^2}}$

18._____

19. $\dfrac{\dfrac{16r^2}{11s^3}}{\dfrac{8r^4}{22s}}$

20. $\dfrac{2x - y^2}{x + \dfrac{y^2}{x}}$

21. $\dfrac{r + \dfrac{3}{r}}{\dfrac{5}{r} + rt}$

22. $\dfrac{\dfrac{x-2}{x+2}}{\dfrac{x}{x-2}}$

23. $\dfrac{2s + \dfrac{3}{s}}{\dfrac{1}{s} - 3s}$

24. $\dfrac{\dfrac{15}{10k+10}}{\dfrac{5}{3k+3}}$

25. $\dfrac{\dfrac{1}{h} - 4}{\dfrac{1}{2} + 2h}$

26. $\dfrac{\dfrac{4}{x} - \dfrac{1}{2}}{\dfrac{5}{x} + \dfrac{1}{3}}$

19._____

20._____

21._____

22._____

23._____

24._____

25._____

26._____

27. $\dfrac{\dfrac{1}{m-1}+4}{\dfrac{2}{m-1}-4}$

27._____

28. $\dfrac{\dfrac{4}{x+4}}{\dfrac{3}{x^2-16}}$

28._____

29. $\dfrac{\dfrac{6}{k+1}-\dfrac{5}{k-3}}{\dfrac{3}{k-3}+\dfrac{2}{k+2}}$

29._____

30. $\dfrac{\dfrac{4}{s+3}-\dfrac{2}{s-3}}{\dfrac{5}{s^2-9}}$

30._____

Chapter 12 RATIONAL EXPRESSIONS AND APPLICATIONS

12.6   Solving Equations with Rational Expressions

Identify each of the following as an operation or an equation. If it is an operation, perform it. If it is an equation, solve it.

1. $\dfrac{2x}{3} + \dfrac{2x}{5} = \dfrac{64}{15}$ 1._____

2. $\dfrac{3x}{5} - \dfrac{4x}{3} = \dfrac{22}{15}$ 2._____

3. $\dfrac{4x}{5} - \dfrac{5x}{10}$ 3._____

4. $\dfrac{2x}{4} - \dfrac{3x}{2}$ 4._____

5. $\dfrac{2x}{5} + \dfrac{7x}{3}$ 5._____

Solve each equation and check your answers.

6. $\dfrac{p}{p-2} = \dfrac{2}{p-2} + 1$ 6._____

7. $\dfrac{4}{m} - \dfrac{2}{3m} = \dfrac{10}{9}$ 7._____

8. $\dfrac{4}{5x} + \dfrac{3}{2x} = \dfrac{23}{50}$

8. _____

9. $\dfrac{x-4}{5} = \dfrac{x+2}{3}$

9. _____

10. $\dfrac{4}{n+2} - \dfrac{2}{n} = \dfrac{1}{6}$

10. _____

11. $\dfrac{x}{3x+16} = \dfrac{4}{x}$

11. _____

12. $\dfrac{2p+3}{3} = \dfrac{4p+2}{15}$

12. _____

13. $\dfrac{4+x}{6} + \dfrac{x}{4} = \dfrac{19}{6}$

13. _____

14. $\dfrac{8}{2m+4} + \dfrac{2}{3m+6} = \dfrac{7}{9}$

14. _____

15. $\dfrac{2}{z-1} + \dfrac{3}{z+1} - \dfrac{17}{24} = 0$

15. _____

16. $\dfrac{2}{m-3} + \dfrac{12}{9-m^2} = \dfrac{3}{m+3}$

16. _____

17. $\dfrac{9}{x^2 - x - 12} = \dfrac{3}{x-4} - \dfrac{x}{x+3}$

17._____

18. $\dfrac{-16}{n^2 - 8n + 12} = \dfrac{3}{n-2} + \dfrac{n}{n-6}$

18._____

19. $\dfrac{4}{y+2} - \dfrac{3}{y+3} = \dfrac{8}{y^2 + 5y + 6}$

19._____

Solve each formula for the specified variable.

20. $P = \dfrac{I}{rt}$ for t

20._____

21. $F = \dfrac{k}{d-D}$ for D

21._____

22. $S = \dfrac{a_1}{1-r}$ for r

22._____

23. $h = \dfrac{2A}{B+b}$ for B

23._____

24. $F = \dfrac{GmM}{d^2}$ for G

24._____

25. $\dfrac{1}{R} = \dfrac{1}{R_1} + \dfrac{1}{R_2}$ for R

25._____

26. $\dfrac{V_1 P_1}{T_1} = \dfrac{V_2 P_2}{T_2}$ for T

26._____

27. $\dfrac{1}{f} = \dfrac{1}{d_0} + \dfrac{1}{d_1}$ for f

27._____

28. $F = \dfrac{G m_1 m_2}{d^2}$ for G

28._____

29. $S_n = \dfrac{n}{2}\left(a_1 + a_n\right)$ for a_1

29._____

30. $A = \dfrac{1}{2} h \left(b_1 + b_2\right)$ for b_2

30._____

Chapter 12 RATIONAL EXPRESSIONS AND APPLICATIONS

12.7   Applications of Rational Expressions

Solve each problem.

1. One-fifth of a number is two less than one-third of the same number. What is the number?

    1._____

2. If the same number is added to the numerator and denominator of the fraction $\frac{5}{9}$, the value of the resulting fraction is $\frac{2}{3}$. Find the number.

    2._____

3. If two times a number is added to one-half of its reciprocal, the result is $\frac{13}{6}$. Find the number.

    3._____

4. If a certain number is added to the numerator and twice that number is subtracted from the denominator of the fraction $\frac{3}{5}$, the result is equal to 5. Find the number.

    4._____

5. In a certain fraction, the numerator is 4 less than the denominator. If 5 is added to both the numerator and the denominator, the resulting fraction is equal to $\frac{7}{9}$. Find the original fraction.

    5._____

6. The sum of a number and its reciprocal is $\frac{13}{6}$. Find the number.

    6._____

7. If twice the reciprocal of a number is added to the number, the result is $\frac{9}{2}$. Find the number.

7._____

8. If three times a number is subtracted from twice its reciprocal, the result is −1. Find the number.

8._____

9. Sharon and Elaine worked as computer analysts. Last year, Sharon earned $\frac{3}{5}$ as much as Elaine. If they earned a total of $152,000, how much did each of them earn?

9._____

10. Lauren takes $\frac{4}{5}$ the number of pills that David takes for the same illness. Together they use 45 pills. Find the number used by David.

10._____

Solve each problem.

11. Mark can row 5 miles per hour in still water. It takes him as long to row 4 miles upstream as 16 miles downstream. How fast is the current?

11._____

12. John flew from City A to City B at 200 miles per hour and from City B to City A at 180 miles per hour. The trip at the slower speed took $\frac{1}{2}$ hour longer. Find the distance between the two cities. (Assume there is no wind in either direction.)

12._____

13. Yohannes traveled to his destination at an average speed   **13.**_____
of 70 miles per hour. Coming home, his average speed
was 50 miles per hour and the trip took 2 hours longer.
How far did he travel each way?

14. Dipti flew her plane 600 miles against the wind in the   **14.**_____
same time it took her to fly 900 miles with the wind. If
the speed of the wind was 30 miles per hour, what was
the speed of the plane?

15. Wendy drove a distance of 250 miles, at a speed that was   **15.**_____
10 miles per hour faster than her speed on her return trip.
If it took Wendy $\frac{5}{6}$ hour longer on the return trip, what
was her speed on the return trip?

16. A boat goes 6 miles per hour in still water. It takes as   **16.**_____
long to go 40 miles upstream as 80 miles downstream.
Find the speed of the current.

17. A ship goes 120 miles downriver in $2\frac{2}{3}$ hours less than it   **17.**_____
takes to go the same distance upriver. If the speed of the
current is 6 miles per hour, find the speed of the ship.

18. A plane traveling 450 miles per hour can go 1000 miles with the wind in $\frac{1}{2}$ hour less than when traveling against the wind. Find the speed of the wind.

18._____

19. On Saturday, Pablo jogged 6 miles. On Monday, jogging at the same speed, it took him 30 minutes longer to cover 10 miles. How fast did Pablo jog?

19._____

20. A plane made the trip from Redding to Los Angeles, a distance of 560 miles, in 1.5 hours. Less than it took to fly from Los Angeles to Portland, a distance of 1130 miles. Find the rate of the plane. (Assume there is no wind in either direction.)

20._____

Solve each problem.

21. Kelly can clean the house in 6 hours, but it takes Linda 4 hours. How long would it take them to clean the house if they worked together?

21._____

22. Nina can wash the walls in a certain room in 2 hours and Mark can wash these walls in 5 hours. How long would it take them to complete the task if they work together?

22._____

23. Phil can install the carpet in a room in 3 hours, but Lil needs 8 hours. How long will it take them to complete this task if they work together?

23._____

24. One pipe can fill a swimming pool in 8 hours and another pipe can fill the pool in 12 hours. How long will it take to fill the pool if both pipes are open?

24._____

25. Chuck can weed the garden in $\frac{1}{2}$ hour, but David takes 2 hours. How long does it take them to weed the garden if they work together?

25._____

26. Jack can paint a certain room in $1\frac{1}{2}$ hours, but Joe needs 4 hours to paint the same room. How long does it take them to paint the room if they work together?

26._____

27. Michael can type twice as fast as Sharon. Together they can type a certain job in 2 hours. How long would it take Michael to type the entire job by himself?

27._____

28. Working together, Ethel and Al can balance the books for a certain company in 3 hours. Working alone, it would take Ethel $\frac{2}{3}$ as long as Al to balance the books. How long would it take Al to do the job alone?

28._____

29. Judy and Tony can mow the lawn together in 4 hours. It takes Tony twice as long as Judy to do the job alone. How long would it take Judy working alone?

29._____

30. Fred can seal an asphalt driveway in $\frac{1}{3}$ the time it takes John. Working together, it takes them $1\frac{1}{2}$ hours. How long would it have taken Fred working alone?

30._____

Chapter 12 RATIONAL EXPRESSIONS AND APPLICATIONS

12.8   Variation

Solve each problem involving direct variation.

1. If m varies directly as p, and $m = 40$ when $p = 5$, find m when p is 9.

   1._____

2. If y varies directly as x, and $x = 14$ when $y = 42$, find y when $x = 4$.

   2._____

3. If a varies directly as b, and $a = 24$ when $b = 16$, find b when $a = 34$.

   3._____

4. If c varies directly as d, and $c = 100$ when $d = 5$, find c when $d = 3$.

   4._____

5. If h varies directly as m, and $h = 9$ when $m = 6$, find h when $m = 10$.

   5._____

6. If x varies directly as y, and $x = 24$ when $y = -2$, find x when $y = 4$.

   6._____

7. If f varies directly as g, and $f = \frac{7}{2}$ when $g = 7$, find f when $g = 18$.

7._____

8. If w varies directly as v, and $w = 24$ when $v = 20$, find w when $v = 25$.

8._____

9. If a varies directly as b, and $a = 61.5$ when $b = 82$, find a when $b = 224$.

9._____

10. If y varies directly as x, and $y = 21$ when $x = 35$, find y when $x = 75$.

10._____

Solve each problem involving inverse variation.

11. If y varies inversely as x, and $y = 20$ when $x = 4$, find y when $x = 12$.

11._____

12. If y varies inversely as x, and $y = 5$ when $x = 3$, find y when $x = 0.5$.

12._____

13. If g varies inversely as f, and $g = 6$ when $f = 12$, find g when $f = 18$.

13._____

14. If d varies inversely as c, and $d = 18$ when $c = \dfrac{1}{3}$,

  find d when $c = \dfrac{2}{5}$.

14._____

15. If n varies inversely as m, and $n = 10.5$ when $m = 1.2$, find n when $m = 5.6$.

15._____

16. If y varies inversely as x, and $y = 10$ when $x = 2$, find y when $x = 4$.

16._____

17. If a varies inversely as b, and $a = 36$ when $b = \dfrac{1}{2}$, find a when $b = 3$.

17._____

18. If y varies inversely as x, and $y = 9$ when $x = 2$, find y when $x = 4$.

18._____

19. If c varies inversely as d, and $c = 11$ when $d = 3$, find c when $d = 11$.

19._____

20. If t varies inversely as s, and $t = 100$ when $s = 0.4$, find t when $s = 6$.

20._____

Solve each problem.

21. For a given period of time, the interest earned on an investment varies directly as the interest rate. If the interest is $125 when the rate is 5%, find the interest when the rate is $6\frac{1}{2}\%$.

21._____

22. For a specified distance, time varies inversely with speed. If Ramona walks a certain distance on a treadmill in 40 minutes at 4.2 miles per hour, how long will it take her to walk the same distance at 3.5 miles per hour?

22._____

23. If the temperature is constant, the pressure of a gas in a container varies inversely as the volume of the container. If the pressure is 140 pounds per square foot when a gas is in a container of 1000 cubic feet, what is the volume of the container when the gas exerts a pressure of 700 pounds per square feet?

23._____

24. If the temperature is constant, the pressure of a gas in a container varies inversely as the volume of the container. If the pressure is 9 pounds per square foot in a container of 6 cubic feet, what is the pressure in a container of 7.5 cubic feet?

24._____

25. The circumference of a circle varies directly as the radius. A circle with a radius of 7 centimeters has a circumference of 43.96 centimeters. Find the circumference if the radius changes to 11 centimeters.

25._____

26. For a constant area, the length of a rectangle varies inversely as the width. If the length of a rectangle is 16 feet when the width is 3 feet, find the length of a rectangle with the same area where the width is 9 feet.

26._____

27. For a given height, the area of a triangle varies directly as its base. Find the area of a triangle with a base of 4 centimeters, if the area is 9.6 square centimeters when the base is 3 centimeters.

27._____

28. The force required to compress a spring varies directly as the change in the length of the spring. If a force of 25 pounds is required to compress a spring 4 inches, how much force is required to compress the spring 8 inches?

28._____

29. For a given rate, the distance that an object travels varies directly with time. Find the distance an object travels in 5 hours if the object travels 165 miles in 3 hours.

29._____

30. The length of a violin string varies inversely with the frequency of its vibrations. A 10-inch violin string vibrates at a frequency of 512 cycles per second. Find the frequency of an 8-inch string.

30._____

Chapter 13 SYSTEMS OF LINEAR EQUATIONS AND INEQUALITIES

13.1   Solving Systems of Linear Equations by Graphing

Decide whether the given ordered pair is a solution of the given system.

1. $(4,1)$

$2x + 3y = 11$

$3x - 2y = 9$

1._____

2. $(2, -4)$

$2x + 3y = 6$

$3x - 2y = 14$

2._____

3. $(-3, -1)$

$5x - 3y = -12$

$2x + 3y = -9$

3._____

4. $(4, 0)$

$4x + 3y = 16$

$x - 4y = -4$

4._____

5. $(-5, -4)$

$x - y = -1$

$4x + y = -24$

5._____

6. $(3, -7)$

$5x + y = 8$

$2x - 3y = 26$

6._____

7. $(-1, -7)$

$x - y = -6$

$-2x + 3y = -19$

7._____

8. $(1, -4)$

$-3x + y = -7$

$4x - 3y = 16$

8._____

9. $(-4,-1)$

    $5x-2y=6$

    $y=-3x-11$

9. _____

10. $(-1,5)$

    $3x+2y=7$

    $y=-2x+3$

10. _____

11. $(4,2)$

    $9x-2y=32$

    $8x-y=30$

11. _____

12. $(6,-2)$

    $x-y=8$

    $2x+3y=6$

12. _____

Solve each system by graphing both equations on the same axes.

13. $x-2y=6$

     $2x+y=2$

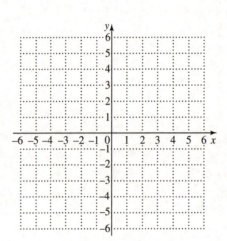

13. _____

14. $2x + 3y = 5$
        $3x - y = 13$

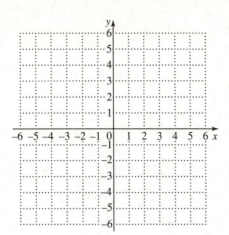

14. _____

15. $6x - 5y = 4$
        $2x - 5y = 8$

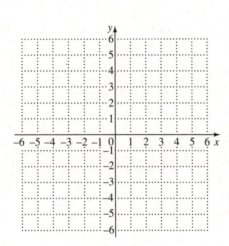

15. _____

16. $3x - y = -7$
        $2x + y = -3$

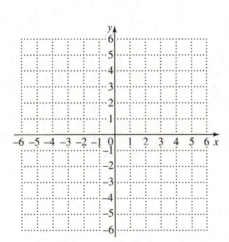

16. _____

17. $2x = y$

$5x + 3y = 0$

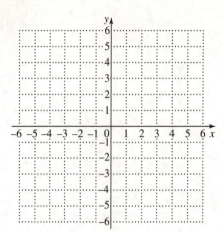

17. _____

18. $y - 2 = 0$

$3x + 4y = -19$

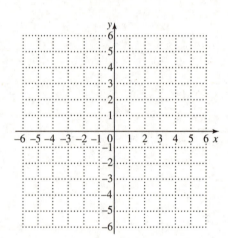

18. _____

19. $8x - 5y = -8$

$2x - y = 0$

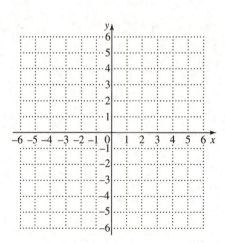

19. _____

20. $x - y = -7$

$x + 11 = 2y$

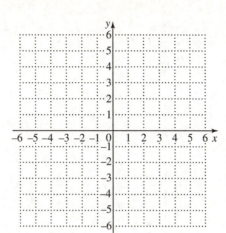

20. _____

21. $3x - 2y = 8$

$7x + 2y = 12$

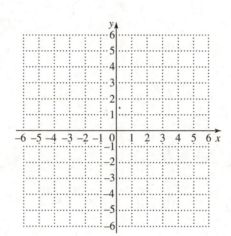

21. _____

22. $2x - 5y = 18$

$3x = 5y + 22$

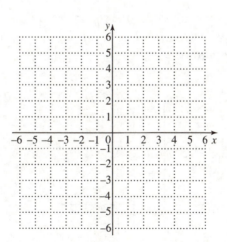

22. _____

23. $x + y = 5$

$3x - y = -1$

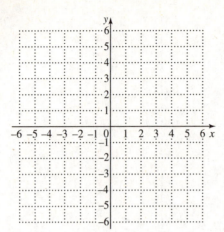

23. _____

24. $2x - y = -10$

$3x + 2y = 6$

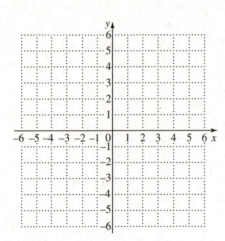

24. _____

Solve each system of equations by graphing both equations on the same axes. If the two equations produce parallel lines, write no solution. If the two equations produce the same line, write infinite number of solutions.

25. $8x + 4y = -1$

$4x + 2y = 3$

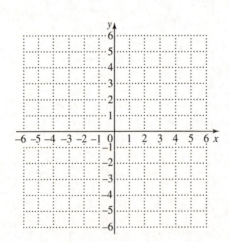

25. _____

26. $x + 2y = 4$

$8y = -4x + 16$

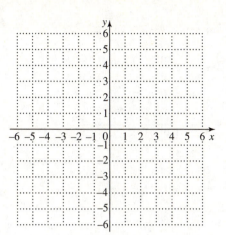

26. _____

27. $4x + 3y = 12$

$6y + 8x = -24$

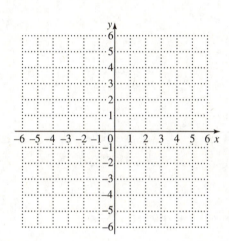

27. _____

28. $2x + 3y = 0$

$6x = -9y$

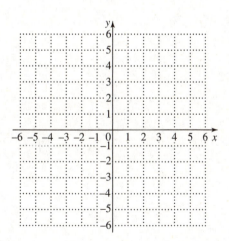

28. _____

29. $-3x + 2y = 6$

      $-6x + 4y = 12$

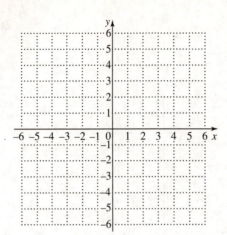

29. _____

30. $3x + 3y = 8$

      $x = 4 - y$

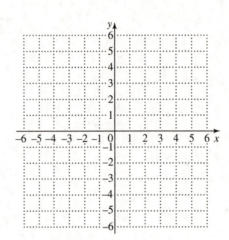

30. _____

Chapter 13 SYSTEMS OF LINEAR EQUATIONS AND INEQUALITIES

13.2   Solving Systems of Linear Equations by Substitution

Solve each system by the substitution method. Check each solution.

1. $x + y = 7$

    $y = 6x$

          1._____

2. $3x + 2y = 14$

    $y = x + 2$

          2._____

3. $x + 3y = 1$

    $x = -5 - 6y$

          3._____

4. $x + y = 9$

    $5x - 2y = -4$

          4._____

5. $x - 4y = 17$

    $3x - 4y = 11$

          5._____

6. $-8x + 5y = 11$

    $x - y = -1$

          6._____

7. $5x + 3y = 11$

    $x + y = 3$

          7._____

8. $3x - 2y = 6$

    $x - 5y = -11$

8._____

9. $3x + 4y = 2$

    $2x + 3y = 2$

9._____

10. $5x + 2y = 14$

    $y + 2 = -1$

10._____

11. $x + 6y = -1$

    $-2x - 9y = 0$

11._____

12. $6x + 8y = 10$

    $4y = 5 - 3x$

12._____

Solve each system by either the addition method or the substitution method. First simplify equations where necessary. Check each solution.

13. $3x + 4y = 2x + 2y + 11$

    $4x - 7y = -16$

13._____

14. $7x + 2y = 2x - y + 19$

       $4x + 3y = 2x - 2y + 0$

14._____

15. $2x + 8y - 4 = 4y - 4x$

       $5x = 5 - 3y$

15._____

16. $6x - 7y = 4x - 3y - 11$

       $16x + 2y = 4x + 12$

16._____

17. $10x + 4y = -4x - 3y$

       $6x = 8y$

17._____

18. $10x - y = 6x + 2y + 1$

       $-3x + 3y = 3x - 2y + 1$

18._____

19. $7x - 3y - 8 = 3x - 5y$

$y = 2x$

19._____

20. $2x + 7y = 5y - 3x + 16$

$2y = -x + y + 2$

20._____

Solve each system by either the addition method or the substitution method. First clear all fractions. Check each solution.

21. $3x + 5y = 7$

$6x + 10y = 3$

21._____

22. $4x + 3y = 2$

$8x + 6y = 4$

22._____

23. $\dfrac{x}{6} + \dfrac{y}{6} = 1$

$\dfrac{x}{2} + \dfrac{y}{2} = 2$

23._____

24. $\dfrac{5}{4}x - y = -\dfrac{1}{4}$

$-\dfrac{7}{8}x + \dfrac{5}{8}y = 1$

24._____

25. $x - \dfrac{7}{5}y = \dfrac{6}{5}$

$\dfrac{1}{4}x - \dfrac{1}{2}y = \dfrac{1}{6}$

25._____

26. $\dfrac{1}{4}x + \dfrac{3}{8}y = -3$

$\dfrac{5}{6}x - \dfrac{3}{7}y = -10$

26. _____

27. $\dfrac{5x}{4} + \dfrac{2y}{3} = \dfrac{8}{3}$

$\dfrac{2x}{3} - \dfrac{3y}{2} = -6$

27. _____

28. $\dfrac{x}{2} - \dfrac{y}{3} = -8$

$\dfrac{x}{4} - \dfrac{y}{6} = -4$

28. _____

29. $\dfrac{9}{2}x - \dfrac{3}{4}y = 3$

$-\dfrac{3}{4}x + \dfrac{1}{8}y = -\dfrac{1}{2}$

29. _____

30. $\dfrac{3}{14}x - \dfrac{1}{7}y = 1$

$-\dfrac{1}{2}x = \dfrac{1}{4}y$

30. _____

Chapter 13 SYSTEMS OF LINEAR EQUATIONS AND INEQUALITIES

13.3   Solving Systems of Linear Equations by Elimination

Solve each system by the elimination method. Check your answers.

1. $x + y = 5$
    $x - y = -3$

    1._____

2. $x - 4y = -4$
    $-x + y = -5$

    2._____

3. $4x + 3y = -4$
    $2x - 3y = 16$

    3._____

4. $8x + 2y = 14$
    $3x - 2y = -14$

    4._____

5. $x - 3y = 5$
    $-x + 4y = -5$

    5._____

6. $5x + 8y = 12$
    $3x - 8y = 20$

    6._____

7. $5x - 6y = 9$
    $3x + 6y = 7$

    7._____

8. $15x - 3y = 8$
    $21x + 3y = 10$

    8._____

Solve each system by the elimination method. Check your answers.

9. $6x + 7y = 10$

$2x - 3y = 14$

9._____

10. $3x + 2y = 5$

$2x - 3y = 12$

10._____

11. $x - 4y = 10$

$x + 6y = -10$

11._____

12. $2x + y = 6$

$-3x + y = -19$

12._____

13. $4x - 9y = 7$

$3x + 2y = 14$

13._____

14. $3x - 7y = 12$

$5x + 3y = -2$

14._____

15. $4x - 3y = 7$

$5x + 4y = 1$

15._____

16. $3x + 2y = 16$

   $4x - 3y = -7$

16._____

Solve each system.

17. $5x - 3y = 23$

   $10 + 2y = 2x$

17._____

18. $4y = 2x - 2$

   $-9 + 3y = 5x$

18._____

19. $4x - 3y - 20 = 0$

   $6x + 5y + 8 = 0$

19._____

20. $6x = 16 - 7y$

   $4x = 3y + 26$

20._____

21. $7 - y = 2x$

   $4x = 19 + 3y$

21._____

22. $2x = 21 - 3y$

$\frac{1}{3}x + \frac{2}{5}y = 3$

22.＿＿＿＿＿＿＿＿＿

23. $3x + 1 = 2y$

$2y - 2 = 2x$

23.＿＿＿＿＿＿＿＿＿

Solve each system by the elimination method.

24. $12x - 8y = 3$

$6x - 4y = 6$

24.＿＿＿＿＿＿＿＿＿

25. $2x + 4y = -6$

$-x - 2y = 3$

25.＿＿＿＿＿＿＿＿＿

26. $6x - 12y = 3$

$2x - 4y = 1$

26.＿＿＿＿＿＿＿＿＿

27. $4x - 2y = -8$

$2x - y = 4$

27.＿＿＿＿＿＿＿＿＿

28. $15x + 6y = 9$ **28.** _____

    $10x + 4y = 6$

29. $12x - 18y = 10$ **29.** _____

    $4x - 6y = 2$

30. $x - 3y = 12$ **30.** _____

    $3x - 9y = 4$

Chapter 13 SYSTEMS OF LINEAR EQUATIONS AND INEQUALITIES

13.4   Applications of Linear Systems

Use a system of equations to solve each problem.

1. The sum of two numbers is 64. Their difference is 18.
   Find the numbers.

   1._____

2. Find two numbers whose sum is –66 and whose
   difference is –116.

   2._____

3. The difference between two numbers is 27. If the larger
   is four more than twice the smaller, find the numbers.

   3._____

4. The sum of two numbers is 20. Three times the smaller
   is equal to twice the larger. Find the numbers.

   4._____

5. The difference between two numbers is 14. If two times
   the smaller is added to one-half the larger, the result is
   52. Find the numbers.

   5._____

6. Two towns have a combined population of 9045. There
   are 2249 more people living in one than in the other.
   Find the population in each town.

   6._____

7. A rope 82 centimeters long is cut into two pieces with one piece four more than twice as long as the other. Find the length of each piece.

7._____

8. The perimeter of a rectangular room is 50 feet. The length is three feet greater than the width. Find the dimensions of the rectangle.

8._____

Use a system of equations to solve each problem.

9. Admission prices at a football game were $12 for adults and $9 for children. The total receipts for the game were $87,000. Tickets were sold to 8000 people. How many adults and how many children attended the game?

9._____

10. The receipts from a concert were $2100. The price for a regular ticket was $6 and the student tickets were half the regular price. If 400 tickets were sold, how many of each type were there?

10._____

11. Charlie has only $5 bills and $20 bills and has a total of $130. If there is a total of 11 bills, how many of each type are there?

11._____

12. There were 411 tickets sold for a soccer game, some for students and some for nonstudents. Student tickets cost $4.25 and nonstudent tickets cost $8.50 each. The total receipts were $3021.75. How many of each type were sold?

12._____

13. A cashier has some $5 bills and some $10 bills. The total value of the money is $750. If the number of tens is equal to twice the number of fives, how many of each type are there?

13._____

14. Twice as many general admission tickets to a basketball game were sold as reserved seat tickets. General admission tickets cost $10 and reserved seat tickets cost $15. If the total value of both kinds of tickets was $26,250, how many tickets of each kind were sold?

14._____

15. Carla has $12,000 to invest at 7% and 9%. She wants the income from simple interest on the two investments to total $1000 yearly. How much should she invest at each rate?

15._____

Use a system of equations to solve each problem.

16. Steve wishes to mix coffee worth $6 a pound with coffee worth $9 a pound to get 45 pounds of a mixture worth $8 a pound. How many pounds of the $6 and the $9 coffee will be needed?

16._____

17. Ben wishes to blend candy selling for $1.60 a pound with candy selling for $2.50 a pound to get a mixture that will be sold for $1.90 a pound. How many pounds of the $1.60 and the $2.50 candy should be used to get 30 pounds of the mixture?

17._____

18. How many bags of coffee worth $90 a bag must be mixed with coffee worth $75 a bag to get 50 bags worth $87 a bag?

18._____

19. How many pounds of walnuts that sell for $8 a pound should be mixed with peanuts that sell for $6 a pound to produce a 10-pound mix that sells for $7 a pound?

19._____

20. Nina sells caramels that cost $3.65 per pound mixed with creams that cost $3.25 per pound. How much of each kind of candy is in a pound of the mixture if it costs $3.49?

20._____

21. How many liters of 75% solution should be mixed with a 55% solution to get 70 liters of 63% solution? How many liters of the 90% and 75% solutions should be used?

21._____

22. A 90% antifreeze solution is to be mixed with a 75% solution ot make 30 liters of an 80% solution. How many liters of the 90% and 75% solutions should be used?

22._____

23. Milton needs 45 liters of 20% solution. He has only 15% alcohol solution and 30% alcohol solution on hand to make the mixture. How many liters of each solution should he combine to make the mixture?

23._____

Use a system of equations to solve each problem.

24. Rick and Hilary drive from positions 378 miles apart and race toward each other. They meet after 3 hours. Find the average speed of each if Hilary travels 30 miles per hour faster than Rick.

24._____

25. Two trains start from positions 1242 miles apart and travel toward each other. They meet after $4\frac{1}{2}$ hours. Find the average speed of each train if one train travels 20 miles per hour faster than the other.

25._____

26. At the beginning of a fund-raising walk, Steve and Vic are 30 miles apart. If they leave at the same time and walk in the same direction, Steve would overtake Vic in 15 hours. If they walked toward each other, they would meet in 3 hours. What are their speeds?

26._____

27. Two bicyclists leave from Washington DC and ride in opposite directions. One travels $1\frac{1}{2}$ times as fast as the other. After 2 hours, they are 40 miles apart. Find the speed of each bicyclist.

27._____

28. John left Louisville at noon on the same day that Mike left Louisville at 1 P.M. Both were traveling in the same direction. At 5 P.M., Mike was 62 miles behind John. If John was traveling 2 miles per hour faster than Mike, what were their speeds?

28._____

29. A plane can travel 300 miles per hour with the wind and
    230 miles per hour against the wind. Find the speed of
    the wind and the speed of the plane in still air.

29._____

30. Two planes left Philadelphia traveling in opposite
    directions. Plane A left 15 minutes before plane B.
    After plane B had been flying for 1 hour, the planes were
    860 miles apart. What were the speeds of the two planes
    if plane A was flying 40 miles per hour faster than plane
    B?

30._____

Name: Date:

Instructor: Section:

Chapter 13 SYSTEMS OF LINEAR EQUATIONS AND INEQUALITIES

13.5   Solving Systems of Linear Inequalities

Graph the solution of each system of linear inequalities.

1. $7x + 3y \geq 21$

$x - y \leq 6$

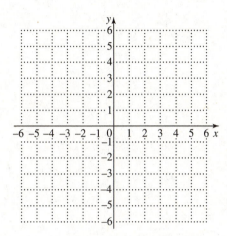

1._____

2. $3x - y \leq 3$

$x + y \leq 0$

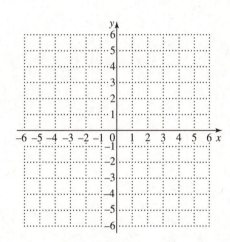

2._____

3. $3x - y \leq 6$

$3y - 6 \leq 2x$

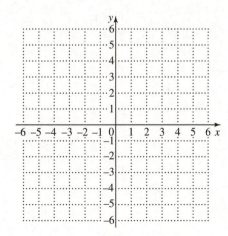

3._____

4. $3x + 5y \geq 15$

$y \geq x - 2$

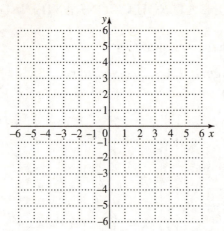

4. _____

5. $x + y \leq 3$

$5x - y \geq 5$

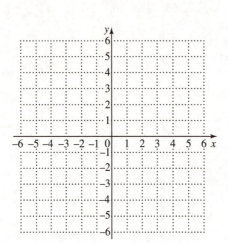

5. _____

6. $x + 2y \geq -4$

$5x \leq 10 - 2y$

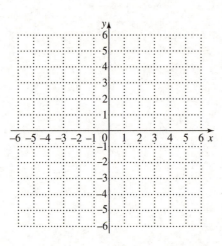

6. _____

7. $3x - y > 3$

$4x + 3y < 12$

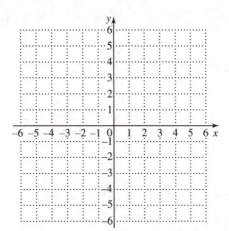

7. _____

8. $4x + 5y \leq 20$

$y \leq x + 3$

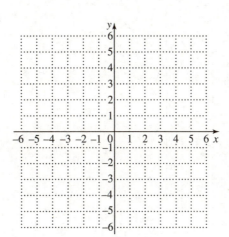

8. _____

9. $2x - y \geq 4$

$5y + 15 \geq -3x$

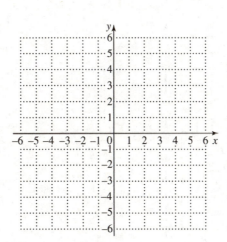

9. _____

10. $3x - 2y < 8$

$x < 4$

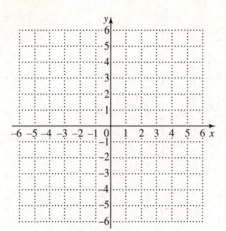

10._____

11. $x - y \leq 2$

$y \leq 2$

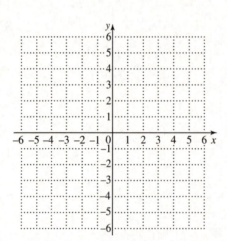

11._____

12. $x + y \geq 3$

$x - 2y \leq 4$

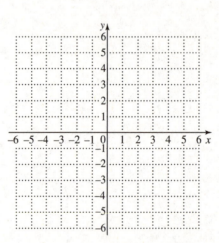

12._____

13. $x < 2y + 3$

  $0 < x + y$

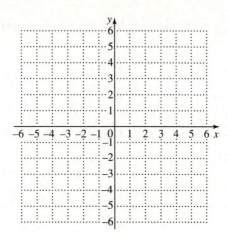

13. _____

14. $6x - y > 6$

  $2x + 5y < 10$

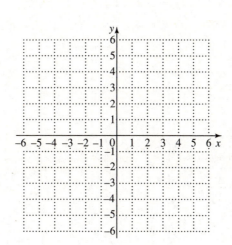

14. _____

15. $3x - 4y < 12$

  $y > -4$

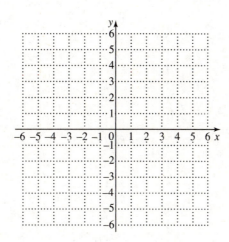

15. _____

16. $x - 2y \leq 4$

$x + 2y \leq 4$

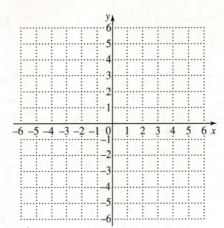

16. _____

17. $y > 2$

$4x - 3y < 9$

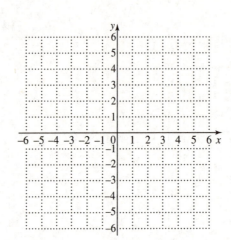

17. _____

18. $4x - 3y > 12$

$x < 4$

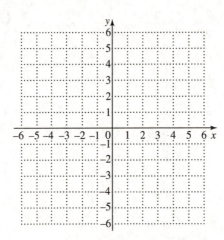

18. _____

19. $4x - y \le 4$

   $7y + 14 \ge -2x$

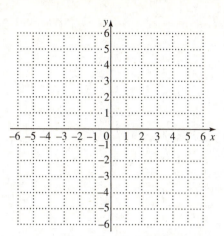

19. _____

20. $5x - 2y \le 10$

   $y \le -2$

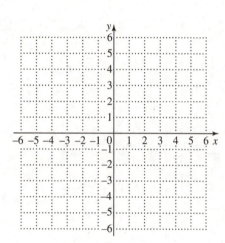

20. _____

21. $x - 2y \le 3$

   $2x + y \le -4$

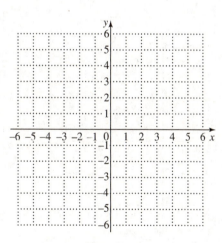

21. _____

22. $x + y > -3$

$2x - 3y \le -2$

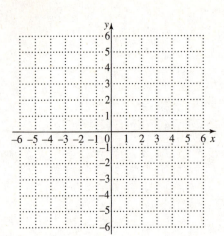

22._____

23. $y < 4$

$x \ge -3$

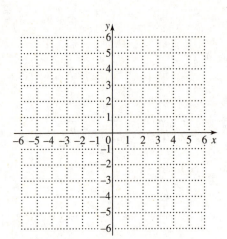

23._____

24. $x - 2y \ge -7$

$x - 2y < 2$

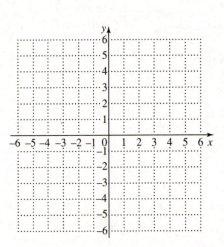

24._____

25. $4x - y > 2$

   $y > -x - 2$

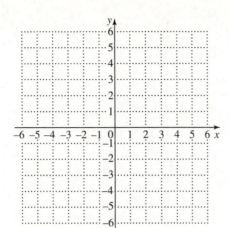

25._____

26. $3x + 2y < 10$

   $5x - 2y \leq 6$

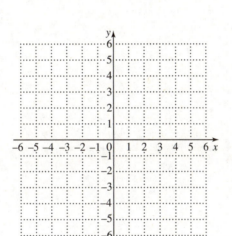

26._____

27. $y \geq -1$

   $2x - y > -1$

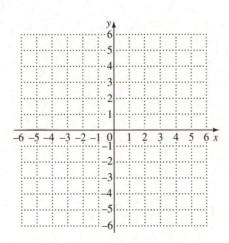

27._____

28. $x - 3y \leq -7$

    $x < 2$

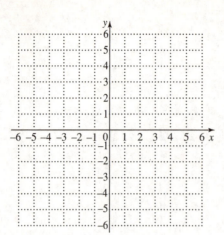

28. _____

29. $2x - y > -4$

    $2x + y > 0$

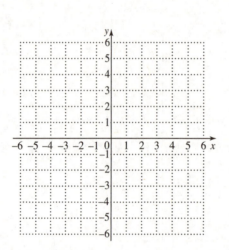

29. _____

30. $x - 3y \geq -9$

    $x + 3y < 3$

30. _____

Chapter 14 ROOTS AND RADICALS

14.1   Evaluating Roots

Find all square roots of each number.

1. 81

1._____

2. 256

2._____

3. 576

3._____

4. $\frac{49}{144}$

4._____

Find each square root.

5. $\sqrt{64}$

5._____

6. $-\sqrt{900}$

6._____

7. $-\sqrt{529}$

7._____

8. $\sqrt{\frac{81}{25}}$

8._____

Tell whether each square root is rational, irrational, *or* not a real number.

9. $\sqrt{72}$

9._____

10. $-\sqrt{81}$

10._____

11. $\sqrt{-49}$

11._____

12. $\sqrt{6400}$

12._____

13. $\sqrt{\frac{9}{49}}$

13._____

Use a calculator to find a decimal approximation for each square root. Round answers to the nearest thousandth.

14. $\sqrt{29}$ 14._____

15. $-\sqrt{18}$ 15._____

16. $-\sqrt{120}$ 16._____

17. $-\sqrt{131}$ 17._____

18. $\sqrt{640}$ 18._____

Find the length of the unknown side of each right triangle with sides a, b, and c, where c is the hypotenuse. If necessary, round your answer to the nearest thousandth.

19. $a = 5, \ b = 9$ 19._____

20. $c = 15, \ a = 12$ 20._____

21. $c = 17, \ a = 16$ 21._____

Use the Pythagorean formula to solve each problem. If necessary, round your answer to the nearest thousandth.

22. The hypotenuse of a right triangle measures 18   22._____
centimeters and one leg measures 4 centimeters. How
long is the other leg?

23. Two sides of a rectangle measure 9 centimeters and 17   23._____
centimeters. How long are the diagonals of the
rectangle?

24. Susan started to drive due south at the same time John started to drive due west. John drove 21 miles in the same time that Susan drove 28 miles. How far apart were they at that time?

24._____

25. Stan is flying a kite on 60 feet of string. How high is the kite above his head (vertically) if the horizontal distance between Stan and the kite is 36 feet?

25._____

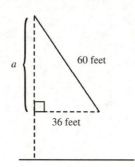

Find each root that is a real number.

26. $\sqrt[3]{-64}$

26._____

27. $\sqrt[4]{-625}$

27._____

28. $-\sqrt[3]{-8}$

28._____

29. $-\sqrt[5]{243}$

29._____

30. $\sqrt[4]{-81}$

30._____

Chapter 14 ROOTS AND RADICALS

14.2   Multiplying, Dividing, and Simplifying Radicals

Use the product rule for radicals to find each product.

1. $\sqrt{7} \cdot \sqrt{5}$ 1._____

2. $\sqrt{3} \cdot \sqrt{3}$ 2._____

3. $\sqrt{7} \cdot \sqrt{2}$ 3._____

4. $\sqrt{7} \cdot \sqrt{7}$ 4._____

5. $\sqrt{13} \cdot \sqrt{8}$ 5._____

6. $\sqrt{3x} \cdot \sqrt{7},\; x > 0$ 6._____

Simplify each radical. Assume that all variables represent nonnegative real numbers.

7. $\sqrt{75}$ 7._____

8. $\sqrt{250}$ 8._____

9. $-\sqrt{88}$ 9._____

10. $-2\sqrt{54}$ 10._____

Find each product and simplify.

11. $\sqrt{6} \cdot \sqrt{15}$

11._____

12. $\sqrt{6} \cdot \sqrt{8}$

12._____

13. $\sqrt{11} \cdot \sqrt{33}$

13._____

Use the quotient rule and product rule, as necessary to simplify each expression.

14. $\sqrt{\dfrac{16}{9}}$

14._____

15. $\sqrt{\dfrac{144}{49}}$

15._____

16. $\dfrac{9\sqrt{75}}{3\sqrt{3}}$

16._____

17. $\dfrac{5\sqrt{48}}{10\sqrt{8}}$

17._____

18. $\dfrac{\sqrt{24}}{2\sqrt{6}}$

18._____

19. $\sqrt{\dfrac{15}{8}} \cdot \sqrt{\dfrac{1}{2}}$

19._____

Name: Date:
Instructor: Section:

Simplify each radical. Assume that all variables represent positive real numbers.

20. $\sqrt{x^{10}}$ 20._____

21. $\sqrt{49x^3}$ 21._____

22. $\sqrt{48x^7}$ 22._____

23. $\sqrt{\dfrac{6}{x^2}}$ 23._____

24. $\sqrt{\dfrac{81}{25x^6}}$ 24._____

Simplify each expression.

25. $\sqrt[4]{32}$ 25._____

26. $\sqrt[3]{128}$ 26._____

27. $\sqrt[5]{-64}$ 27._____

28. $-\sqrt[3]{\dfrac{64}{27}}$ 28._____

29. $-\sqrt[5]{-\dfrac{1}{32}}$ 29._____

30. $\sqrt[3]{3} \cdot \sqrt[3]{9}$ 30._____

Chapter 14 ROOTS AND RADICALS

14.3   Adding and Subtracting Radicals

Add or subtract as indicated.

1. $6\sqrt{2} - 4\sqrt{2}$

    1._____

2. $9\sqrt{11} - 3\sqrt{11}$

    2._____

3. $5\sqrt[3]{10} + \sqrt[3]{10}$

    3._____

4. $5\sqrt{2} - 4\sqrt{2}$

    4._____

5. $5\sqrt{2} + \sqrt{3}$

    5._____

6. $6\sqrt{3} - 2\sqrt{3} + 4\sqrt{3}$

    6._____

7. $3\sqrt{5} - 9\sqrt{5} + \sqrt{5}$

    7._____

8. $3\sqrt[4]{12} + 5\sqrt[4]{12} - 8\sqrt[4]{12}$

    8._____

Simplify and add or subtract terms wherever possible. Assume that all variables represent nonnegative real numbers.

9. $\sqrt{80} + \sqrt{45}$

    9._____

10. $4\sqrt{128} - 2\sqrt{32}$

    10._____

11. $9\sqrt{300} - 4\sqrt{75}$

11._____

12. $2\sqrt{27} - 8\sqrt{12}$

12._____

13. $5\sqrt{32} - 8\sqrt{18} + 2\sqrt{20}$

13._____

14. $-7\sqrt{63} - 9\sqrt{28} + 6\sqrt{28}$

14._____

15. $\dfrac{1}{3}\sqrt{27} - \dfrac{3}{4}\sqrt{48}$

15._____

16. $\dfrac{\sqrt{8}}{2} - \dfrac{7}{2}\sqrt{32}$

16._____

17. $\dfrac{5}{6}\sqrt{72} - \dfrac{3}{2}\sqrt{24}$

17._____

18. $\dfrac{5}{8}\sqrt{128} - \dfrac{3}{2}\sqrt{8}$

18._____

19. $3\sqrt{72z} - 6\sqrt{50z}$

20. $6\sqrt{54w} - 4\sqrt{6w}$

21. $3\sqrt{125x} - \sqrt{80x} + 2\sqrt{45x}$

22. $\sqrt[4]{162} - \sqrt[4]{32}$

Perform the indicated operations. Assume that all variables represent nonnegative real numbers.

23. $\sqrt{5} \cdot \sqrt{7} + 3\sqrt{35}$

24. $4\sqrt{7} \cdot \sqrt{3} - 2\sqrt{21}$

25. $7\sqrt{14} - 9\sqrt{7} \cdot \sqrt{2}$

26. $\sqrt{6} \cdot \sqrt{10} - 4\sqrt{15}$

27. $7\sqrt{x} \cdot \sqrt{9} + \sqrt{9x}$

27._____

28. $3\sqrt{2y} \cdot \sqrt{6} - 5\sqrt{6y}$

28._____

29. $4x\sqrt{5} - 3\sqrt{25x} \cdot \sqrt{3x}$

29._____

30. $11\sqrt{5w} \cdot \sqrt{30w} - 8w\sqrt{24}$

30._____

Chapter 14 ROOTS AND RADICALS

14.4   Rationalizing the Denominator

Rationalize each denominator.

1. $\dfrac{5}{\sqrt{5}}$ 1._____

2. $\dfrac{-2}{\sqrt{7}}$ 2._____

3. $\dfrac{-3}{\sqrt{3}}$ 3._____

4. $\dfrac{10}{\sqrt{15}}$ 4._____

5. $\dfrac{1}{\sqrt{5}}$ 5._____

6. $\dfrac{-44}{\sqrt{11}}$ 6._____

7. $\dfrac{\sqrt{3}}{\sqrt{5}}$ 7._____

8. $\dfrac{\sqrt{3}}{\sqrt{24}}$ 8._____

9. $\dfrac{\sqrt{4}}{\sqrt{24}}$ 9._____

10. $\dfrac{25\sqrt{5}}{\sqrt{250}}$ 10._____

Perform the indicated operations and write all answers in simplest form. Rationalize all denominators. Assume that all variables represent positive real numbers.

11. $\sqrt{\dfrac{1}{5}}$

11._____

12. $\sqrt{\dfrac{5}{7}}$

12._____

13. $\sqrt{\dfrac{5}{8}}$

13._____

14. $\sqrt{\dfrac{32}{27}}$

14._____

15. $\sqrt{\dfrac{63}{50}}$

15._____

16. $\sqrt{\dfrac{5}{21}} \cdot \sqrt{7}$

16._____

17. $\sqrt{\dfrac{1}{5}} \cdot \sqrt{\dfrac{9}{35}}$

17._____

18. $\sqrt{\dfrac{3x}{y}}$

18._____

19. $\dfrac{\sqrt{k^2 m^4}}{\sqrt{k^5}}$

19._____

20. $\sqrt{\dfrac{20a^3 b^4}{6a^2}}$

20._____

Rationalize each denominator. Assume that all variables in the denominator represent nonzero real numbers.

21. $\sqrt[3]{\dfrac{2}{5}}$

21._____

22. $\dfrac{\sqrt[3]{9}}{\sqrt[3]{4}}$

22._____

23. $\sqrt[3]{\dfrac{1}{25}}$

23._____

24. $\sqrt[3]{\dfrac{5}{3}}$

24._____

25. $\sqrt[3]{\dfrac{7}{2}}$

25._____

26. $\dfrac{\sqrt[3]{6}}{\sqrt[3]{9}}$

26._____

27. $\dfrac{3}{\sqrt[3]{49}}$

27._____

28. $\sqrt[3]{\dfrac{3}{80}}$

28. _____

29. $\sqrt[3]{\dfrac{2}{5r^2}}$

29. _____

30. $\sqrt[3]{\dfrac{x^2}{25y}}$

30. _____

Chapter 14 ROOTS AND RADICALS

14.5   More Simplifying and Operations with Radicals

Simplify each expression.

1. $\sqrt{2}\left(\sqrt{2}-\sqrt{5}\right)$ 1._____

2. $\sqrt{10}\left(\sqrt{6}+\sqrt{3}\right)$ 2._____

3. $\sqrt{5}\left(\sqrt{12}+4\sqrt{7}\right)$ 3._____

4. $\sqrt{7}\left(2\sqrt{8}-9\sqrt{7}\right)$ 4._____

5. $\left(4\sqrt{5}+\sqrt{3}\right)\left(\sqrt{2}-\sqrt{7}\right)$ 5._____

6. $\left(\sqrt{5}-\sqrt{8}\right)\left(\sqrt{3}+\sqrt{2}\right)$ 6._____

7. $\left(\sqrt{6}-2\sqrt{5}\right)\left(4\sqrt{6}+\sqrt{10}\right)$ 7._____

8. $\left(4\sqrt{2}+3\sqrt{3}\right)\left(\sqrt{2}-7\sqrt{3}\right)$ 8._____

9. $\left(2\sqrt{3}-\sqrt{2}\right)^2$

9._____

10. $\left(\sqrt{5}-\sqrt{8}\right)^2$

10._____

Rationalize each denominator.

11. $\dfrac{2}{7+\sqrt{3}}$

11._____

12. $\dfrac{4}{\sqrt{3}+2}$

12._____

13. $\dfrac{5}{3-\sqrt{7}}$

13._____

14. $\dfrac{4+\sqrt{5}}{\sqrt{5}}$

14._____

15. $\dfrac{\sqrt{2}}{\sqrt{5}-2}$

15._____

16. $\dfrac{3}{\sqrt{2}-\sqrt{5}}$

16._____

17. $\dfrac{4}{\sqrt{5}+\sqrt{2}}$

18. $\dfrac{5}{\sqrt{3}-\sqrt{10}}$

19. $\dfrac{\sqrt{3}+\sqrt{2}}{\sqrt{3}-\sqrt{2}}$

20. $\dfrac{\sqrt{6}+2}{\sqrt{2}-4}$

Write each quotient in lowest terms.

21. $\dfrac{6-3\sqrt{3}}{3}$

22. $\dfrac{6+2\sqrt{5}}{2}$

23. $\dfrac{2\sqrt{7}-4\sqrt{2}}{6}$

24. $\dfrac{8\sqrt{7}+12}{14}$

25. $\dfrac{8+6\sqrt{7}}{4}$

25.＿＿＿＿＿＿＿＿＿＿

26. $\dfrac{4\sqrt{2}-6}{6}$

26.＿＿＿＿＿＿＿＿＿＿

27. $\dfrac{6+\sqrt{8}}{2}$

27.＿＿＿＿＿＿＿＿＿＿

28. $\dfrac{3+\sqrt{27}}{9}$

28.＿＿＿＿＿＿＿＿＿＿

29. $\dfrac{8\sqrt{5}-12}{20}$

29.＿＿＿＿＿＿＿＿＿＿

30. $\dfrac{12+6\sqrt{6}}{8}$

30.＿＿＿＿＿＿＿＿＿＿

Chapter 14 ROOTS AND RADICALS

14.6   Solving Equations with Radicals

Solve each equation.

1. $\sqrt{x} = 7$ 1._____

2. $\sqrt{x} - 9 = 0$ 2._____

3. $8 - \sqrt{y} = 2$ 3._____

4. $\sqrt{b + 16} = 4$ 4._____

5. $\sqrt{3x + 1} = 3$ 5._____

6. $\sqrt{2z + 7} = 0$ 6._____

7. $\sqrt{2 + 4k} = 3\sqrt{k}$ 7._____

8. $\sqrt{3x + 3} = 2\sqrt{3x}$ 8._____

9. $\sqrt{5t + 2} = \sqrt{6t - 1}$ 9._____

Solve each equation that has a solution.

10. $\sqrt{y} = -9$

10._____

11. $6 - \sqrt{p} = 0$

11._____

12. $\sqrt{z} + 9 = 0$

12._____

13. $0 = 7 - \sqrt{r}$

13._____

14. $\sqrt{y+3} = -4$

14._____

15. $\sqrt{x+2} + 7 = 0$

15._____

16. $\sqrt{x+1} + 9 = 0$

16._____

17. $\sqrt{4-3n} = 2\sqrt{-5n}$

17._____

18. $r = \sqrt{r^2 - 6r + 12}$

18._____

19. $b = \sqrt{b^2 - 3b + 15}$

19._____

Find all solutions for each equation.

20. $\sqrt{5x+1} = x+1$

20._____

21. $\sqrt{x+3} = x-3$

21._____

22. $\sqrt{x-5}+5 = x-2$

22._____

23. $\sqrt{x+11}-1 = x-2$

23._____

24. $t+1 = \sqrt{t^2+4}-1$

24._____

25. $q-1 = \sqrt{q^2-4q+7}$

25._____

26. $5\sqrt{p} - 4 = p + 2$ **26.**_____

27. $4\sqrt{a} - 1 = a + 2$ **27.**_____

28. $\sqrt{x} + 2 = x - 4$ **28.**_____

29. $2\sqrt{c + 2} = c + 3$ **29.**_____

30. $x - 1 = \sqrt{x^2 + 8x + 11}$ **30.**_____

Chapter 15 QUADRATIC EQUATIONS

15.1   Solving Equations by the Square Root Property

Solve each equation by using the square root property. Express all radicals in simplest form.

1. $x^2 = 49$

1._____

2. $y^2 = 121$

2._____

3. $r^2 = 900$

3._____

4. $s^2 = 81$

4._____

5. $a^2 = 24$

5._____

6. $b^2 = -4$

6._____

7. $k^2 = 45$

7._____

8. $q^2 = 50$

8._____

9. $x^2 = 72$

9._____

10. $w^2 = 98$

10._____

11. $c^2 + 36 = 0$

11._____

12. $d^2 - 250 = 0$ 12._____

13. $x^2 = \frac{9}{25}$ 13._____

14. $x^2 = -\frac{36}{16}$ 14._____

15. $t^2 - 12.25 = 0$ 15._____

Solve each equation by using the square root property. Express all radicals in simplest form.

16. $(x+4)^2 = 25$ 16._____

17. $(x-3)^2 = 81$ 17._____

18. $(y+2)^2 = 16$ 18._____

19. $(y-9)^2 = 49$ 19._____

20. $(x-7)^2 = 0$ 20._____

21. $(y+8)^2 = 0$ 21._____

22. $(m+7)^2 = 4$ 22._____

23. $(n+3)^2 = 18$ **23.** _____

24. $(p-9)^2 = 28$ **24.** _____

25. $(q-2)^2 = 27$ **25.** _____

26. $(r+7)^2 = -25$ **26.** _____

27. $(2x+5)^2 = 32$ **27.** _____

28. $(3m+2)^2 = 27$ **28.** _____

29. $(2p+9)^2 - 4 = 0$ **29.** _____

30. $\left(\frac{1}{4}x+4\right)^2 = 16$ **30.** _____

Chapter 15 QUADRATIC EQUATIONS

15.2   Solving Quadratic Equations by Completing the Square

Solve each equation by completing the square.

1. $x^2 - 2x = 15$

1._____

2. $x^2 - 5x = 14$

2._____

3. $p^2 + 6p = 0$

3._____

4. $m^2 - 6m = -12$

4._____

5. $x^2 - 10x = -21$

5._____

6. $x^2 + 4x - 2 = 0$

6._____

7. $b^2 + 5b - 5 = 0$

7._____

8. $x^2 - 8x + 16 = 0$

8._____

9. $z^2 + 3z - \frac{7}{4} = 0$

9._____

Solve each equation by completing the square.

10. $4x^2 + 8x = 0$

10._____

11. $3m^2 - 15m = 42$

11._____

12. $2x^2 - 13x + 20 = 0$

12._____

13. $3z^2 + 3z - 4 = 0$

13._____

14. $6a^2 - 4a = -5$

14._____

15. $2x^2 + 13x - 7 = 0$

15._____

16. $-r^2 + 3r = -1$

16._____

17. $3r^2 - 6r - 2 = 0$

17._____

18. $3x^2 - 2x + 4 = 0$

18._____

Simplify each of the following equations and then solve by completing the square.

19. $3p^2 = 3p + 5$

19._____

20. $2x - 4 = x^2 - 2x$

20._____

21. $4y^2 + 6y = 2y + 3$

21._____

22. $(x-2)(x+3) = 10$

22._____

23. $(c+3)(c+7) = 5$

23._____

24. $(x-1)(x+2) = 4$

24._____

Solve each problem. Round answers to the nearest tenth if necessary.

25. A certain projectile is located at a distance of

$$d = 3t^2 - 6t + 1$$

feet from its starting point after t seconds. How many seconds will it take the projectile to travel 10 feet?

25._____

26. The time t in seconds for a car to skid 48 feet is given (approximately) by

$$48 = 64t - 16t^2$$

Solve this equation for t. Are both answers reasonable?

26._____

27. The amount A that P dollars invested at a rate of interest r will amount to in two years is

$$A = P(1+r)^2.$$

At what interest rate will $100 grow to $110.25 in two years?

27._____

28. If Dipti throws a ball into the air from ground level with an initial velocity of 32 feet per second, its height s (in feet) after t seconds is given by the formula

$$s = -16t^2 + 32t.$$

After how many seconds will the ball return to the ground?

28._____

29. If Pablo throws an object upward from a height of 32   **29.**_____
feet with an initial velocity of 48 feet per second, then its
height h (in feet) after t seconds is given by the formula

$$h = -16t^2 + 48t + 32$$

At what times will it be 50 feet above the ground?

30. The commodities market is very unstable; money can be   **30.**_____
made or lost quickly on investments in soybeans, wheat,
pork bellies, and so on. Suppose that an investor kept
track of his total profit, P (in thousands of dollars), at
time t (in months), after he began investing, and found
that his profit was given by the formula

$$P = 4t^2 - 24t + 32$$

Find the times at which he broke even on his investment.

Chapter 15 QUADRATIC EQUATIONS

15.3   Solving Quadratic Equations by the Quadratic Formula

For each of the following quadratic equations, write in the form $ax^2 + bx + c = 0$*, if necessary, and then identify the values of a, b, and c. Do not actually solve the equation.*

1. $3x^2 - 4x + 2 = 0$ 1._____

2. $5x^2 - 2x + 2 = 0$ 2._____

3. $2x^2 + 7x = 0$ 3._____

4. $3q^2 + 4q - 5 = 0$ 4._____

5. $10x^2 = -4x$ 5._____

6. $3p^2 = 12$ 6._____

7. $x(x-3) = 5$ 7._____

8. $(z-2)(z+2) = 9z$ 8._____

Use the quadratic formula to solve each equation.

9. $x^2 + 2x - 4 = 0$ 9._____

10. $3x^2 - 7x - 6 = 0$

10._____

11. $n^2 + 4n - 5 = 0$

11._____

12. $y^2 = -29$

12._____

13. $x^2 - 3x + 1 = -8$

13._____

14. $-6x^2 - 12x = 0$

14._____

15. $g^2 = 2g + 4$

15._____

16. $(3a - 2)^2 = 5$

16._____

Use the quadratic formula to solve each equation.

17. $x^2 + 6x + 9 = 0$

17._____

18. $y^2 + 12y + 36 = 0$

18._____

19. $16a^2 - 8a + 1 = 0$

19._____

20. $x^2 + 9 = -6x$

20._____

21. $n^2 + 81 = 18n$

21._____

22. $9r^2 = 6r - 1$

22._____

23. $25x^2 - 80x + 64 = 0$

23._____

Use the quadratic formula to solve each equation.

24. $\dfrac{1}{2}x^2 + 2x - 3 = 0$

24._____

25. $\dfrac{1}{6}y^2 + \dfrac{1}{2}y = \dfrac{2}{3}$

25._____

26. $\dfrac{1}{4}q^2 - \dfrac{1}{3}q - \dfrac{2}{3} = 0$

26._____

27. $\dfrac{1}{4}t^2 - \dfrac{1}{3}t + \dfrac{5}{12} = 0$

27._____

28. $r^2 + \dfrac{4}{3}r - \dfrac{1}{3} = 0$

28._____

29. $\dfrac{1}{3}y^2 + \dfrac{2}{3}y = \dfrac{5}{2}$

29._____

30. $\dfrac{1}{3}x^2 = \dfrac{2}{15}x + \dfrac{2}{5}$

30._____

Chapter 15 QUADRATIC EQUATIONS

15.4   Graphing Quadratic Equations and Inequalities

Sketch the graph of each equation. Give the coordinates of the vertex in each case.

1. $y = -x^2$

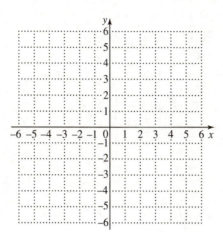

1._____

2. $y = -x^2 - 1$

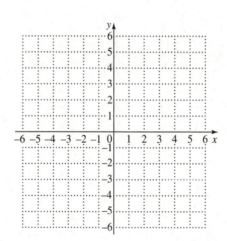

2._____

3. $y = x^2 - 3$

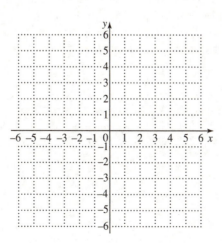

3._____

4. $y = \frac{1}{3}x^2$

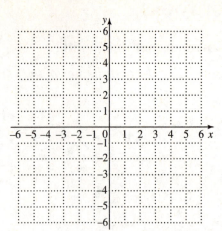

4._____

5. $y = x^2 + 2$

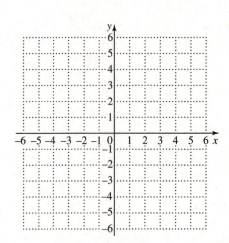

5._____

6. $y = 2x^2 - 4$

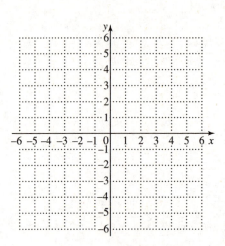

6._____

7. $y = 1 - 2x^2$

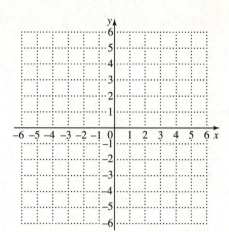

7. _____

8. $y = 9 - x^2$

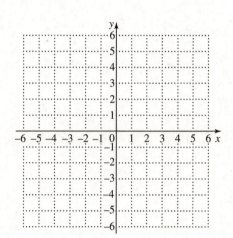

8. _____

9. $y = (x - 2)^2$

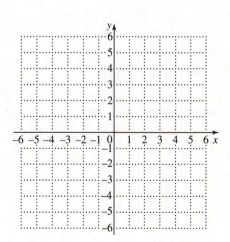

9. _____

10. $y = (x+4)^2$

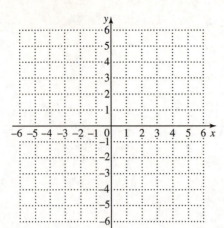

10. _____

11. $y = (x-3)^2$

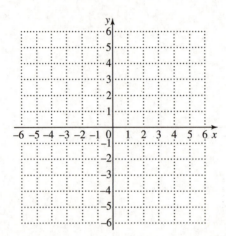

11. _____

12. $y = -(x+1)^2$

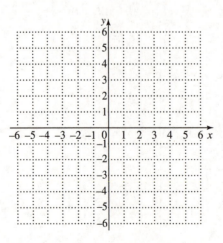

12. _____

13. $y = 2x^2 + 4x$

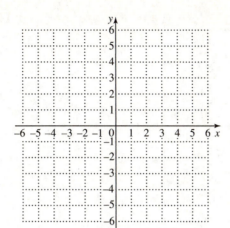

13. _____

14. $y = x^2 + x - 2$

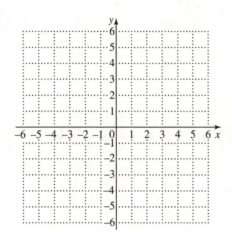

14. _____

15. $y = x^2 - 6x + 11$

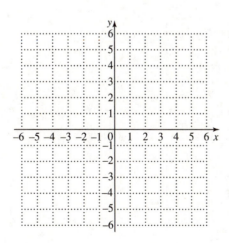

15. _____

16. $y = -x^2 + 6x - 9$

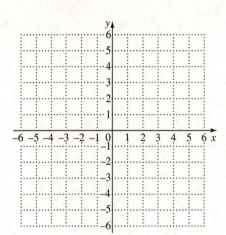

16._____

17. $y = -x^2 + 4x - 1$

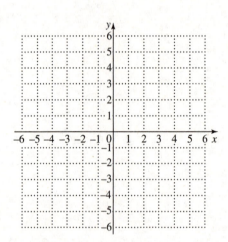

17._____

18. $y = x^2 + 8x + 14$

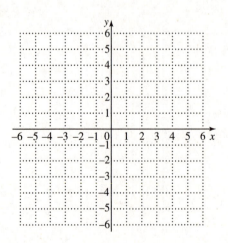

18._____

19. $y = -x^2 + 6x - 13$

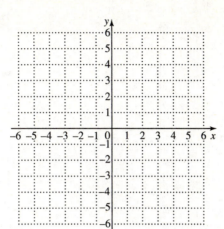

19. _____

20. $y = x^2 + 2x - 2$

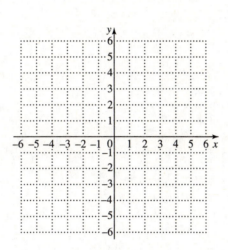

20. _____

21. $y = -x^2 + 5x$

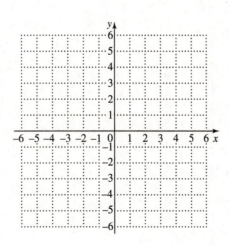

21. _____

22. $y = -x^2 - 3x + 1$

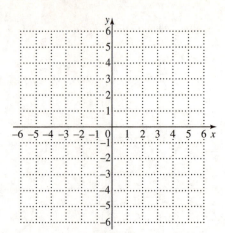

22. _____

Graph each quadratic inequality.

23. $y < x^2 - 2$

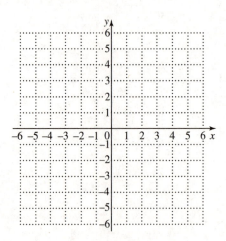

23. _____

24. $y \geq (x-3)^2$

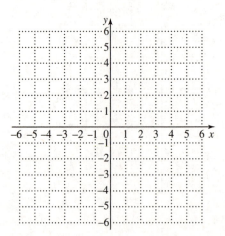

24. _____

25. $y \le x^2 - 5x$

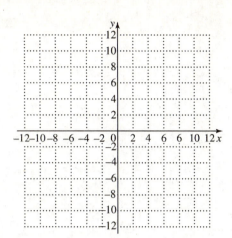

25. _____

26. $y < x^2 + 4x - 5$

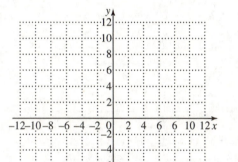

26. _____

27. $y \ge -x^2 + 3$

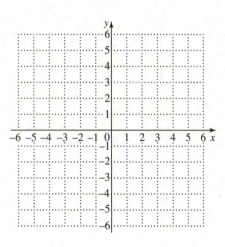

27. _____

28. $y > -x^2 - 3x + 10$

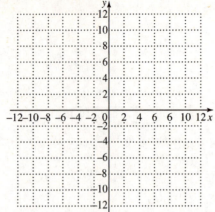

28. _____

29. $y < \frac{1}{2}(x+2)^2$

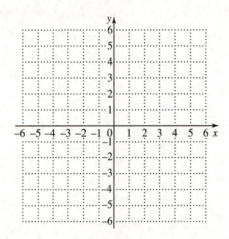

29. _____

30. $y \geq x^2 - x - 12$

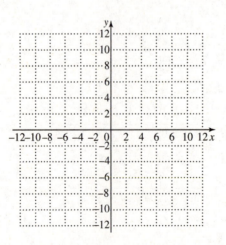

30. _____

Chapter 15 QUADRATIC EQUATIONS

15.5   Introduction to Functions

Decide whether the given relation is a function and give the domain and range of each.

1. $\{(2,7),(5,-4),(-3,-1),(0,-8),(5,2)\}$

1._____

2. $\{(1,3),(5,7),(11,9),(8,-2),(6,-7),(-4,-3)\}$

2._____

3. $\{(0,1),(2,6),(-3,7),(2,9),(-7,1),(-3,4)\}$

3._____

4. $\{(3,5),(3,8),(3,-4),(3,1),(3,0)\}$

4._____

5. $\{(1,4),(3,4),(7,4),(-2,4),(-5,4)\}$

5._____

6. $\{(-3,5),(-2,5),(-1,0),(0,-5),(1,5)\}$

6._____

7. $\{(-1.2,4),(1.8,-2.5),(3.7,-3.8),(3.7,3.8)\}$

7._____

Decide whether or not the relations graphed or defined are functions.

8. $y = 3x + 2$

8._____

9. $y = 7$

9._____

10. $x = -4$

10._____

11. $3x + 2y = 8$

11._____

12. $x^2 + y^2 = 4$

12._____

13. $x = y^2 - 4$

13._____

14. $y = x$

14._____

15. $y = \sqrt{x - 5}$

15._____

16.

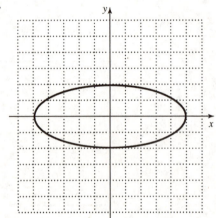

16._____

17.

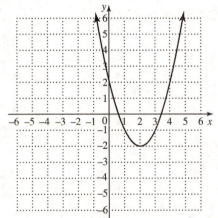

17._____

18.

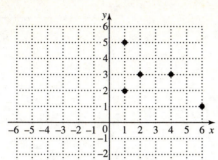

18. _____

For each function f, find (a) $f(-2)$, (b) $f(0)$, and (c) $f(4)$.

19. $f(x) = 3x - 7$

19. _____

20. $f(x) = x^2 - 3x + 2$

20. _____

21. $f(x) = 2x^2 + x - 5$

21. _____

22. $f(x) = |2x + 3|$

22. _____

23. $f(x) = x^3 - 2x^2 + 4$

23. _____

24. $f(x) = 9$

24. _____

25. $f(x) = 3x^2 - 2x + 1$ **25.**_____

For each function f, find (a) $f(-5)$, (b) $f\left(\frac{1}{2}\right)$, and (c) $f(3.2)$.

26. $f(x) = 2x - 7$ **26.**_____

27. $f(x) = -x^2 + 3x - 4$ **27.**_____

28. $f(x) = \sqrt{x} + 1$ **28.**_____

29. $f(x) = x^2 - 5x - 3$ **29.**_____

30. $f(x) = 2|x + 4| - 1$ **30.**_____

Chapter 1

WHOLE NUMBERS

1.1 Reading and Writing Whole Numbers

1. Whole number
3. Whole number
5. 9, 4

7. 8, 1
9. 2, 0
11. 75, 229, 301

13. 300, 459, 200, 5
15. Thirty-nine thousand, fifteen

17. Two million, fifteen thousand, one hundred two

19. 4127
21. 685,000,259
23. 7210

25. 15,313
27. 177 calories
29. 315 calories

1.2 Adding Whole Numbers

1. 12
3. 10
5. 25
7. 31
9. 99

11. 98,977
13. 112
15. 15,815
17. 38
19. 44

21. 625 tickets
23. 310 feet
25. 1044 yards

27. Incorrect; 17,280
29. Correct

1.3 Subtracting Whole Numbers

1. $15 - 6 = 9$, $15 - 9 = 6$
3. $187 - 149 = 38$, $187 - 38 = 149$

5. $785 + 426 = 1211$
7. $2196 + 3721 = 5917$

9. 5, 3, 2
11. 98, 36, 62
13. 31.

15. 5151
17. Incorrect; 153
19. Incorrect; 2980

21. Incorrect; 78,087
23. 39
25. 1918

27. 25 boxes
29. $180

1.4 Multiplying Whole Numbers

1. Factors: 5, 3; product: 15   3. Factors: 13, 3; product: 39

5. 32   7. 192   9. 40   11. 216

13. 2835   15. 820   17. 8400   19. 7,740,000

21. 966   23. 12,236   25. 492,119   27. 560 yd

29. $384

1.5 Dividing Whole Numbers

1. $\dfrac{15}{3} = 5, \; 3\overset{5}{\overline{)15}}$ 3. $\dfrac{32}{16} = 2, \; 32 \div 16 = 2$ 5. 30, 5, 6

7. 0   9. 0   11. Undefined   13. 1   15. 1

17. 128   19. 12   21. 72 R3   23. Incorrect; 27

25. Correct   27. Incorrect; 3389 R7   29. Divisible by 3; not 2, 5, 10

1.6 Long Division

1. 82   3. 77   5. 85 R84   7. 98,532 R6

9. 654 R22   11. 9   13. 7   15. 16   17. 70

19. 84   21. Correct   23. Correct   25. Incorrect; 45 R23

27. Incorrect; 296 R79   29. Correct

1.7 Rounding Whole Numbers

1. 8<u>5</u>3
3. 6<u>4</u>5,371
5. 257,3<u>0</u>1
7. 7900

9. 810
11. 53,600
13. 16,700

15. $40 + 20 + 60 + 90 = 210; 210$
17. $70 - 40 = 30; 27$

19. $300 + 300 + 200 + 900 = 1700; 1698$
21. $1000 - 400 = 600; 589$

23. $900 \times 800 = 720,000; 715,008$
25. $600 + 40 + 200 + 2000 = 2840; 3280$

27. $900 - 40 = 860; 833$
29. $1000 \times 40 = 40,000; 36,260$

1.8 Exponents, Roots, and Order of Operations

1. 2, 7; 49
3. 2, 9; 81
5. 5, 1; 1
7. 7, 2; 128

9. 3
11. 11
13. 10
15. 324; 324

17. 2500; 2500
19. 225; 225
21. 2704; 2704
23. Undefined

25. 54
27. 45
29. 28

1.9 Reading Pictographs, Bar Graphs, and Line Graphs

1. Georgia
3. Minnesota
5. $1\frac{1}{2}$
7. 3

9. Spanish
11. 50 pints
13. Department 4

15. Department 1 and Department 5
17. 2900
19. Freshmen

21. The nets dales are increasing every year.
23. 2000
25. 1997 and 1998

27. 1998
29. $1.5 million

1.10   Solving Application Problems

1. +   3. ×   5. ×   7. ÷

9. +   11. 4 toys   13. $3008   15. 828 salmon

17. $352   19. $600 \div 60 = 10$ hr; 11 hr

21. $50,000 + 500 = \$50,500$; $47,541   23. $2000 - 1000 = \$1000$; $949

25. $200,000 \div 30 = 6667$ people;   27. $10 + 200 + 400 = 610$ deer; 589 deer
    6850 people

29. $3000 - 300 - 700 - 200 = \$1800$; $1854

Chapter 2

FRACTIONS AND MIXED NUMBERS

2.1 Basics of Fractions

1. $\frac{3}{8}$ 3. $\frac{5}{6}$ 5. $\frac{1}{3}$ 7. $\frac{5}{3}$ 9. $\frac{5}{8}$

11. N: 4; D: 3   13. N: 2; D: 5   15. N: 8; D: 11   17. N: 112; D: 5

19. N: 7; D: 15   21. Improper   23. Proper   25. Proper

27. Improper   29. Improper

2.2 Mixed Numbers

1. $2\frac{1}{2}, 1\frac{1}{6}$ 3. None   5. $\frac{23}{8}$ 7. $\frac{14}{5}$

9. $\frac{7}{4}$ 11. $\frac{14}{3}$ 13. $\frac{29}{11}$ 15. $\frac{20}{3}$

17. $\frac{34}{3}$ 19. $1\frac{3}{5}$ 21. $3\frac{3}{10}$ 23. $2\frac{6}{7}$

25. $3\frac{5}{7}$ 27. $4\frac{5}{9}$ 29. $7\frac{1}{4}$

2.3 Factors

1. 1, 7   3. 1, 7, 49   5. 1, 2, 5, 10   7. 1, 5, 25

9. 1, 2, 3, 6, 9, 18   11. Neither   13. Prime   15. Composite

17. Prime   19. Prime   21. $2^2 \cdot 3$ 23. $3 \cdot 5$

25. $2^2 \cdot 7$ 27. 2^5 29. $3^2 \cdot 7$

2.4   Writing a Fraction in Lowest Terms

1. No   3. No   5. No   7. Yes   9. $\frac{1}{4}$

11. $\frac{2}{7}$ 13. $\frac{5}{6}$ 15. $\frac{2}{7}$ 17. $\frac{1}{2}$ 19. $\frac{4}{5}$

21. $\frac{1}{2}$ 23. Not equivalent   25. Not equivalent

27. Equivalent   29. Not equivalent

2.5   Multiplying Fractions

1. $\frac{5}{24}$ 3. $\frac{2}{15}$ 5. $\frac{55}{24}$ 7. $\frac{7}{40}$ 9. $\frac{2}{3}$

11. $\frac{3}{16}$ 13. $\frac{5}{12}$ 15. $\frac{2}{3}$ 17. $\frac{2}{21}$ 19. $\frac{1}{4}$

21. $\frac{1}{6}$ 23. 42   25. 4   27. $3\frac{1}{2}$ 29. $\frac{7}{8}$

2.6   Applications of Multiplication

1. 1680 paperbacks   3. $1500   5. 133 employees

7. $500   9. $104   11. 320 muffins

13. 160 pages   15. 90 games   17. 1688 votes

19. 312 sq ft   21. 6 gallons   23. $12,000

25. $2250   27. $16,000   29. $36,000

2.7   Dividing Fractions

1. $\frac{4}{3}$ 3. 3   5. $\frac{1}{10}$ 7. $\frac{5}{8}$ 9. $2\frac{2}{15}$ 11. $\frac{3}{4}$

13. $3\frac{1}{3}$ 15. $37\frac{1}{2}$ 17. $\frac{1}{33}$ 19. $2\frac{1}{3}$ 21. 75

23. 54 dresses   25. 24 Brownies   27. 32 guests   29. 24 patties

2.8 Multiplying and Dividing Mixed Numbers

1. $5 \cdot 3 = 15$; $13\frac{1}{3}$ 3. $5 \cdot 3 = 15$; $16\frac{4}{5}$ 5. $4 \cdot 2 = 8$; $10\frac{2}{3}$

7. $2 \cdot 4 = 8$; $8\frac{1}{8}$ 9. $6 \cdot 4 = 24$; 21 11. $1 \cdot 3 \cdot 2 = 6$; 5

13. $6 \div 5 = 1\frac{1}{5}$; $1\frac{1}{9}$ 15. $5 \div 1 = 5$; $3\frac{7}{10}$ 17. $6 \div 1 = 6$; $4\frac{4}{5}$

19. $5 \div 4 = 1\frac{1}{4}$; $1\frac{1}{3}$ 21. $7 \div 6 = 1\frac{1}{6}$; $1\frac{2}{9}$ 23. $8 \div 1 = 8$; $11\frac{1}{4}$

25. $20 \cdot 3 = 60$ yards; 65 yards   27. $41 \cdot 6 = 246$ yards; 248 yards

29. $29 \cdot 2 = 58$ ounces; $50\frac{2}{5}$ ounces

2.9 Adding and Subtracting Like Fractions

1. Like   3. Like   5. Unlike   7. 1   9. $\frac{4}{5}$

11. $1\frac{1}{8}$ 13. 1   15. $1\frac{3}{10}$ 17. $\frac{1}{2}$ mile   19. $\frac{8}{13}$

21. 1   23. $\frac{2}{3}$ 25. $\frac{4}{5}$ 27. $\frac{5}{9}$ 29. $\frac{2}{5}$

2.10 Least Common Multiples

1. 14   3. 84   5. 200   7. 80   9. 70

11. 28   13. 480   15. 420   17. 110   19. 108

21. 140   23. 4   25. 36   27. 33   29. 35

2.11 Adding and Subtracting Unlike Fractions

1. $\frac{5}{6}$ 3. $\frac{33}{40}$ 5. $\frac{23}{30}$ 7. $\frac{1}{3}$

9. $\frac{37}{45}$ 11. $\frac{19}{24}$ 13. $\frac{5}{6}$ 15. $\frac{5}{6}$

17. $\frac{13}{21}$ 19. $\frac{3}{4}$ 21. $\frac{3}{8}$ 23. $\frac{1}{3}$

25. $\frac{5}{24}$ 27. $\frac{1}{15}$ 30. $\frac{11}{24}$

2.12   Adding and Subtracting Mixed Numbers

1. $5+4=9$; $9\frac{4}{7}$

3. $18+12+6=36$; $35\frac{17}{24}$

5. $6-2=4$; $3\frac{1}{2}$

7. $9+2=11$ gallons; $11\frac{1}{12}$ gallons

9. $13-6=7$ hours; $6\frac{3}{8}$ hours

11. $5+9+3+1=18$ tons; $17\frac{1}{3}$ tons

13. $6-6=0$; $\frac{3}{4}$

15. $12-12=0$; $\frac{35}{48}$

17. $42-20=22$; $22\frac{1}{4}$

19. $146-35-43-33=35$ yards; $34\frac{5}{8}$ yards

21. $15-3-5-4=3$ yards; $3\frac{1}{4}$ yards

23. $15\frac{1}{20}$

25. $8\frac{1}{6}$

27. $2\frac{3}{8}$

29. $2\frac{7}{8}$

2.13   Order Relations and the Order of Operations

1. $<$

3. $>$

5. $<$

7. $<$

9. $>$

11. $\frac{1}{4}$

13. $\frac{1}{25}$

15. $\frac{1}{125}$

17. $5\frac{1}{16}$

19. $\frac{27}{64}$

21. $\frac{4}{15}$

23. $\frac{1}{36}$

25. $1\frac{5}{12}$

27. $\frac{5}{28}$

29. $\frac{1}{2}$

Chapter 3

DECIMALS

3.1 Reading and Writing Decimals

1. $\frac{5}{10}$; 0.5; five tenths

3. $\frac{35}{100}$; 0.35; thirty-five hundredths

5. $\frac{47}{100}$; 0.47; forty-seven hundredths

7. 6, 3

9. 7, 1

11. 3, 6

13. Tenths, hundredths

15. Tenths, hundredths thousandths

17. Four and six hundredths

19. Two and three hundred four thousandths

21. 5.04

23. 38.00052

25. $\frac{4}{5}$

27. $3\frac{3}{5}$

29. $\frac{19}{20}$

3.2 Rounding Decimals

1. 17.9

3. 785.498

5. 53.33

7. 486

9. 21.77; 21.8

11. 1.44; 1.4

13. 114.04; 114.0

15. 78.70; 78.7

17. $79

19. $226

21. $11,840

23. $276

25. $1.09

27. $134.21

29. $2096.01

3.3 Adding and Subtracting Decimals

1. 92.49

3. 178

5. 48.35

7. 123.6802 inches

9. 1186.1162 feet

11. 115.8

13. 16.124

15. 24.016 feet

17. 0.862 inch

19. $30 + 40 + 8 = 78$; 82.91

21. $10 - 5 = 5$; 4.838

23. $9 - 3 = 6$; 5.53

25. $400 + 90 = \$490$; 452.88

27. $20 - 10 = \$10$; $8.71

29. $9 + 10 + 30 = \$49$; $47.92

3.4   Multiplying Decimals

1. 0.2279   3. 90.71   5. 1.5548   7. 0.0037

9. 0.00000963   11. $163.08   13. $200.20   15. $9.44

17. $20.64   19. $50 \times 6 = 300$; 288.26   21. $60 \times 4 = 240$; 218.4756

23. $80 \times 1 = 80$; 43.548   25. $400 \times 8 = 3200$;   27. $494.40

                                       3033.306

29. $2105.99

3.5   Dividing Decimals

1. 14.65   3. 7.994   5. 20.9   7. $1.99 per sock

9. $3.16 per book   11. 129.467   13. 17.660   15. 67.231

17. 33.3 miles per gallon   19. $0.32 per brick   21. Reasonable

23. Reasonable   25. Unreasonable   27. 54.02   29. 7.91

3.6   Writing Fractions and Decimals

1. 6.5   3. 2.667   5. 0.091   7. 0.6

9. 4.111   11. 0.15   13. 0.611   15. $<$

17. $>$ 19. $>$ 21. $>$ 23. $>$

25. $\frac{3}{11}$, 0.29, $\frac{1}{3}$ 27. 0.166, 0.1666, $\frac{1}{6}$ 29. $\frac{1}{7}$, 0.187, $\frac{3}{16}$

Chapter 4

RATIO, PROPORTION, AND PERCENT

4.1 Ratios

1. $\dfrac{7}{8}$ 3. $\dfrac{76}{101}$ 5. $\dfrac{19}{25}$ 7. $\dfrac{17}{27}$ 9. $\dfrac{3}{64}$

11. $\dfrac{13}{4}$ 13. $\dfrac{6}{5}$ 15. $\dfrac{5}{6}$ 17. $\dfrac{3}{4}$ 19. $\dfrac{35}{1}$

21. $\dfrac{3}{2}$ 23. $\dfrac{16}{5}$ 25. $\dfrac{7}{2}$ 27. $\dfrac{4}{15}$ 29. $\dfrac{2}{1}$

4.2 Rates

1. $\dfrac{3 \text{ miles}}{1 \text{ minute}}$ 3. $\dfrac{7 \text{ dresses}}{1 \text{ person}}$ 5. $\dfrac{15 \text{ gallons}}{1 \text{ hour}}$ 7. $\dfrac{7 \text{ pills}}{1 \text{ patient}}$

9. $\dfrac{32 \text{ pages}}{1 \text{ chapter}}$ 11. \$15/hour   13. \$110/day   15. \$13.64/hour

17. 27.3 miles/gallons   19. \$8.75/hour   21. $\frac{1}{2}$ acre/hour; 2 hour/acre

23. \$5.90/yard   25. \$11.50/share   27. 16 ounces for \$0.89

29. 5 cans for \$2.75

4.3 Proportions

1. $\frac{11}{15} = \frac{22}{30}$ 3. $\frac{24}{30} = \frac{8}{10}$ 5. $\frac{14}{21} = \frac{10}{15}$ 7. $\frac{26}{4} = \frac{39}{6}$ 9. $\dfrac{1\frac{1}{2}}{4} = \frac{21}{56}$

11. True   13. False   15. False   17. True   19. False

21. True   23. False   25. True   27. False   29. True

4.4 Solving Proportions

1. 9	3. 36	5. 21	7. 5
9. 45	11. 77	13. 18	15. 55
17. $11\frac{2}{3}$	19. 13	21. $4\frac{1}{2}$	23. 0
25. 21	27. 14	29. $\frac{1}{15}$	

4.5 Solving Application Problems with Proportions

1. $112.50	3. $15	5. 8 pounds	7. $818.40
9. 960 miles	11. $440	13. $4399.50	15. $122.50
17. $22.50	19. $2160	21. 480 minutes	23. $62\frac{1}{2}$ minutes
25. 76.8 inches	27. $2.56	29. about 11 days	

4.6 Basics of Percent

1. 43%	3. 29%	5. 38%	7. 0.42
9. 0.09	11. 0.325	13. 30%	15. 20%
17. 42%	19. 98.6%	21. $19	23. $228
25. 87 days	27. 492 televisions	29. 125 signs	

4.7 Percents and Fractions

1. $\frac{7}{20}$	3. $\frac{14}{25}$	5. $\frac{5}{8}$	7. $\frac{49}{300}$	9. $\frac{1}{15}$
11. 70%	13. 81%	15. 85.3%	17. 94%	
19. 380%	21. 0.5, 50%	23. 0.25, 25%	25. $\frac{7}{8}$, 0.875	
27. $\frac{1}{3}$, 0.333	29. $\frac{13}{40}$, 32.5%			

4.8 Using the Percent Proportion and identifying the Components in a Percent Problem

1. whole = 120	3. whole = 17.5	5. part = 12	7. part = 3.5
9. percent = 22	11. 71%	13. Unknown	15. 42%

17. 48   19. 78   21. 60   23. Unknown

25. 29   27. Unknown   29. Unknown

4.9 Using Proportions to Solve Percent Problems

1. 146   3. 3.8   5. 118.2   7. 66 children

9. $210   11. 205.6   13. 400   15. 70

17. 75 members   19. 2325 customers   21. 26%   23. 3%

25. 0.2%   27. 2.5%   29. 22%

4.10 Using the Percent Equation

1. 192   3. 845   5. 83.2   7. 21.6   9. 14 clients

11. 400   13. 1500   15. 2160   17. 600   19. 640 gallons

21. 50%   23. 52%   25. 35%   27. 32%   29. 25%

4.11 Solving Application Problems with Percent

1. $3; $103   3. $3.50; $53.50   5. $0.30; $15.30

7. $6.03; $73.03   9. $931.25   11. $110

13. $1024   15. $3000   17. $3750

19. $30; $170   21. $15.20; $22.80   23. $6.75; $15.75

25. $14.97; $9.98   27. 25%   29. 44.0%

4.12 Simple Interest

1. $20   3. $144   5. $4   7. $270   9. $8

11. $41.25   13. $36.90   15. $19.50   17. $95.20   19. $1170

21. $3075   23. $1224   25. $5145.83   27. $6169.20   29. $9240

4.13 Compound Interest

1. $8103.38   3. $1273.45   5. $4867.20

7. $20,145.60   9. $1262.50   11. $6205.20

Answers to Worksheets

13. $66.78

15. $47.24

17. $10,976.01

19. $1302.30

21. $1628.90; $628.90

23. $17,103.70; $8603.70

25. $16,935.74; $7785.74

27. $34,730.06; $13,330.06

29. $14,282.10; $5282.10

Chapter 5

GEOMETRY

5.1   Lines and Angles

1. ray named $\overrightarrow{ST}$ 3. intersecting   5. parallel   7. ∠PRT or ∠TRP

9. obtuse   11. intersecting   13. ∠AOR and ∠ROM;   15. ∠WMT and ∠TMV;
∠MOT and TOW   ∠TMV and ∠VMN;
∠VMN and ∠NMW;
∠NMW and ∠WMT

17. 52°   19. 79°   21. 168°   23. 20°

25. ∠LOP ∠MOQ; ∠POQ ∠LOM   27. ∠1, ∠4, ∠5, ∠8 all measure 100°;
∠2, ∠3, ∠6, ∠7 all measure 80°

29. ∠3, ∠4, ∠5, ∠6 all measure 125°;
∠1, ∠2, ∠7, ∠8 all measure 55°

5.2   Rectangles and Squares

1. 24 cm; 32 cm^2 3. 36 cm; 17 cm^2 5. 22 yd; $29\frac{1}{4}$ yd^2

7. 281.2 cm; 3859.68 cm^2 9. 7248 ft^2 11. 36 m; 81 m^2

13. 31.2 ft; 60.84 ft^2 15. $5\frac{3}{5}$ in; $1\frac{24}{25}$ in^2 17. 12.4 cm; 9.61 cm^2

19. $18\frac{2}{3}$ mi; $21\frac{7}{9}$ mi^2 21. 30 ft; 18 ft^2 23. 30 m; 30 m^2

25. 32 in; 28 in^2 27. 76 mm; 192 mm^2 29. 42 yd; 54 yd^2

5.3   Parallelograms and Trapezoids

1. 168 m   3. 30 ft   5. 26.0 in   7. 713 yd^2

9. $11\frac{1}{4}$ m^2 11. 27 in^2 13. 14.72 m^2 15. $587.52

17. 159 in   19. $53\frac{3}{4}$ yd   21. 1943.7 cm^2 23. $43\frac{3}{4}$ ft^2

25. 32.5 cm^2 27. 3515.4 cm^2 29. $700

5.4   Triangles

1. 25 yd   3. 34 yd   5. 37.2 ft   7. $10\frac{3}{4}$ ft

9. 17.7 m   11. 1260 m^2 13. $21\frac{3}{4} \text{ ft}^2$ 15. 15.81 m^2

17. 42 ft^2 19. 1008 m^2 21. 50° 23. 50°

25. 20° 27. 59° 30. 45°

5.5   Circles

1. 86 m   3. 13.25 m   5. 5.8 ft   7. 6.3 in   9. $4\frac{7}{8}$ in

11. 69.1 ft   13. 94.2 m   15. 106.8 ft   17. 62.8 cm

19. 4.7 mi   21. 78.5 in^2 23. 1519.8 yd^2 25. 22.3 yd^2

27. 57 cm^2 29. 2101.3 m^2

5.6   Volume and Surface Area

1. 468 cm^3 3. 400 m^3 5. 95.2 cm^3 7. 4.2 m^3

9. 16.7 cm^3 11. 2.6 cm^3 13. 471 ft^3 15. 141.3 ft^3

17. 1433.5 in^3 19. 121.3 m^3 21. 6 km^3 23. $22{,}344 \text{ m}^3$

25. $SA = 558 \text{ in.}^2$ 27. $SA \approx 1507.2 \text{ in.}^2$ 29. $V \approx 13{,}564.8 \text{ in.}^3$; $SA \approx 3165.1 \text{ in.}^2$

5.7   Pythagorean Theorem

1. 4.123   3. 7.874   5. 4.899   7. 1.414

9. 3.606   11. 2.646 in   13. 9.849 in   15. 2.236 cm

17. 10.198 m   19. 13.229 km   21. 8.062 ft   23. 18.028 mi

25. 8 ft   27. 467.039 cm   29. 7.616 cm

5.8   Congruent and Similar Triangles

1. $m\angle 1 \cong m\angle 4$; $m\angle 2 \cong m\angle 5$, $m\angle 3 \cong m\angle 6$; $AB \cong DE$, $BC \cong EF$, $AC \cong DF$

3. $m\angle 1 \cong m\angle 5$; $m\angle 2 \cong m\angle 6$, $m\angle 3 \cong m\angle 4$; $DE \cong JG$, $EF \cong GH$, $DF \cong JH$

5. SSS   7. SAS   9. ASA   11. ASS

13. $\dfrac{3}{5}$; $\dfrac{3}{5}$; $\dfrac{3}{5}$ 15. $\dfrac{5}{12}$; $\dfrac{5}{12}$; $\dfrac{5}{12}$ 17. $a = 72$ in.; $b = 30$ in.

19. $a = 12$ ft; $b = 6$ ft   21. 115.2 yd; 89.6 yd

23. 24 in.; 60 in.   25. 54 in.; 67.5 in.   27. 21 ft   29. 42 m

Chapter 6

STATISTICS

6.1   Circle Graphs

1. 896 students
3. 608 students
5. 288 students
7. Liberal Arts

9. 3:1
11. $10,400
13. $\frac{3}{8}$
15. $\frac{3}{14}$

17. $\frac{10}{29}$
19. $\frac{5}{9}$
21. $237,500
23. $95,000

25. 15%, 54°
27. 15%
29. $160, 10%

6.2   Bar Graphs and Line Graphs

1. 1000
3. 1800
5. 2002
7. 300
9. 350

11. 10:7
13. 17:11
15. September
17. $20

19. 3:1
21. $30
23. $1,000,000
25. $2,500,000

27. $3,000,000
29. 2000

6.3   Frequency Distributions and Histograms

1. ‖; 2
3. |||| |; 6
5. ||||; 4
7. |; 1

9. |||| ‖; 7
11. |; 1
13. 80–89 and 120–129
15. |||| |||; 8

17. |; 1
19. |; 1
21. 150–169 and 190–209

23.

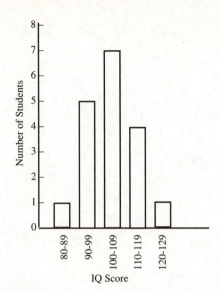

25. 16–20 27. 120 members 29. 30 members

6.4 Mean, Median, and Mode

1. 7.2 3. 60.3 5. 6.2 7. 27,955 9. 14.8

11. 40.2 13. 3.1 15. 2.5 17. 298 19. 29.4

21. 1.7 23. 57 25. 5 27. No mode 29. 24, 35, and 39

Chapter 7

THE REAL NUMBER SYSTEM

7.1 Exponents, Order of Operations, and Inequality

1. 27

3. $\frac{9}{16}$

5. 0.000027

7. 9

9. $\frac{7}{12}$

11. $\frac{29}{16}$

13. 52

15. 68

17. 176

19. False

21. False

23. $7 = 13 - 5$

25. $30 - 7 > 20$

27. $12 > 9$

29. $.19 < .21$

7.2 Variables, Expressions, and Equations

1. 5

3. 0

5. 0

7. $\frac{28}{13}$

9. $3 - x$

11. $10x + 21$ or $21 + 10x$

13. $\frac{2}{3}x - \frac{1}{2}x$

15. No

17. No

19. Yes

21. $x - 5 = 9$

23. $\frac{y}{9} = 17$

25. $6x = 18$

27. Equation

29. Expression

7.3 Real Numbers and the Number Line

1. -273; 103

3. -72

5. $-42,500$

7.

9.

11. -15

13. -2

15. True

17. False

19. 25

21. -4

23. 0

25. -10

27. $\frac{3}{4}$

29. $-|5|$

7.4 Adding Real Numbers

1. 15   3. –10   5. 2   7. $-\frac{1}{6}$ 9. 0

11. False   13. False   15. True   17. –23   19. –6

21. $\frac{3}{5}$ 23. $-2+16$; 14   25. $10+(-20+9)$; –1   27. 8 yd lost

29. 2556 ft

7.5 Subtracting Real Numbers

1. 5   3. –12   5. –6   7. 22   9. 4.4

11. $-\frac{1}{30}$ 13. 7   15. –6   17. $\frac{4}{5}$ 19. –2

21. $-6-(-2)$; –4   23. $-8-(-1-2)$; –5   25. $\left[4+(-7)\right]-(-6)$; 3

27. –50.2°C   29. 37°F

7.6 Multiplying and Dividing Real Numbers

1. –28   3. $-\frac{3}{13}$ 5. 40   7. $-\frac{1}{2}$ 9. –8

11. 6   13. –11   15. 70   17. 5   19. $\frac{37}{10}$

21. –3   23. $\frac{3}{4}$ 25. $-2+2\left[14+(-4)\right]$; 18

27. $\frac{(40)(-3)}{5-(-10)}$; -8 29. $x-(-7)=12$

7.7 Properties of Real Numbers

1. $y+4=4+y$ 3. $7m=m\cdot 7$

5. $2+\left[10+(-9)\right]=\left[10+(-9)\right]+2$ 7. $\left[-4+(-2)\right]+y=-4+(-2+y)$

9. $(-12x)(-y)=(-12)\left[x(-y)\right]$ 11. $\left[x+(-4)\right]+3y=x+\left[(-4)+3y\right]$

13. -7 15. 7 17. $-\frac{1}{7}+\frac{1}{7}=0$; inverse   19. $-\frac{3}{5}\cdot-\frac{5}{3}=1$; inverse

21. $0+0=0$; either property   23. $y(6+7)$; $13y$ 25. $4(r+p)$

27. $2an-4bn+6cn$ 29. $2k-7$

7.8   Simplifying Expressions

1. $6+3y$ 3. $-17+4b$ 5. $10x+3$ 7. -2

9. 125 11. $\frac{7}{9}$ 13. Like   15. Unlike

17. Like   19. $5.9r+9.7$ 21. $6x+10$ 23. $5q+30$

25. $-1.5y+16$ 27. $(6x+12)+4x$; $10x+12$

29. $(2-9x)-(10x+7)$; $-19x-5$

Chapter 8

EQUATIONS, INEQUALITIES, AND APPLICATIONS

8.1   The Addition Property of Equality

1. Yes   3. No   5. No   7. 20   9. −5

11. $\frac{3}{2}$ 13. −5   15. −12.8   17. −4   19. 7

21. −8   23. $\frac{5}{4}$ 25. 0   27. $\frac{1}{3}$ 29. 7.2

8.2   The Multiplication Property of Equality

1. 3   3. 3   5. −14   7. −9   9. $\frac{4}{3}$

11. 6   13. 5.8   15. 6.4   17. 9   19. 9

21. −8   23. 8   25. 26   27. −8   29. −8

8.3   More on Solving Linear Equations

1. $\frac{5}{2}$ 3. 10   5. $-\frac{1}{4}$ 7. 4   9. 1

11. $\frac{3}{5}$ 13. $\frac{53}{11}$ 15. 30   17. −3   19. No solution

21. No solution   23. No solution   25. No solution   27. No solution

29. $\dfrac{17}{p}$

8.4   An Introduction to Applications of Linear Equations

1. $4+3x=7$; 1   3. $-2(4-x)=24$; 16

5. $-3(x-4)=-5x+2$; −5   7.   27 inches

9. 52   11. $4.25   13. Mark 3; Pablo 14; Faustino 12

15. 133°   17. 66°   19. 27°   21. 76°   23. 55°

25. 76, 78   27. 27, 28   29. 13, 15

8.5   Formulas and Applications from Geometry

1. 24   3. 10   5. 7   7. 95   9. 36 ft

11. 12 ft   13. 12 m   15. 4 cm   17. $(3x+5)^o = 35^o$; $(6x-25)^o = 35^o$

19. $(6x)^o = 72^o$; $(10x-48)^o = 72^o$ 21. $(2x+16)^o = 48^o$; $(7x+20)^o = 132^o$

23. $H = \dfrac{V}{LW}$ 25. $h = \dfrac{2A}{b}$ 27. $A = \dfrac{P}{1-rt}$ 29. $F = \frac{9}{5}C + 32$

8.6   Solving Linear Inequalities

1.

3.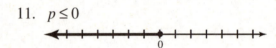

5.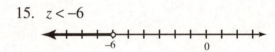

7. $j \le 5$

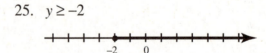

9. $y > -5$

11. $p \le 0$

13. $r > -4$

15. $z < -6$

17. $t \ge 7$

19. $y > 4$ 21. $z > -1$

23. $p < -1$

25. $y \ge -2$

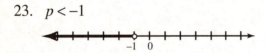

Chapter 9

GRAPHS OF LINEAR EQUATIONS AND INEQUALITIES IN TWO VARIABLES

9.1   Reading Graphs: Linear Equations in Two Variables

1. 1991 – 1992   3. 40   5. 280   7. $(-4, 0)$

9. $\left(0, \frac{1}{3}\right)$ 11. No   13. No

15. (a) $(-4, -5)$ (b) $(2, 7)$ (c) $\left(-\frac{3}{2}, 0\right)$ (d) $(-2, -1)$ (e) $(-5, -7)$

17.

x	2	$\frac{1}{2}$	1
y	-2	4	2

19.

x	0	3	-1
y	9	0	12

21.

x	y
0	$\frac{3}{2}$
-2	0
2	3

23 and 25.   27. IV   29. III

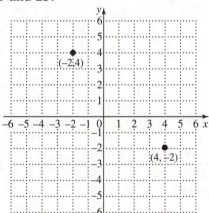

9.2 Graphing Linear Equations in Two Variables

1. $(0,3)$, $(3,0)$, $(2,1)$

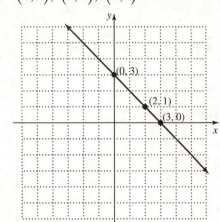

3. $(0,2)$, $(-4,0)$, $(-2,1)$

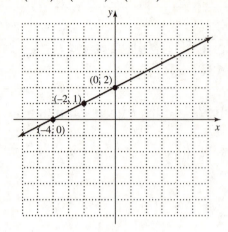

5. $(0,1)$, $(2,0)$, $(-2,2)$

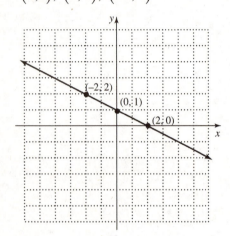

7. $(0,-3)$, $\left(\frac{3}{2},0\right)$, $(1,-1)$

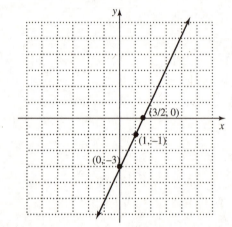

9. x-intercept: $(4,0)$;

   y-intercept: $(0,6)$

11. x-intercept: $(2,0)$;

    y-intercept: $\left(0,\frac{8}{5}\right)$

13. x-intercept: $\left(\frac{5}{2},0\right)$

    y-intercept: $(0,3)$

15. x-intercept: $(-6,0)$

    y-intercept: $(0,-5)$

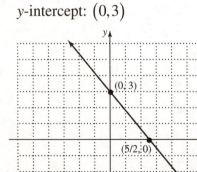

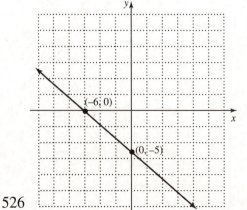

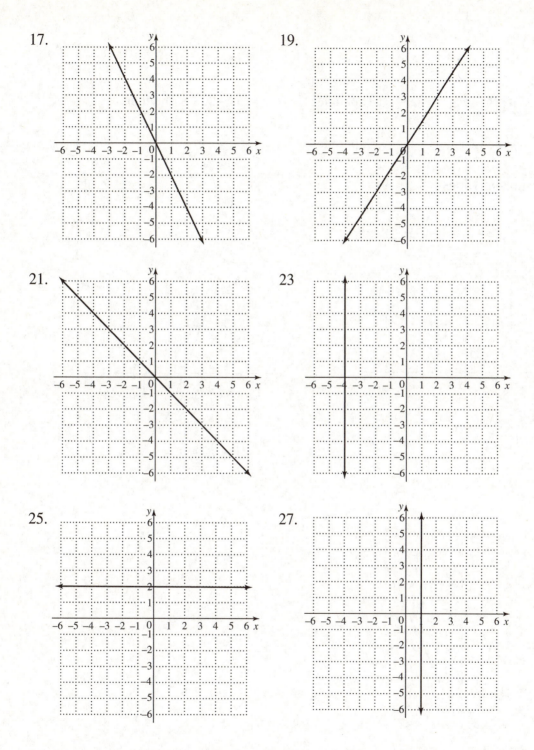

17.

19.

21.

23

25.

27.

29. 1994: $4.9 million;   1995: $5.53 million;   1996: $6.16 million   1997: $6.79 million

9.3 Slope of a Line

1. -2 3. $\frac{1}{5}$ 5. $-\frac{3}{5}$ 7. 1 9. 1

11. -5 13. $-\frac{2}{5}$ 15. $\frac{3}{4}$ 17. $-\frac{2}{7}$ 19. $\frac{4}{3}$

21. -5; 5; neither   23. 1; 1; parallel   25. -2; $-\frac{1}{4}$; neither

27. -2; $-\frac{5}{3}$; neither   29. 0; 0; parallel

9.4 Equations of Lines

1. $y = \frac{2}{3}x - 4$ 3. $y = \frac{1}{3}x - \frac{1}{2}$ 5. $y = -\frac{1}{2}x - 3$ 7. $y = -4$

9.
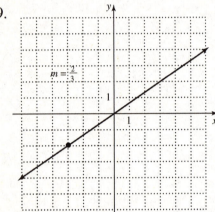

11.
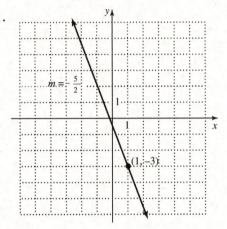

13.
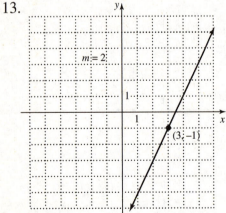

15. $x - 3y = -7$ 17. $2x + 3y = -9$ 19. $2x - 3y = -8$

21. $5x - 2y = 29$ 23. $2x - 5y = -11$ 25. $2x + 11y = 8$

27. $2x - y = -5$ 29. $3x - 2y = 0$

9.5   Graphing Inequalities in Two Variables

1.
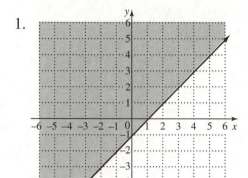

3.
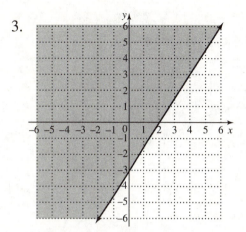

5.
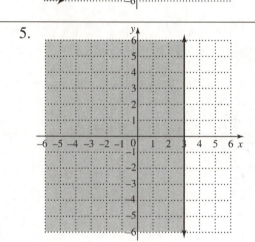

7.
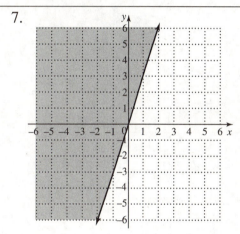

9.
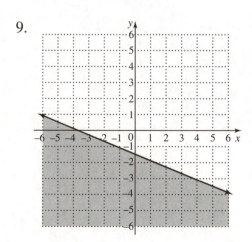

11.
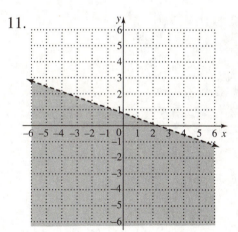

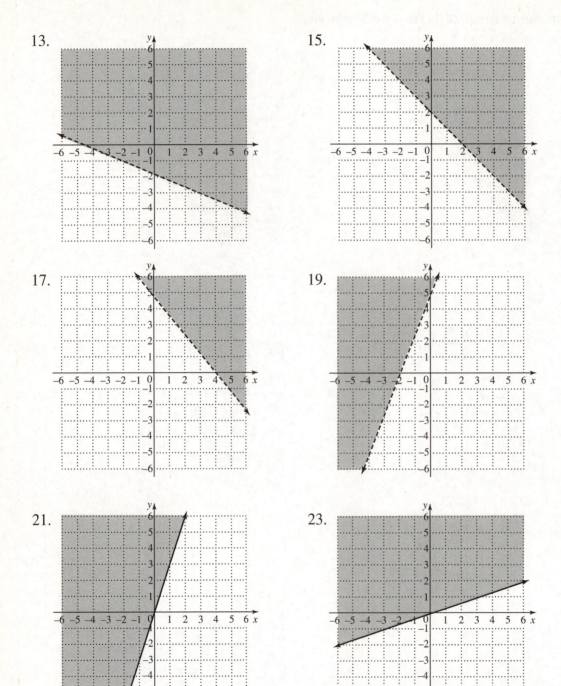

25.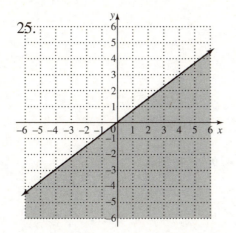

27.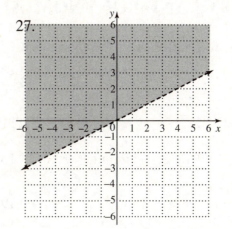

29.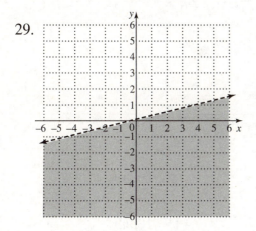

Chapter 10

EXPONENTS AND POLYNOMIALS

10.1   Adding and Subtracting Polynomials

1. $1; -7$

3. $2; \frac{1}{2}, -\frac{3}{4}$

5. $-3z^3$

7. $-\frac{1}{4}r^3$

9. (c)

11. $3z^4 + 3z^3$; degree 4; binomial

13. $\frac{1}{2}x^2 - \frac{1}{2}x$; degree 2; binomial

15. $5m^3 - 2m^2 - 4m + 4$

17. $-3x^4 + 8x^3 - 3x^2 - 3$

19. $6x^2 - 2x$

21. $19y^4 - 8y^3 + 4y^2 - 2y$

23. $3p^2 + 10p - 6$

25. $-8m^2n - 4n - 8m$; degree 3

27. $2x^3y - 2xy^3$

29. $-3a^3b + 6a^2b^2 + 9ab^3 + 11$

10.2   The Product Rule and Power Rules for Exponents

1. $2^6 = 64$

3. $\left(\frac{1}{3}\right)^5 = \frac{1}{243}$

5. $(-2y)^5$

7. -81; base, 3; exponent, 4

9. $(-3)^{13}$

11. $100n^{12}$

13. $16y^2$; $63y^4$

15. 3^{12}

17. $(-3)^{21}$

19. y^3z^{12}

21. $2^4 w^{12} z^{28}$

23. $\frac{3^3}{5^3}$

25. $\frac{x^4 y^4}{z^8}$

27. $7^9 a^9$

29. x^{26}

10.3   Multiplying Polynomials

1. $12y^5$

3. $-6x^4 - 12x^5 - 4x^6$

5. $12k^2 + 8k^5 + 24k^6$

7. $32m^3n + 16m^2n^2 + 56mn^3$

9. $3p^2 + 10p + 8$

11. $6x^2 + 23x + 20$

13. $y^3 + 64$

15. $2z^6 + 4z^5 - z^4 + z^2 - 9z - 6$

17. $6y^4 - 5y^3 - 8y^2 + 7y - 6$

19. $8x^5 + 12x^4 + 12x^3 + 12x^2 + 13x + 6$

21. $-4m^6 - 14m^4 + 8m^3 - 12m^2 + 15m - 3$

23. $4m^2 - 25m - 21$

25. $20a^2 + 11ab - 3b^2$

27. $4x^2 + 5xy - 6y^2$

29. $4y^2 - .8y - .05$

10.4 Special Products

1. $z^2 + 6z + 9$

3. $a^2 + 4ab + 4b^2$

5. $4m^2 + 20m + 25$

7. $49 + 14x + x^2$

9. $16y^2 - 5.6y + .49$

11. $z^2 - 36$

13. $49x^2 - 9y^2$

15. $16p^2 - 49q^2$

17. $81 - 16y^2$

19. $4a^2 - \frac{16}{9}b^2$

21. $x^3 + 6x^2 + 12x + 8$

23. $y^3 + 12y^2 + 48y + 64$

25. $8x^3 + 12x^2 + 6x + 1$

27. $t^4 - 12t^3 + 54t^2 - 108t + 81$

29. $81b^4 - 216b^3 + 216b^2 - 96b + 16$

10.5 Integer Exponents and the Quotient Rule

1. 1

3. -1

5. 1

7. -1

9. $-\frac{1}{27}$

11. $\frac{3}{8}$

13. $2r^7$

15. 16

16. 9^5

17. $(-2)^5$

19. $\frac{x^6}{12^3 y^2}$

21. $\frac{p^8}{3^5 m^3}$

23. 9

25. $\frac{1}{2^{16}}$

27. $\frac{3^2}{2^4 y^2}$

29. $\frac{1}{x^4 y^{12}}$

10.6 Dividing a Polynomial by a Monomial

1. $4m+2$

3. $7m^3-1$

5. $7m+12-\dfrac{4}{m}$

7. $-21m+9+\dfrac{2}{m}$

9. $8m-\dfrac{12}{m}-\dfrac{3}{m^2}$

11. $1+3p^3$

13. $5a^2-\frac{9}{4}$

15. $10x^3-5x$

17. $\dfrac{1}{2}m+\dfrac{7}{2}-\dfrac{21}{m}$

19. $7q^2-4+\dfrac{1}{q}$

21. $3z^2+7z-2+\dfrac{3}{4z^2}$

23. $-5u^2+4uv-9v^2$

25. $-13m^2+4m-\dfrac{5}{m^2}$

27. $-3x^4+2x^3+2x-\dfrac{4}{5x}$

29. $4y^2+5y-\dfrac{2}{3y}$

10.7 Dividing a Polynomial by a Polynomial

1. $x+2$

3. $6a-5$

5. $x+8$

7. $a-7+\dfrac{37}{2a+3}$

9. $5w-2+\dfrac{-4}{w-4}$

11. $3m-2+\dfrac{8}{3m-4}$

13. $2x+5$

15. $z^2-5z+9+\dfrac{-25}{2z+3}$

17. $9p^3-2p+6$

19. $3x^2-6x+2+\dfrac{13x-7}{2x^2+3}$

21. a^2+1

23. y^2-y+1

25. $2x^4+x^3-3x^3+2x+1+\dfrac{2}{3x+2}$

27. $-3x+4$

29. $3y^3+4y^2-4y+\dfrac{3}{2y^3-1}$

10.8   An Application of Exponents: Scientific Notation

1. 3.25×10^2 3. 2.3651×10^4 5. 7.42×10^0

7. 2.57×10^{-2} 9. 4.13×10^{-6} 11. 25,000

13. $-2,450,000$ 15. 23,000   17. .04007

19. -4.02 21. 210,000,000   23. 253

25. .02   27. 200   29. 2.1

FACTORING AND APPLICATIONS

11.1   The Greatest Common Factor

1. 3 3. 15 5. 32 7. $9b^3$ 9. $6a^6$

11. $15a^2y$ 13. $6x$ 15. $6(3r+4t)$ 17. $10x(2x+4xy-7y^2)$

19. $5a^3(3a^4-5-8a)$ 21. $8xy^2(7xy^2-3y+4)$ 23. $(x+2)(x+5)$

25. $(x-4)(x+5)$ 27. $(x-2)(y-2)$ 29. $(2a-3b)(a^2+b^2)$

11.2   Factoring Trinomials

1. 1 and 42, –1 and –42, 2 and 21, -2 and –21, 7 and 6, –7 and –6, 14 and 3, –14 and –3; the pair with sum of 17 is 3 and 14.

3. –8 and 8, 1 and –64, –1 and 64, 2 and –32, –2 and 32, 16 and –4, –16 and 4; the pair with a sum of 12 is –4 and 16.

5. $x+4$ 7. $x+2$ 9. $(x+2)(x+9)$

11. $(x-2)(x+1)$ 13. $(x-7)(x+5)$ 15. $(x+5)(x+1)$

17. $(x-7y)(x+3y)$ 19. $2(x-2)(x+7)$ 21. $3hk(h+2)(h-9)$

23. Prime   25. $2ab(a-3b)(a-2b)$

27. $5(r+4)(r+3)$ 29. $10k^4(k+2)(k+5)$

11.3   Factoring Trinomials by Grouping

1. $x+3$ 3. $4x-2$ 5. $(4b+3)(2b+3)$

7. $(5a+2)(3a+2)$ 9. $(3b+2)(b+2)$ 11. $p(3p+2)(p+2)$

13. $b(7a+4)(a+2)$ 15. $3(c+2d)(3c+2d)$ 17. $(2c-3t)(5c-7t)$

19. $(2x+7y)(6x-5y)$ 21. $2m(m-n)(3m+4n)$ 23. $(6f-g)(3f+5g)$

25. $2a(5a-4)(4a-5)$ 27. $4(x-y)(2x+y)$ 29. $(5c+12d)(2c+3d)$

11.4 Factoring Trinomials using FOIL

1. $(5x+2)(2x+3)$ 3. $(a+6)(2a+1)$ 5. $(2q+1)(4q+3)$

7. $(2b+1)(7b-2)$ 9. $(3a+2b)(a+2b)$ 11. $(5c+2d)(2c-d)$

13. $(6x-y)(3x-4y)$ 15. $(3x+1)(x-4)$ 17. $(3y-1)(2y+1)$

19. $(p+5)(3p+2)$ 21. $(x+4)(7x-1)$ 23. $x^2(2x-3)(x+4)$

25. $(9r-4t)(3r+2t)$ 27. $(2x^2y-3)(3x^2y+5)$ 29. $x(2x-y)(4x-3y)$

11.5 Special Factoring Techniques

1. Prime   3. $(10r+3s)(10r-3s)$ 5. $(5a+6)(5a-6)$

7. $(6+11d)(6-11d)$ 9. $(x^2+9)(x+3)(x-3)$ 11. $(3x^2+4)(3x^2-4)$

13. Prime   15. $(3p+11)(3p-11)$ 17. $(q+7)^2$

19. $(c+11)^2$ 21. $(2w+3)^2$ 23. $(3j+2)^2$

25. $\left(10p-\frac{5}{8}r\right)^2$ 27. $-4(2x+3)^2$ 29. $2(3x+7y)^2$

11.6 Solving Quadratic Equations by Factoring

1. $-9, \frac{3}{2}$ 3. $-2, -\frac{1}{3}$ 5. $-\frac{5}{2}, 4$ 7. $0, 3$

9. $-1, \frac{3}{8}$ 11. $0, \frac{4}{5}$ 13. $-\frac{4}{3}, \frac{3}{4}$ 15. $-4, \frac{3}{5}$

17. $0, -\frac{3}{2}, 5$ 19. $-7, 0, 7$ 21. $-4, 0, 2$ 23. $-\frac{3}{2}, \frac{3}{2}, 2$

25. $-\frac{4}{3}, 0, 1$ 27. $-6, 2, 3, 6$ 29. $7, -5, \frac{3}{2}$

11.7 Applications of Quadratic Equations

1. Length, 17 cm; width, 9 cm

3. Length, 18 ft; width, 14 ft

5. First square, 5 m; second square, 3 m

7. Length, 9 cm; width, 3 cm

9. 6 ft   11. 0, 1 or 7, 8   13. 7, 9   15. −4, −3, or 3, 4

17. 7, 9 or −7, −5   19. 6, 8   21. 6 inches

23. 7 m, 24 m, 25 m   25. 15 ft   27. 18 ft

29. 20 mi

Chapter 12

RATIONAL EXPRESSIONS AND APPLICATIONS

12.1 The Fundamental Property of Rational Expressions

1. 0
3. 0, 4
5. Never undefined
7. $-4, 2$

9. (a) 5
   (b) $-\frac{11}{2}$
11. (a) $\frac{64}{7}$
    (b) Undefined
13. (a) $\frac{24}{7}$
    (b) $-\frac{27}{7}$
15. (a) 5
    (b) $\frac{19}{8}$

17. $\dfrac{3}{8k}$
19. $\dfrac{-5b}{8c}$
21. $\dfrac{a-1}{3a-1}$
23. $4k$

25. $\dfrac{5x+3y}{x-3y}$

27. $\dfrac{-(2x-3)}{x+2}$; $\dfrac{-2x+3}{x+2}$; $\dfrac{2x-3}{-(x+2)}$; $\dfrac{2x-3}{-x-2}$

29. $\dfrac{-(2x-1)}{3x+5}$; $\dfrac{-2x+1}{3x+5}$; $\dfrac{2x-1}{-(3x+5)}$; $\dfrac{2x-1}{-3x-5}$

12.2 Multiplying and Dividing Rational Expressions

1. $\dfrac{10m^3n}{3}$
3. $-\frac{3}{2}$
5. $\dfrac{a-1}{a+1}$
7. $\dfrac{3-x}{2x+3}$

9. $\dfrac{-1}{8+2x}$
11. $\dfrac{x+4}{2x-8}$
13. $\dfrac{2y}{7}$
15. $\dfrac{1}{4r^2+2r+3}$

17. $\dfrac{2x^2+9}{x^2-2x+3}$
19. $\dfrac{12m^3}{5}$
21. $-\frac{1}{2}$
23. $\dfrac{2(m-5)}{m+3}$

25. $\dfrac{(m-1)(m+n)}{m(m-n)}$
27. $\dfrac{3k+1}{3k-1}$
29. $\dfrac{y^2-9yz+18z^2}{y^2+yz-20z^2}$

12.3 Least Common Denominator

1. 48
3. 180
5. $24a^2b^2$
7. $15r^3(r-5)$

9. $x-y$ or $y-x$
11. $m(2m-1)(m+5)$
13. $z^2(z-2)(z+4)^2$

15. $(2q-5)(q+2)(q-2)$ 17. 15   19. $21mn^5$

21. $-3y(y+3)$ or $-3y^2-9y$ 23. $5(r^2+2r-8)$ 25. $4p^2$

27. $3(k+7)$ or $3k+21$ 29. $20a^2$

12.4   Adding and Subtracting Rational Expressions

1. $\dfrac{6x}{x-2}$ 3. $\dfrac{8x}{x^2-4}$ 5. $\dfrac{1}{b-2}$ 7. $\dfrac{3}{2x+1}$

9. $\dfrac{5x+6}{15}$ 11. $\dfrac{3a-4}{a^2-4}$ 13. $\dfrac{3y^2+10y-12}{(y+4)(y+2)(y-2)}$

15. $\dfrac{19z-1}{15(z+4)}$ 17. $\dfrac{16p^2+14p+17}{(3p+2)(2p+3)(p+3)}$ 19. 5

21. $\dfrac{1}{x+5}$ 23. $\dfrac{16}{3x+12}$ 25. $\dfrac{5-20s}{18}$ 27. $\dfrac{10+y}{x-y}$

29. $\dfrac{q-22}{(2q+1)(q+2)(q-2)}$

12.5   Complex Fractions

1. $-\frac{2}{3}$ 3. $\dfrac{7m^2}{2n^3}$ 5. $\dfrac{2y-5}{3y-8}$ 7. $\dfrac{3p-2}{2p+1}$

9. $\dfrac{9s+12}{6s^2+2s}$ 11. $(a+2)^2$ 13. $\dfrac{24}{w-3}$ 15. $\dfrac{4(2a+3)}{7a}$

17. $-\frac{7}{3}$ 19. $\dfrac{4}{r^2s^2}$ 21. $\dfrac{r^2+3}{5+r^2t}$ 23. $\dfrac{2s^2+3}{1-3s^2}$

25. $\dfrac{2(1-4h)}{h(1+4h)}$ 27. $\dfrac{4m-3}{6-4m}$ 29. $\dfrac{(k-23)(k+2)}{5k(k+1)}$

12.6 Solving Equations with Rational Expressions

1. equation; 4

3. operation; $\dfrac{3x}{10}$

5. operation; $\dfrac{41x}{15}$

7. 3

9. -11

11. $-4, 16$

13. 6

15. $7, \frac{1}{17}$

17. $0, 7$

19. 2

21. $D = \dfrac{dF - k}{F}$

23. $B = \dfrac{2A - hb}{h}$ or $B = \dfrac{2A}{h} - b$

25. $R = \dfrac{R_1 R_2}{R_1 + R_2}$

27. $f = \dfrac{d_0 d_1}{d_1 + d_0}$

29. $a_1 = \dfrac{2S_n}{n} - a_n$

12.7 Applications of Rational Expressions

1. 15

3. $\frac{3}{4}$ or $\frac{1}{3}$

5. $\frac{9}{13}$

7. $\frac{1}{2}$ or 4

9. Sharon, \$57,000; Elaine, \$95,000

11. 3 mph

13. 350 mi

15. 60 mph

17. 24 mph

19. 8 mph

21. $\frac{12}{5}$ or $2\frac{2}{5}$

23. $\frac{24}{11}$ or $2\frac{2}{11}$

25. $\frac{2}{5}$ hr

27. 3 hr

29. 6 hr

12.8 Variation

1. 72

3. $\frac{68}{3}$

5. 15

7. 9

9. 168

11. $\frac{20}{3}$

13. 4

15. 2.25

17. 6

19. 3

21. \$162.50

23. 200 cu ft

25. 69.08 cm

27. 12.8 sq cm

29. 275 miles

Chapter 13

SYSTEMS OF LINEAR EQUATIONS AND INEQUALITIES

13.1   Solving Systems of Linear Equations by Graphing

1. No   3. Yes   5. Yes   7. No   9. No

11. Yes   13. $(2,-2)$ 15. $(-1,-2)$ 17. $(0,0)$ 19. $(4,8)$

21. $(2,-1)$ 23. $(1,4)$ 25. No solution   27. No solution

29. Infinite number of solutions

13.2   Solving Systems of Linear Equations by Substitution

1. $(1,6)$ 3. $(7,-2)$ 5. $(-3,-5)$ 7. $(1,2)$

7. $(1,2)$ 9. $(-2,2)$ 11. $\left(3,-\frac{2}{3}\right)$ 13. $(3,4)$

15. $(4,-5)$ 17. $(0,0)$ 19. $(1,2)$ 21. No solution

23. No solution   25. $\left(\frac{22}{9},\frac{8}{9}\right)$ 27. $(0,4)$ 29. Infinite number
                                                                          of solutions

13.3   Solving Systems of Linear Equations by Elimination

1. $(1,4)$ 3. $(2,-4)$ 5. $(5,0)$ 7. $\left(2,\frac{1}{6}\right)$

9. $(4,-2)$ 11. $(2,-2)$ 13. $(4,1)$ 15. $(1,-1)$

17. $(4,-1)$ 19. $(2,-4)$ 21. $(4,-1)$ 23. $(1,2)$

25. Infinite number of solutions   27. No solution   29. No solution

13.4 Applications of Linear Systems

1. 41 and 23   3. 23 and 50   5. 18 and 32

7. 26 cm; 56 cm   9. 5000 adults; 3000 children   11. 6 fives; 5 twenties

13. 30 fives; 60 tens   15. $4000 at 7%; $8000 at 9%

17. 20 pounds at $1.60 per pound; 10 pounds at $2.50 per pound   19. 5 pounds of walnuts; 5 pounds of peanuts

21. 28 liters of 75% solution; 42 liters of 55% solution   23. 30 liters of 15% solution; 15 liters of 30% solution

25. 128 mph; 148 mph   27. 8 mph; 12 mph   29. Wind, 35 mph; plane, 265 mph

13.5 Solving Systems of Linear Inequalities

1.
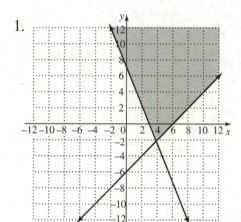

3.
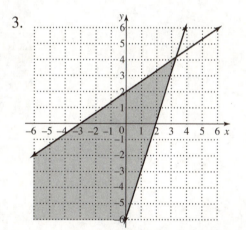

5.
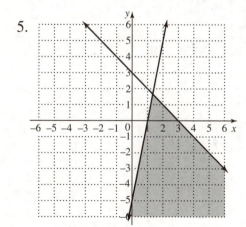

7.
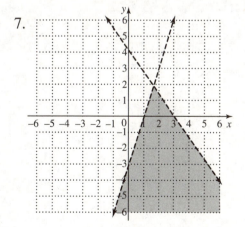

9.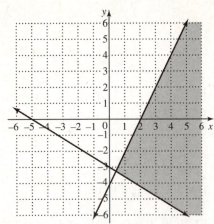

11.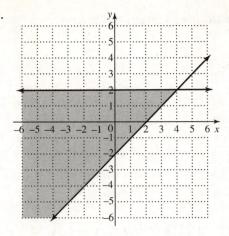

13.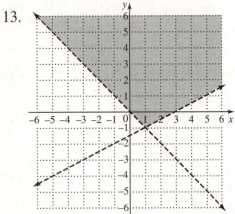

15.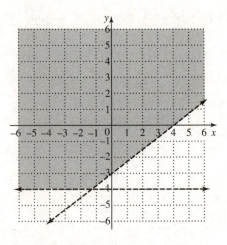

17.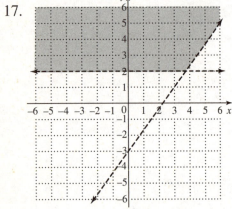

19.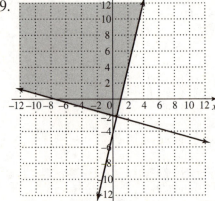

21.

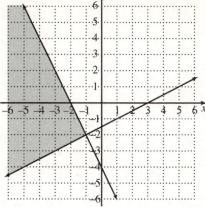

23.

25.

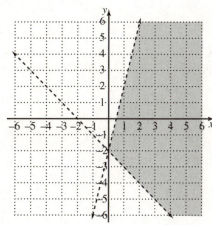

27.

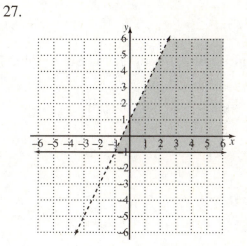

29.

ROOTS AND RADICALS

14.1 Evaluating Roots

1. $-9, 9$
3. $-24, 24$
5. 8
7. -23

9. Irrational
11. Not a real number
13. Rational
15. -4.243

17. -11.446
19. $c = 10.296$
21. $b = 5.745$
23. 19.235 cm

25. 48 ft
27. Not a real number
29. -3

14.2 Multiplying, Dividing, and Simplifying Radicals

1. $\sqrt{35}$
3. $\sqrt{14}$
5. $\sqrt{104}$
7. $5\sqrt{3}$

9. $-2\sqrt{22}$
11. $3\sqrt{10}$
13. $11\sqrt{3}$
15. $\frac{12}{7}$

17. $\frac{\sqrt{6}}{2}$
19. $\frac{\sqrt{15}}{4}$
21. $7x\sqrt{x}$
23. $\frac{\sqrt{6}}{x}$

25. $2\sqrt[4]{2}$
27. $-2\sqrt[5]{2}$
29. $\frac{1}{2}$

14.3 Adding and Subtracting Radicals

1. $2\sqrt{2}$
3. $6\sqrt[3]{10}$
5. Cannot be simplified further

7. $-5\sqrt{5}$
9. $7\sqrt{5}$
11. $70\sqrt{3}$
13. $4\sqrt{5} - 4\sqrt{2}$

15. $-2\sqrt{3}$
17. $5\sqrt{2} - 3\sqrt{6}$
19. $-12\sqrt{2z}$
21. $17\sqrt{5x}$

23. $4\sqrt{35}$
25. $-2\sqrt{14}$
27. $24\sqrt{x}$
29. $4x\sqrt{5} - 15x\sqrt{3}$

14.4 Rationalizing the Denominator

1. $\sqrt{5}$
3. $-\sqrt{3}$
5. $\frac{\sqrt{5}}{5}$
7. $\frac{\sqrt{15}}{5}$

9. $\dfrac{\sqrt{6}}{6}$ 11. $\dfrac{\sqrt{5}}{5}$ 13. $\dfrac{\sqrt{10}}{4}$ 15. $\dfrac{3\sqrt{14}}{10}$

17. $\dfrac{3\sqrt{7}}{35}$ 19. $\dfrac{m^2\sqrt{k}}{k^2}$ 21. $\dfrac{\sqrt[3]{50}}{5}$ 23. $\dfrac{\sqrt[3]{5}}{5}$

25. $\dfrac{\sqrt[3]{28}}{2}$ 27. $\dfrac{3\sqrt[3]{7}}{7}$ 29. $\dfrac{\sqrt[3]{50r}}{5r}$

14.5 More Simplifying and Operations with Radicals

1. $2 - \sqrt{10}$ 3. $2\sqrt{15} + 4\sqrt{35}$ 5. $4\sqrt{10} - 4\sqrt{35} + \sqrt{6} - \sqrt{21}$

7. $24 + 2\sqrt{15} - 8\sqrt{30} - 10\sqrt{2}$ 9. $14 - 4\sqrt{6}$ 11. $\dfrac{7 - \sqrt{3}}{23}$

13. $\dfrac{5\left(3 + \sqrt{7}\right)}{2}$ 15. $\sqrt{10} + 2\sqrt{2}$ 17. $\dfrac{3\left(\sqrt{5} - \sqrt{2}\right)}{3}$

19. $5 + 2\sqrt{6}$ 21. $2 - \sqrt{3}$ 23. $\dfrac{\sqrt{7} - 2\sqrt{2}}{3}$

25. $\dfrac{4 + 3\sqrt{7}}{2}$ 27. $3 + \sqrt{2}$ 29. $\dfrac{2\sqrt{5} - 3}{5}$

14.6 Solving Equations with Radicals

1. 49 3. 36 5. $\frac{8}{3}$ 7. $\frac{2}{5}$

9. 3 11. 36 13. 49 15. No solution

17. $-\frac{4}{17}$ 19. 5 21. 6 23. 5

25. 3 27. $1, 9$ 29. -1

Chapter 15

QUADRATIC EQUATIONS

15.1 Solving Equations by the Square Root Property

1. $-7, 7$ 3. $-30, 30$ 5. $-2\sqrt{6}, 2\sqrt{6}$ 7. $-3\sqrt{5}, 3\sqrt{5}$

9. $-6\sqrt{2}, 6\sqrt{2}$ 11. No real number solutions 13. $-\frac{3}{5}, \frac{3}{5}$ 15. $-3.5, 3.5$

17. $12, -6$ 19. $2, 16$ 21. -8 23. $-3-3\sqrt{2}, -3+3\sqrt{2}$

25. $2-3\sqrt{3}, 2+3\sqrt{3}$ 27. $\dfrac{-5-4\sqrt{2}}{2}, \dfrac{-5+4\sqrt{2}}{2}$ 29. $-\frac{7}{2}, -\frac{11}{2}$

15.2 Solving Quadratic Equations by Completing the Square

1. $-3, 5$ 3. $-6, 0$ 5. $3, 7$

7. $\dfrac{-5-3\sqrt{5}}{2}, \dfrac{-5+3\sqrt{5}}{2}$ 9. $-\frac{7}{2}, \frac{1}{2}$ 11. $-2, 7$

13. $\dfrac{-3-\sqrt{57}}{6}, \dfrac{-3+\sqrt{57}}{6}$ 15. $-7, \frac{1}{2}$ 17. $\dfrac{3-\sqrt{15}}{3}, \dfrac{3+\sqrt{15}}{3}$

19. $\dfrac{3-\sqrt{69}}{6}, \dfrac{3+\sqrt{69}}{6}$ 21. $-\frac{3}{2}, \frac{1}{2}$ 23. $-8, -2$

24. $-3, 2$ 25. 3 sec 27. 5%

29. .4 sec, 2.6 sec

15.3 Solving Quadratic Equations by the Quadratic Formula

1. $a=3, \ b=-4, \ c=2$ 3. $a=2, \ b=7, \ c=0$

548

5. $a = 10$, $b = 4$, $c = 0$ 7. $a = 1$, $b = -3$, $c = -5$

9. $-1 - \sqrt{5}$, $-1 + \sqrt{5}$ 11. -5, 1 13. No real number solution

15. $1 - \sqrt{5}$, $1 + \sqrt{5}$ 17. -3 19. $\frac{1}{4}$ 21. 9

23. $\frac{8}{5}$ 25. -4, 1 27. No real number solution   29. -3, $\frac{5}{3}$

15.4 Graphing Quadratic Equations and Inequalities

1. Vertex: $(0, 0)$

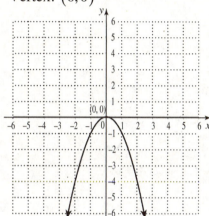

3. Vertex: $(0, -3)$

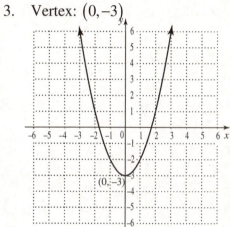

5. Vertex: $(0, 2)$

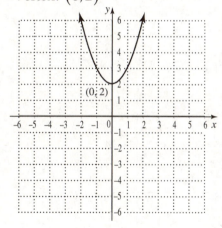

7. Vertex: $(0, 1)$

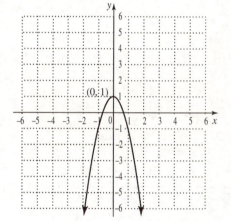

9. Vertex: $(2,0)$

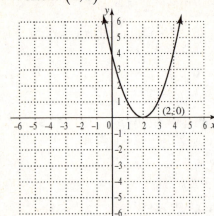

11. Vertex: $(3,0)$

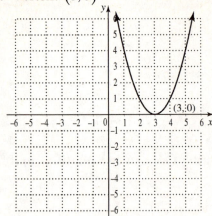

13. Vertex: $(-1,-2)$

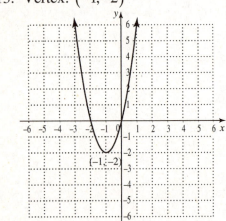

15. Vertex: $(3,2)$

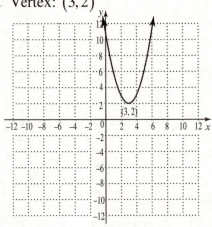

17. Vertex: $(2,3)$

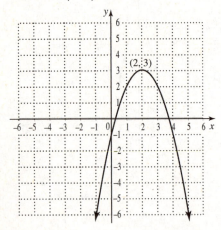

19. Vertex: $(3,-4)$

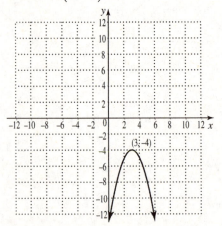

21. Vertex: $\left(\frac{5}{2}, \frac{25}{4}\right)$

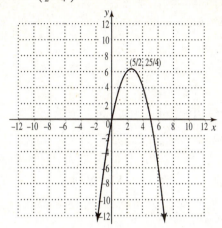

23.

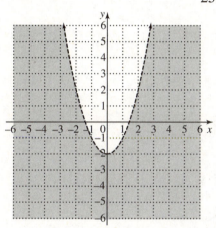

25.

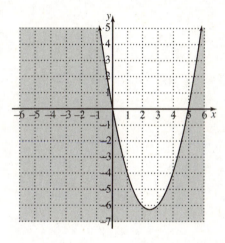

27.

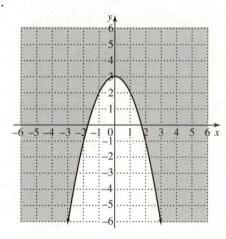

29.

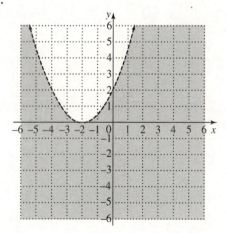

15.5   Introduction to Functions

1. Not a function;
   domain: $\{-3,0,2,5\}$
   range: $\{-8,-4,-1,2,7\}$

3. Not a function;
   domain: $\{-7,-3,0,2\}$
   range: $\{1,4,6,7,9\}$

5. Function;
   domain: $\{-5,-2,1,3,7\}$
   range: $\{4\}$

7. Not a Function;
   domain: $\{-1.2,1.8,3.7\}$
   range: $\{-3.8,-2.5,3.8,4\}$

9. Yes   11. Yes   13. No   15. Yes

17. Yes   19. (a) -13 (b) -7 (c) 5

21. (a) 1 (b) -5 (c) 31 23. (a) -12 (b) 4 (c) 36

25. (a) 17 (b) 1 (c) 41 27. (a) -44 (b) $-2\frac{3}{4}$ (c) -4.64

29. (a) 47 (b) $-5\frac{1}{4}$ (c) -8.76 4